# Lecture Notes in Computer Scie

T0238156

*Commenced Publication in 1973*
Founding and Former Series Editors:
Gerhard Goos, Juris Hartmanis, and Jan van Leeuwen

Sung Hyon Myaeng   Ming Zhou
Kam-Fai Wong   Hong-Jiang Zhang (Eds.)

# Information Retrieval Technology

Asia Information Retrieval Symposium, AIRS 2004
Beijing, China, October 18-20, 2004
Revised Selected Papers

Volume Editors

Sung Hyon Myaeng
Information and Communications University (ICU)
119 Munji-Ro, Yuseong-Gu, Daejeon, 305-714, South Korea
E-mail: myaeng@icu.ac.kr

Ming Zhou
Hong-Jiang Zhang
Microsoft Research Asia
5F, Beijing Sigma Center
No. 49 Zhichun Road Haidian District, Beijing 100080, China
E-mail: {mingzhou,hjzhang}@microsoft.com

Kam-Fai Wong
The Chinese University of Hong Kong
Shatin, N.T., Hong Kong, China
E-mail: kfwong@se.cuhk.edu.hk

Library of Congress Control Number: 2005921104


CR Subject Classification (1998): H.3, H.4, F.2.2, E.1, E.2

ISSN 0302-9743
ISBN 3-540-25065-4 Springer Berlin Heidelberg New York

Springer is a part of Springer Science+Business Media

springeronline.com

© Springer-Verlag Berlin Heidelberg 2005
Printed in Germany

Typesetting: Camera-ready by author, data conversion by Scientific Publishing Services, Chennai, India
Printed on acid-free paper       SPIN: 11398479       06/3142       5 4 3 2 1 0

# Preface

The Asia Information Retrieval Symposium (AIRS) was established by the Asian information retrieval community after the successful series of Information Retrieval with Asian Languages (IRAL) workshops held in six different locations in Asia, starting from 1996. While the IRAL workshops had their focus on information retrieval problems involving Asian languages, AIRS covers a wider scope of applications, systems, technologies and theory aspects of information retrieval in text, audio, image, video and multimedia data. This extension of the scope reflects and fosters increasing research activities in information retrieval in this region and the growing need for collaborations across subdisciplines.

We are very pleased to report that we saw a sharp increase in the number of submissions and their quality, compared to the IRAL workshops. We received 106 papers from nine countries in Asia and North America, from which 28 papers (26%) were presented in oral sessions and 38 papers in poster sessions (36%). It was a great challenge for the Program Committee to select the best among the excellent papers. The low acceptance rates witness the success of this year's conference.

After a long discussion between the AIRS 2004 Steering Committee and Springer, the publisher agreed to publish our proceedings in the Lecture Notes in Computer Science (LNCS) series, which is SCI-indexed. We feel that this strongly attests to the excellent quality of the papers.

The attendees were cordially invited to participate in and take advantage of all the technical programs at this conference. A tutorial was given on the first day to introduce the state of the art in Web mining, an important application of Web document retrieval. Two keynote speeches covered two main areas of the conference: video retrieval and language issues. There were a total of eight oral sessions run, with two in parallel at a time, and two poster/demo sessions.

The technical and social programs, which we are proud of, were made possible by the hard-working people behind the scenes. In addition to the Program Committee members, we are thankful to the Organizing Committee (Shao-Ping Ma and Jianfeng Gao, Co-chairs), Interactive Posters/Demo Chair (Gary G. Lee), and the Special Session and Tutorials Chair (Wei-Ying Ma). We also thank the sponsoring organizations: Microsoft Research Asia, the Department of Systems Engineering and Engineering Management at the Chinese University of Hong Kong, and LexisNexis for their financial support, the Department of Computer Science and Technology, Tsinghua University for local arrangements, the Chinese NewsML Community for website design and administration, Ling Huang for the logistics, Weiwei Sun for the conference webpage management, EONSOLUTION for the conference management, and Springer for the postconference

LNCS publication. We believe that this conference set a very high standard for a regionally oriented conference, especially in Asia, and we hope that it continues as a tradition in the upcoming years.

Sung Hyon Myaeng and Ming Zhou (PC Co-chairs)
Kam-Fai Wong and Hong-Jiang Zhang (Conference Co-chairs)

# Organization

## General Conference Co-chairs

Kam-Fai Wong, Chinese University of Hong Kong, China
Hong-Jiang Zhang, Microsoft Research Asia, China

## Program Co-chairs

Sung Hyon Myaeng, Information and Communications University (ICU),
    South Korea
Ming Zhou, Microsoft Research Asia, China

## Organization Committee Co-chairs

Jianfeng Gao, Microsoft Research Asia, China
Shao-Ping Ma, Tsinghua University, China

## Special Session and Tutorials Chair

Wei-Ying Ma, Microsoft Research Asia, China

## Interactive Posters/Demo Chair

Gary Geunbae Lee, POSTECH, South Korea

## Steering Committee

Jun Adachi, National Institute of Informatics, Japan
Hsin-Hsi Chen, National Taiwan University, Taiwan
Lee-Feng Chien, Academia Sinica, Taiwan
Tetsuya Ishikawa, University of Tsukuba, Japan
Gary Geunbae Lee, POSTECH, South Korea
Mun-Kew Leong, Institute for Infocomm Research, Singapore
Helen Meng, Chinese University of Hong Kong, China
Sung-Hyon Myaeng, Information and Communications University, South Korea

Hwee Tou Ng, National University of Singapore, Singapore
Kam-Fai Wong, Chinese University of Hong Kong, China

## Organizing Committee

Lee-Feng Chien, Academia Sinica, Taiwan (Publicity, Asia)
Susan Dumais, Microsoft, USA (Publicity, North America)
Jianfeng Gao, Microsoft Research Asia, China (Publication, Co-chair)
Mun-Kew Leong, Institute for Infocomm Research, Singapore (Finance)
Shao-Ping Ma, Tsinghua University, China (Local Organization, Co-chair)
Ricardo Baeza-Yates, University of Chile, Chile (Publicity, South America)
Shucai Shi, BITI, China (Local Organization)
Dawei Song, DSTC, Australia (Publicity, Australia)
Ulich Thiel, IPSI, Germany (Publicity, Europe)
Chuanfa Yuan, Tsinghua University, China (Local Organization)

# Program Committee

Peter Anick, Yahoo, USA
Hsin-Hsi Chen, National Taiwan University, Taiwan
Aitao Chen, University of California, Berkeley, USA
Lee-Feng Chien, Academia Sinica, Taiwan
Fabio Crestani, University of Strathclyde, UK
Edward A. Fox, Virginia Tech, USA
Jianfeng Gao, Microsoft Research Asia, China
Hani Abu-Salem, DePaul University, USA
Tetsuya Ishikawa, University of Tsukuba, Japan
Christopher Khoo, Nanyang Technological University, Singapore
Jung Hee Kim, North Carolina A&T University, USA
Minkoo Kim, Ajou University, South Korea
Munchurl Kim, Information and Communication University, South Korea
Kazuaki Kishida, Surugadai University, Japan
Kui-Lam Kwok, Queens College, City University of New York, USA
Wai Lam, Chinese University of Hong Kong, China
Gary Geunbae Lee, POSTECH, South Korea
Mun-Kew Leong, Institute for Infocomm Research, Singapore
Gena-Anne Levow, University of Chicago, USA
Hang Li, Microsoft Research Asia, China
Robert Luk, Hong Kong Polytechnic University, China
Gay Marchionini, University of North Carolina, Chapel Hill, USA
Helen Meng, Chinese University of Hong Kong, China
Hiroshi Nakagawa, University of Tokyo, Japan
Hwee Tou Ng, National University of Singapore, Singapore
Jian-Yun Nie, University of Montreal, Canada
Jon Patrick, University of Sydney, Australia
Ricardo Baeza-Yates, University of Chile, Chile
Hae-Chang Rim, Korea University, South Korea
Tetsuya Sakai, Toshiba Corporate R&D Center, Japan
Padmini Srinivasan, University of Iowa, USA
Tomek Strzalkowski, State University of New York, Albany, USA
Maosong Sun, Tsinghua University, China
Ulrich Thiel, Fraunhofer IPSI, Germany
Takenobu Tokunaga, Tokyo Institute of Technology, Japan
Hsin-Min Wang, Academia Sinica, Taiwan
Ross Wilkinson, CSIRO, Australia
Lide Wu, Fudan University, China
Jinxi Xu, BBN Technologies, USA
ChengXiang Zhai, University of Illinois, Urbana Champaign, USA
Min Zhang, Tsinghua University, China

# Reviewers

| | | |
|---|---|---|
| Peter Anick | Kui-Lam Kwok | Hae-Chang Rim |
| Yunbo Cao | Wai Lam | Tetsuya Sakai |
| Yee Seng Chan | Gary Geunbae Lee | Tomek Strzalkowski |
| Hsin-Hsi Chen | Mun-Kew Leong | Maosong Sun |
| Zheng Chen | Gena-Anne Levow | Ulrich Thiel |
| Aitao Chen | Hang Li | Takenobu Tokunaga |
| Tee Kiah Chia | Mu Li | Hsin-Min Wang |
| Lee-Feng Chien | Hongqiao Li | Haifeng Wang |
| Fabio Crestani | Chin-Yew Lin | Ross Wilkinson |
| Edward A. Fox | Robert Luk | Kam-Fai Wong |
| Jianfeng Gao | Wei-Ying Ma | Lide Wu |
| Hani Abu-Salem | Gay Marchionini | Jinxi Xu |
| Xuanjing Huang | Helen Meng | Peng Yu |
| Tetsuya Ishikawa | Sung Hyon Myaeng | Chunfa Yuan |
| Christopher Khoo | Hiroshi Nakagawa | ChengXiang Zhai |
| Jung Hee Kim | Hwee Tou Ng | Hong-Jiang Zhang |
| Minkoo Kim | Jian-Yun Nie | Min Zhang |
| Munchurl Kim | Jon Patrick | Ming Zhou |
| Kazuaki Kishida | Ricardo Baeza-Yates | Jian-lai Zhou |

# Table of Contents

## Information Organization

## Automatic Summarization

## Alignment/Paraphrasing in IR

## Web Search

## Linguistic Issues in IR

## Document/Query Models

# Enabling Technology

# Mobile Applications

# Automatic Word Clustering for Text Categorization Using Global Information

Chen Wenliang, Chang Xingzhi, Wang Huizhen, Zhu Jingbo, and Yao Tianshun

Natural Language Processing Lab
Northeastern University, Shenyang, China 110004
chenwl@mail.neu.edu.cn

**Abstract.** High dimensionality of feature space and short of training documents are the crucial obstacles for text categorization. In order to overcome these obstacles, this paper presents a cluster-based text categorization system which uses class distributional clustering of words. We propose a new clustering model which considers the global information over all the clusters. The model can be understood as the balance of all the clusters according to the number of words in them. It can group words into clusters based on the distribution of class labels associated with each word. Using these learned clusters as features, we develop a cluster-based classifier. We present several experimental results to show that our proposed method performs better than the other three text classifiers. The proposed model has better results than the model which only considers the information of the two related clusters. Specially, it can maintain good performance when the number of features is small and the size of training corpus is small.

## 1 Introduction

The goal of text categorization is to classify documents into a certain number of predefined categories. A variety of techniques for supervised learning algorithms have demonstrated reasonable performance for text categorization[5][11][12]. A common and overwhelming characteristic of text data is its extremely high dimensionality. Typically the document vectors are formed using bag-of-words model. It is well known, however, that such count matrices tend to be highly sparse and noisy, especially when the training data is relatively small. So when the text categorization systems are applied, there are two problems to be counted:

- High-dimensional feature space: Documents are usually represented in a high-dimensional sparse feature space, which is far from optimal for classification algorithms.
- Short of training documents: Many applications can't provide so many training documents.

A standard procedure to reduce feature dimensionality is feature selection, such as Document Frequency, $\chi^2$ statistic, Information Gain, Term Strength, and

S. H. Myaeng et al. (Eds.): AIRS 2004, LNCS 3411, pp. 1–11, 2005.

Mutual Information[13]. But feature selection is better at removing detrimental, noisy features. The second procedure is cluster-based text categorization[1][2][3] [10]. Word clustering methods can reduce feature spaces by joining similar words into clusters. First they grouped words into the clusters according to their distributions. Then they used these clusters as features for text categorization.

In this paper, we cluster the words according to their class distributions. Based on class distributions of words, Baker[1] proposes a clustering model. In clustering processing, we will select two most similar clusters by comparing the similarities directly. But Baker's model only considers two related clusters, when computing the similarity between the clusters without taking into account the information of other clusters. In order to provide better performance, we should take into account the information of all the clusters when computing the similarities between the clusters. This paper proposes a clustering model which considers the global information over all the clusters. The model can be understood as the balance of all the clusters according to the number of words in them.

Using these learned clusters as features, we develop a cluster-based Classifier. We present experimental results on a Chinese text corpus. We compare our text classifier with the other three classifiers. The results show that the proposed clustering model provides better performance than Baker's model. The results also show that it can perform better than the feature selection based classifiers. It can maintain high performance when the number of features is small and the size of training corpus is small.

In the rest of this paper: Section 2 reviews previous works. Section 3 proposes a global Clustering Model (globalCM). Section 4 describes a globalCM-based text categorization system. Section 5 shows the experimental results. Finally, we draw our conclusions at section 6.

## 2   Related Work

Distributional Clustering has been used to address the problem of sparse data in building statistical language models for natural language processing[7][10]. There are many works[1][2] related with using distributional clustering for text categorization.

Baker and McCallum[1] proposed an approach for text categorization based on word-clusters. First, find word-clusters that preserve the information about the categories as much as possible. Then use these learned clusters to represent the documents in a new feature space. Final, use a supervised classification algorithm to predict the categories of new documents. Specifically, it was shown there that word-clustering can be used to significantly reduce the feature dimensionality with only a small change in classification performance.

# 3    Global Clustering Model Based on Class Distributions of Words

In this section, we simply introduce the class distribution of words[1]. Then we propose the Global Clustering Model, here we name it as globalCM. In our clustering model, we define a similarity measure between the clusters, and add the candidate word into the most similar cluster that no longer distinguishes among the words different.

## 3.1    Class Distribution of Words

Firstly, we define the distribution $P(C|w_t)$ as the random variable over classes C, and its distribution given a particular word $w_t$. When we have two words $w_t$ and $w_s$, they will be put into the same cluster $f$. The distribution of the cluster $f$ is defined

$$
\begin{aligned}
P(C|f) &= P(C|w_t \vee w_s) \\
&= \frac{P(w_t)}{P(w_t) + P(w_s)} \times P(C|w_t) \\
&+ \frac{P(w_s)}{P(w_t) + P(w_s)} \times P(C|w_s) .
\end{aligned} \tag{1}
$$

Now we consider the case that a word $w_t$ and a cluster $f$ will be put into a new cluster $f_{new}$. The distribution of $f_{new}$ is defined

$$
\begin{aligned}
P(C|f_{new}) &= P(C|w_t \vee f) \\
&= \frac{P(w_t)}{P(w_t) + P(f)} \times P(C|w_t) \\
&+ \frac{P(f)}{P(w_t) + P(f)} \times P(C|f) .
\end{aligned} \tag{2}
$$

## 3.2    Similarity Measures

Secondly, we turn to the question of how to measure the difference between two probability distributions. Kullback-Leibler divergence is used to do this. The KL divergence between the class distributions induced by $w_t$ and $w_s$ is written $D(P(C|w_t)||P(C|w_s))$, and is defined

$$
-\sum_{j=1}^{|C|} P(c_j|w_t) \log \frac{P(c_j|w_t)}{P(c_j|w_s)} . \tag{3}
$$

But KL divergence has some odd properties: It is not symmetric, and it is infinite when $p(w_s)$ is zero. In order to resolve these problems, Baker[1] proposes a measure named "KL divergence to the mean" to measure the similarity of two distributions(Here we name it as $S_{mean}$). It is defined

$$\frac{P(w_t)}{P(w_t) + P(w_s)} \times D(P(C|w_t)||P(C|w_s \vee w_t))$$

$$+ \frac{P(w_s)}{P(w_t) + P(w_s)} \times D(P(C|w_s)||P(C|w_s \vee w_t)) . \tag{4}$$

$S_{mean}$ uses a weighted average and resolves the problems of KL divergence. But it only considers the two related clusters without thinking about other clusters. Our experimental results show that the numbers of words in learned clusters, which are generated by Baker's clustering model, are very different. Several clusters include so many words while most clusters include only one or two words.

We study the reasons of these results. When Equation 4 is applied in the clustering algorithm, it can't work well if the numbers of words in the clusters are very different at iterations.

For example, we have a cluster $f$ which include only a word(In Baker's clustering model, a new candidate word will be put into an empty cluster). We will compute the similarities between $f$ and the other two clusters($f_i$ and $f_j$) using Equation 4. Let $f_i$ has many words(ie. 1000 words) and $f_j$ has one or two words. We define:

$$S_i = \frac{P(f)}{P(f) + P(f_i)} \times D(P(C|f)||P(C|f \vee f_i))$$

$$+ \frac{P(f_i)}{P(f) + P(f_i)} \times D(P(C|f_i)||P(C|f \vee f_i))$$

$$= (1 - \alpha_i) \times D_{i1} + \alpha_i \times D_{i2} . \tag{5}$$

$$S_j = \frac{P(f)}{P(f) + P(f_j)} \times D(P(C|f)||P(C|f \vee f_j))$$

$$+ \frac{P(f_j)}{P(f) + P(f_j)} \times D(P(C|f_j)||P(C|f \vee f_j))$$

$$= (1 - \alpha_j) \times D_{j1} + \alpha_j \times D_{j2} . \tag{6}$$

According to Equation 2, if a word is added to a cluster, the word will affect tiny to the cluster which includes many words and affect remarkable to the cluster which includes few words. So the distribution of $f \vee f_i$ is very similar to $f_i$ because $f_i$ has many words and $f$ has only one word. And then $D_{i2}$ is near zero. $\alpha_i$ is near 1 and $(1 - \alpha_i)$ is near zero because the number of $f_i$ is very large than $f$. We know:

$$S_i \approx D_{i2} \approx 0 . \tag{7}$$

So when we compute the similarities between $f$ and the other clusters using Equation 4, $f$ will be more similar to the cluster which includes more words.

**Table 1.** The globalCM Algorithm

---

Input:
  W - the vocabulary includes the candidate words
  M - desired number of clusters
Output:
  F - the learned clusters

Clustering:

1. Sort the vocabulary by $\chi^2$ statistic with the class variable.

2. Initialize the M clusters as singletons with the top M words.

3. Loop until all words have been put into one of the M clusters.
   (a) Compute the similarities between the M clusters(Equation 8).

   (b) Merge the two clusters which are most similar, resulting in M-1 clusters.

   (c) Get the next word from the sorted list.

   (d) Create an new cluster consisting of the new word.

---

The problems of $S_{mean}$ indicate that we should consider the information of all the clusters when computing the similarity between the two clusters. If we only take into account the two related clusters, the system will can't work well. In order to resolve the problems, we propose a new similarity measure that considers the global information over the clusters. The similarity between a cluster $f_i$ and a cluster $f_j$ is defined

$$S_{global} = \frac{N(f_i) + N(f_j)}{2 \sum_{k=1}^{|M|} N(f_k)} \times S_{mean} .$$   (8)

Where $N(f_k)$ denotes the number of words in the cluster $f_k$, M is the list of clusters. Equation 8 can be understood as the balance of all the clusters according to the numbers of words in them. In our experimental results show that it can work well even if the numbers of words in the clusters are very different.

## 3.3    Global Clustering Model(globalCM)

Now we introduce a clustering model which use Equation 8. The model is similar to Baker's clustering model[1]. In this paper, we name Baker's model as BakerCM, and our model as globalCM.

In the algorithm, we set M is the final number of clusters. First, we sort the vocabulary by $\chi^2$ statistic with the class variable. Then the clusters are initialized with the Top M words from the sorted list. Then we will group the rest words into the clusters. We compute the similarities between all the clusters(Equation 8) and then merge the two clusters which are most similar. Now we have M-1 clusters. An empty cluster is created and the next word is added. So the number of clusters is back to M. Table 1 shows the clustering algorithm.

# 4    The globalCM-Based Text Categorization System

This section introduces our globalCM-based Chinese text categorization System. The system includes Preprocessing, Extracting the candidate words, Word Clustering, Cluster-based Text Classifier. Word Clustering has been described at Section 3.

## 4.1    Preprocessing

First, the html tags and special characters in the collected documents are removed. Then we should use a word segmentation tool to segment the documents because there are no word boundaries in Chinese documents.

## 4.2    Extracting the Candidate Words

We extract candidate words from the documents: First we use a stoplist to eliminate no-informative words, and then we remove the words whose frequencies are less than $F_{min}$. Final, we generate the class distributions of words which is described at Section 3.

## 4.3    The Cluster-Based Classifier

Using the learned clusters as features, we develop a cluster-based text classifier. The document vectors are formed using bag-of-clusters model. If the words are included in the same cluster, they will be presented as the single cluster symbol. After representation, we develop a classifier based on these features.

In this paper, we use naïve Bayes for classifying documents. We only describe naïve Bayes briefly since full details have been presented in the paper [9]. The basic idea in naïve Bayes approach is to use the joint probabilities of features and categories to estimate the probabilities of categories when a document is given. Given a document d for classification, we compute the probabilities of each category c as follows:

$$P(c_j|d_i;\hat{\theta}) = \frac{P(c_j|\hat{\theta})P(d_i|c_j;\hat{\theta}_j)}{P(d_i|\hat{\theta})} \approx P(d_i|c_j;\hat{\theta}) . \qquad (9)$$

$$P(d_i|c_j;\hat{\theta}) = P(|d_i|)|d_i|! \prod_{t=1}^{|F|} \frac{P(f_t|c_j;\theta)^{N_{it}}}{N_{it}!} . \qquad (10)$$

Where $P(c_j)$ is the class prior probabilities, $|d_i|$ is length of document $d_i$, $N_{it}$ is the frequency of the feature $f_t$ (Notes that the features are the cluster symbols in this paper.) in document $d_i$, F is the vocabulary and $|F|$ is the size of F, $f_t$

is the $t^{th}$ feature in the vocabulary, and $P(f_t|c_j)$ thus represents the probability that a randomly drawn feature from a randomly drawn document in category $c_j$ will be the feature $f_t$. The probability is estimated by the following formulae:

$$P(f_t|c_j; \theta) = \frac{1 + \sum_{i=1}^{|D|} N_{it} P(c_j|d_i)}{|F| + \sum_{s=1}^{|F|} \sum_{i=1}^{|D|} N_{is} P(c_j|d_i))} \ . \tag{11}$$

## 5   Evaluation

In this section, we provide empirical evidence to prove that the globalCM-based text categorization system is a high-accuracy system.

### 5.1   Performance Measures

In this paper, a document is assigned to only one category. We use the conventional recall, precision and F1 to measure the performance of the system. For evaluating performance average across categories, we use the micro-averaging method. F1 measures is defined by the following formula[6]:

$$F1 = \frac{2rp}{r+p} \ . \tag{12}$$

Where r represents recall and p represents precision. It balances recall and precision in a way that gives them equal weight.

### 5.2   Experimental Setting

The NEU_TC data set contains Chinese web pages collected from web sites. The pages are divided into 37 categories according to "China Library Categorization" [4][1]. It consists of 14,459 documents. We do not use tag information of pages. We use the toolkit CipSegSDK[14] for word segmentation. We removed all words that have less than two occurrences($F_{min} = 2$). The resulting vocabulary has 75480 words.

In experiments, we use 5-fold cross validation where we randomly and uniformly split each category into 5 folds and we take four folds for training and one fold for testing. In the cross-validated experiments we report on the average performance.

### 5.3   Experimental Results

We compare our globalCM-based classifier with the other three clustering and feature selection algorithms: BakerCM-based classifier, $\chi^2$ statistic based clas-

---

[1] China Library Categorization includes 38 categories. We use 37 categories of all, except category Z(/Comprehensive Books).

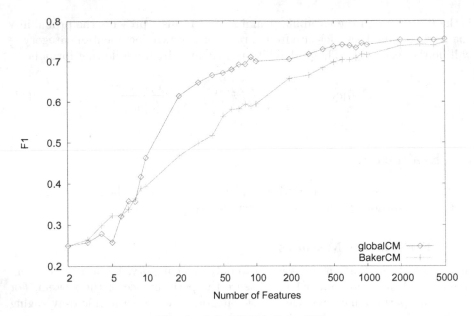

**Fig. 1.** globalCM Vs BakerCM

sifier, and document frequency based classifier. These two feature selection methods are the best of feature selection methods according to Yang's experiments[13].

**Experiment 1: globalCM VS BakerCM.** In this experiment, we provide empirical evidence to prove that the globalCM based text classifier provides better performance than that based on BakerCM. Figure 1 shows the experimental results.

From Figure 1 we can find that globalCM provides better performance than BakerCM in most different features size cases. With 100 features, globalCM provides 10.6% higher than BakerCM. Only when the number of features is less than 7, BakerCM can provide the similar performance to globalCM.

**Experiment 2: globalCM-based classifier VS Feature-Selection-based classifiers.** In this experiment, we use three different size of training corpus: 10%, 50%, 100% of the total training corpus(Here we name them as T10, T50 and T100). And we select two feature selection methods: document frequency and $\chi^2$ statistic for text categorization.

Figure 2 shows the effect of varying the amount of features with 3 different amounts of training dataset, where globalCM denotes our clustering model, fs_x2 denotes $\chi^2$ statistic feature selection method and fs_df denotes document frequency feature selection method. For 3 different quantities of documents for training, we keep the number of features constant, and vary the number of documents in the horizontal axis.

Naturally, the more documents for training are used, the better the performance is. The best result of globalCM with T100 training corpus is 75.57%, 1.73%

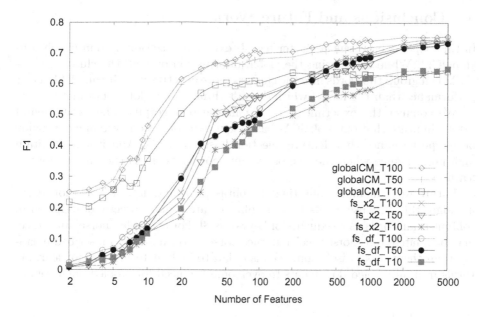

**Fig. 2.** globalCM-based classifier vs Feature-Selection-based classifiers

higher than the best result with T50 and 11.42% higher than the best result with
T10. The best result of fs_x2 with T100 training corpus is 74.37%, higher than
the result of the other two training corpus.

Then, we study the results of the T100 training corpus. Notice that with only
100 features globalCM achieves 70.01%, only 5.6% lower than with 5000 features.
In comparison, fs_x2 provides only 56.15% and fs_df provides only 52.12%. When
with 1000 features, fs_x2 can yields the similar result as globalCM with 100
features. Even with only 50 features, globalCM provides 67.10%. The best of
globalCM is 1.20% higher than the best of fs_x2 and 1.37% higher than fs_df. The
performance indicates that globalCM is providing more accuracy. And it can
maintain near 70% with only 100 or less features while feature selection based
classifiers have fallen into the 50s.

When we study the results of the other two training corpus, we can find that
globalCM can maintain good performance with small training corpus. With T10
training corpus and 50 features, globalCM achieves 60.33%. It is near 20% higher
than fs_x2 and fs_df. To our surprise, it is similar to the results of the feature
selection based classifier using T100 training corpus and 200 features.

Then we study the reasons why our cluster-based text classifier performs
better than the feature selection based classifier. In feature selection method, the
system discards some words that are infrequent. But in our clustering algorithm
merges them into the clusters instead of discards them. So it can preserves
information during merging.

# 6   Conclusions and Future Work

In this paper, we present a cluster-based text categorization system which uses globalCM. While considering the global information over all the clusters, globalCM can group the words into the clusters more effectively. So it can yield better performance than the model which doesn't think about global information.

We conduct the experiments on a Chinese text corpus. The experimental results indicate that our globalCM-based text categorization system can provide better performance than feature selection based systems. And it can maintain high performance when the size of training corpus is small and the number of features is small.

Future work includes collecting the phrases as candidate features for learning algorithm because words forming phrases are a more precise description of content than words as a sequence of keywords[8]. For example, 'horse' and 'race' may be related, but 'horse race' and 'race horse' carry more circumscribed meaning than the words in isolation. We also plan to look at techniques for learning words from unlabeled documents to overcome the need for labeled documents.

## Acknowledgements

This research was supported in part by the National Natural Science Foundation of China & Microsoft Asia Research (No. 60203019) and the Key Project of Chinese Ministry of Education (No. 104065).

## References

1. Baker,L. D. and McCallum,A. K. Distributional clustering of words for text classification. In Croft,W. B. , Moffat,A. , van Rijsbergen,C. J. , Wilkinson,R. , and Zobel,J. , editors, Proceedings of SIGIR-98, 21st ACM International Conference on Research and Development in Information Retrieval, Melbourne, AU (1998) 96-103
2. Bekkerman,R. , El-Yaniv,R. , Tishby,N. , and Winter,Y. On feature distributional clustering for text categorization. In Croft,W. B. , Harper,D. J. , Kraft,D. H. , and Zobel,J. , editors, Proceedings of SIGIR-01, 24th ACM International Conference on Research and Development in Information Retrieval, New Orleans, US (2001) 146-153
3. Bekkerman,R. , El-Yaniv,R. , Tishby,N. , and Winter,Y. Distributional word clusters vs. words for text categorization. Journal of Machine Learning Research, (2003) 1183-1208
4. Board,C. L. C. E. China Library Categorization(The 4th ed.). Beijing Library Press, Beijing(1999)
5. Joachims,T. Text categorization with support vector machines: learning with many relevant features. In Nedellec,C. and Rouveirol,C. , editors, Proceedings of ECML-98, 10th European Conference on Machine Learning, Chemnitz, DE, Springer Verlag, Heidelberg, DE. Published in the "Lecture Notes in Computer Science" series, Vol. 1398 (1998)137-142

6. Ko,Y. and Seo,J. Automatic text categorization by unsupervised learning. In Proceedings of COLING-00, the 18th International Conference on Computational Linguistics, Saarbrucken, DE (2000)

7. Lee,L. Similarity-Based Approaches to Natural Language Processing. PhD thesis, Harvard University, Cambridge, MA (1997)

8. Lee,S. and Shishibori,M. Passage segmentation based on topic matter. Computer Processing of Oriental Languages, 15(3),(2002) 305-340

9. McCallum,A. and Nigam,K. A comparison of event models for naive bayes text classification. In In AAAI-98 Workshop on Learning for Text Categorization (1998)

10. Pereira,F. C. N. , Tishby,N. , and Lee,L. Distributional clustering of english words. In Meeting of the Association for Computational Linguistics, (1993)183-190

11. Sebastiani,F. Machine learning in automated text categorization. ACM Computing Surveys, 34(1), 2002(1-47)

12. Yang,Y. and Liu,X. A re-examination of text categorization methods. In Hearst,M. A. , Gey,F. , and Tong,R. , editors, Proceedings of SIGIR-99, 22nd ACM International Conference on Research and Development in Information Retrieval, Berkeley, US (1999)42-49

13. Yang,Y. and Pedersen,J. O. A comparative study on feature selection in text categorization. In Fisher,D. H. , editor, Proceedings of ICML-97, 14th International Conference on Machine Learning, pages 412-420, Nashville, US, Morgan Kaufmann Publishers, San Francisco, US(1997)

14. Yao,T. , Zhu,J. , Zhang,L. , and Yang,Y. Natural Language Processing - A research of making computers understand human languages. Tsinghua University Press, Beijing (2002)

# Text Classification Using Web Corpora and EM Algorithms

Chen-Ming Hung[1] and Lee-Feng Chien[1,2]

[1] Institute of Information Science, Academia Sinica, Taipei, Taiwan
{rglly, lfchien}@iis.sinica.edu.tw
[2] Department of Information Management, National Taiwan University,
Taipei, Taiwan
lfchien@iis.sinica.edu.tw

**Abstract.** The insufficiency and irrelevancy of training corpora is always the main task to overcome while doing text classification. This paper proposes a Web-based text classification approach to train a text classifier without the pre-request of labeled training data. Under the assumption that each class of concern is associated with several relevant concept classes, the approach first applies a greedy EM algorithm to find a proper number of concept clusters for each class, via clustering the documents retrieved by sending the class name itself to Web search engines. It then retrieves more training data through the keywords generated from the clusters and set the initial parameters of the text classifier. It further refines the initial classifier by an augmented EM algorithm. Experimental results have shown the great potential of the proposed approach in creating text classifiers without the pre-request of labeled training data.

## 1 Introduction

In practical use, the Web has become the largest source of training data [1, 4]; however, it is cost-ineffective to handlabel each extracted documents from the Web. Thus many algorithms using few labeled data and large unlabeled data to train a text classifier were presented [3]; and [2] proposed a method without requiring labeled training data. The purpose of the paper is extended from the previous work and focuses much on using EM techniques to extract more reliable training data.

In [2], the training data of a topic class of concern was the documents retrieved via sending the class name itself to Web search engines. But while using the Web as the training data source, the extended problem is how to check the relevancy of the extracted documents to the class of concern. If the training data contains too many noisy documents, no doubt the classifiers trained by the data might not be reliable enough. In view of this problem, our idea is to retrieve only a small set of documents as the initial data for each class, and generate proper keywords from it as the relevant concepts of the class to retrieve augmented but reliable training data. The approach we proposed in this paper is based on the

S. H. Myaeng et al. (Eds.): AIRS 2004, LNCS 3411, pp. 12–23, 2005.

concept of the Naive Based modeling and EM algorithms. It assumes that the training data of a class is distributed and mixed with some relevant concepts; thus, we could model the distribution of each mixture based on a small set of data retrieved from Web search engines.

There are three problems needed to be dealt with. First is the acquisition of the initial data; second is the keywords extracted from the initial data to describe relevant concepts of the class of concern; third is to refine the initial modeling of each concept (keywords). As for extracting initial documents from the Web, depending on the ranking ability of search engines, like Google, it can be retrieved from the top-ranked documents, since their relevance to the target class might be high. With the initial data, we apply a greedy EM algorithm to determine the proper number of relevant concepts of each class. Then, for each mixture, refine its corresponding distribution depending on more "augmented" training data extracted via sending several queries into the search engine. The queries are those with the highest mutual information in distributions, i.e. keywords. In this step, an augmented EM algorithm is applied to iteratively update parameters in each distribution with the increasing of likelihood. Experimental results have shown the great potential of the proposed approach in creating text classifiers without the pre-request of labeled training data.

The remainder of the paper is organized as follows. Section 2 briefly describes the background assumption, i.e. Naive Bayes, and the modeling based on Naive Bayes. Section 3 describes the main idea of using the Web-based approach to acquiring training data from the Web, and its extended problems will need to be resolved, including document clustering, keyword generation, extracted data verification, and etc. Section 4 presents the augmented EM algorithm for text classification. Section 5 shows the experiments and their result. The summary and our future work are described in Section 6.

## 2  Naive Bayes Assumption

Before introducing our proposed approach, here introduce a well known way of text representation, i.e. Naive Bayes assumption. Naive Bayes assumption is a particular probabilistic generative model for text. First, introduce some notation about text representation. A document, $d$, is considered to be an ordered list of words, $\{w_1, w_2, ..., w_{|d|}\}$, where $w_j$ means the $j^{th}$ word in document $d$ and $|d|$ means the length of $d$. Second, every document is generated by a mixture of components $\{C_k\}$, for k=1 to K. Thus, we can characterize the likelihood of document $d$ with a sum of total probabilities conditioned on all mixture components:

$$p(d|\theta) = \sum_{k=1}^{K} p(C_k|\theta) P(d|C_k, \theta) , \tag{1}$$

where $\theta$ is a set of parameters including the probabilities of all words in class $C_k$ and the prior probability of $C_k$.

Furthermore, given a certain class $C_k$, we can express the probability of a document as:

$$p(d|C_k, \theta) = p(< w_1, w_2, ..., w_{|d|} > |C_k, \theta) = \prod_{j=1}^{|d|} p(w_j|C_k, \theta, w_z, z < j) , \quad (2)$$

where

$$p(w_{d_{i,j}}|C_k, \theta, w_{d_{i,z}}, z < j) = p(w_{d_{i,j}}|C_k, \theta) . \quad (3)$$

Based on standard Naive Bayes assumption, the words of a document are generated independently of context, that is, independently of the other words in the same document given the class model. We further assume that the probability of a word is independent of its position within the document; thus (3) is derived.

Combine (2) and (3),

$$p(d|C_k, \theta) = \prod_{j=1}^{|d|} p(w_j|C_k, \theta) . \quad (4)$$

Thus, the parameters of an individual class are the collection of word probabilities, $\theta_{w_j|C_k} = p(w_j|C_k, \theta)$. The other parameters are the weight of mixture class, $p(C_k|\theta)$, that is, the prior probabilities of class, $C_k$. The set of parameters is $\theta = \{\theta_{w_t|C_k}, \theta_{C_k}\}$.

# 3    The Proposed Approach

In this section we will introduce the framework of the proposed approach to acquiring training data from the Web and creating text classifiers, and in the next section address the detailed descriptions of the proposed text classification algorithm that utilizes EM techniques. Since it is normally lack of sufficient and relevant training data in many practical text classification applications, unlike conventional supervised learning approaches, we pursue to create a text classifier that requires no labeled training data. We use the Web as the unlabeled corpus source and real search engines the agents to retrieve relevant documents as the training data for each subject class of concern. To assure the accuracy of the auto-retrieved training data, the proposed approach is composed of an EM-based document clustering and keyword generation method to acquire more training data without increasing many noses. In addition, an EM-based classification algorithm is presented to make the created classifiers more reliable.

## 3.1    The Approach for Corpus Acquisition

As shown in Figure.1, suppose now we want to train a classifier for K classes $C_1, C_2, ..., C_K$; then as our previous approach proposed in [2] we treat each class name of $C_k$ for k=1 to K as a query sent to search engines to retrieve a number of relevant documents from the Web. Considering the information capacity and convenience, only the short descriptions of the retrieved documents (the snippets of search result pages) are used, which include page titles, short descriptions and URLs. Since most of the search engines, like Google and Yahoo, provide ranking service for the retrieved documents, we could believe that the documents with the higher ranking scores have higher association with the corresponding class, $C_k$, and these highly-ranked documents are then taken the *initial training data* of $C_k$, i.e., $D_k^l$. It is not hard to imagine that the training data might contain some texts describing the relevant concepts of $C_k$. To acquire more relevant training data, the keywords, $T_{k,m}$, describing the relevant concepts of $C_k$ will be generated, from $D_k^l$ where m=1 to $|C_k|$ and $|C_k|$ is the total number of generated keywords in $C_k$. As for $T_{k,m}$'s generation, the documents in $D_k^l$ will be clustered with their similarity into a set of clusters and keywords with the highest *weighted log likelihood ratio* in each cluster will be extracted as the relevant concept of $C_k$, i.e. $\{T_{k,m}\}_{m=1}^{|C_k|}$. More training data will be acquired for $C_k$ via sending the keywords to search engines. Each $T_{k,m}$ is then taken as a *sub-concept class* of $C_k$ and the retrieved documents as the corresponding training corpus in the process of training the text classifier, which will be further described in next section. However, here comes with the difficulties such as how to

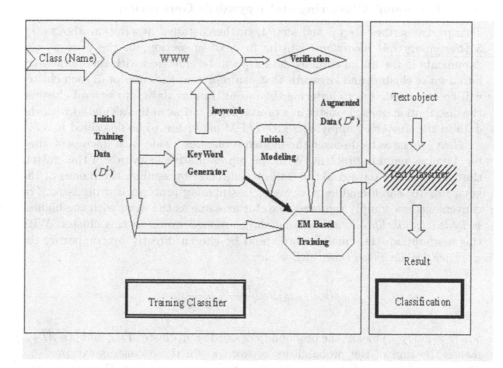

**Fig. 1.** Text classifier training using Web corpora

determine a proper number of clusters when clustering $D_k^l$, and how to extract keywords to represent each auto-created cluster. The complete process of the corpus acquisition approach is thus designed as follows:

Step1. Given a set of $C_1, C_2, ..., C_K$ that require training data from the Web.

Step2. Send $C_1, C_2, ..., C_K$ as the queries into search engines to retrieve the "initial" training data of each class, $D_k^l$.

Step3. Apply an unsupervised clustering method to determine the number of component clusters, $|C_k|$, in $D_k^l$, and separate $D_k^l$ as $D_{k,m}^l$, where m=1 to $|C_k|$.

Step4. Choose those words in $D_{k,m}^l$ with the highest *weighted log likelihood ratio* as $T_{k,m}$.

Step5. Put $T_{k,m}$ as the new queries into search engines and retrieve new "augmented" training data, $D_{k,m}^u$.

Step6. Perform a verification process to filter out the unreliable augmented training data.

Step7. Refine parameter $\theta_{T_{k,m}}$ with $D_k^l$ and $D_{k,m}^u$, for m = 1 to $|C_k|$ and k = 1 to K.

Below are the detailed processes from Step 3 to Step 7.

## 3.2    Document Clustering and Keywords Generation

This part describes step 3 and step 4; further detailed description about step 3 (*the greedy EM algorithm*) will be in next subsection. In the process, the documents in the initial training data $D_k^l$ will be clustered with their similarity into a set of clusters and keywords that can represent the concept of each cluster will be extracted. After clustering the initial training data into several clusters, the distribution of each cluster in a probabilistic form can be calculated with the data in the cluster by applying a greedy EM algorithm to be described in 3.3.

Next, we have to discover the hidden semantics inside each document cluster, i.e class naming problem. With the representative keywords of the cluster, the "augmented" data could be collected directly by sending the names of the keywords to search engines. However, class naming problem is a big issue. For convenience, we simply represent the cluster name as the word with the highest *weighted log likelihood ratio* among the contained words in this cluster. With this assumption, the "main" word could be chosen directly by comparing the *weighted log likelihood ratio*[1] defined as:

$$p(w_j|T_{k,m}) \log(\frac{p(w_j|T_{k,m})}{p(w_j|\overline{T}_{k,m})}) , \tag{5}$$

where $p(w_j|T_{k,m})$ means the probability of word $w_j$ in cluster $T_{k,m}$ and $p(w_j|\overline{T}_{k,m})$ means the sum of the probabilities of word $w_j$ in those clusters except $T_{k,m}$. Those chosen keywords by 5 are used to extract new "augmented" training data, $D_{k,m}^u$ from the Web, where $D^u$ means the "augmented" data.

---

[1] The sum of this quantity over all words is the Kullback-Leibler divergence between the distribution of words in $T_{k,m}$ and the distribution of words in $\overline{T}_{k,m}$, (Cover and Thomas, 1991).

## 3.3    The Greedy EM Algorithm

Because we have no idea about how many concepts strongly associated with each class, thus for each class, $C_k$, it is straightforward to apply the Greedy EM algorithm to clustering the "initial" data from the Web into a certain number of clusters automatically. The algorithm is based on the assumptions of the theoretical evidence developed in [5,6], and the basic idea is to suppose that all the training data belong to one component at the initial stage, then successively adding one more component (concept) and redistributing the training data step by step until the maximal likelihood is approached.

Let $D$ be a document set and $\theta_m$ be the parameter set for $m^{th}$ component in any class, which are mixed with M components. In particular, assume that a new component with parameter set $\theta_{M+1}$ is added to this M-component class; then the mixture of this M-component class will be:

$$f_{M+1}(D) = (1 - \alpha)f_M(D) + \alpha\phi(D, \theta_{M+1}) \qquad (6)$$

with $\alpha$ in (0,1), where $f_M(D)$ means the probability of document set $D$ in this M-component class and $\phi(D, \theta_{M+1})$ means the probability of $D$ in the newly added component. Thereafter, as long as the weight $\alpha$ and the parameter set $\theta_{M+1}$ of $M + 1$ component are optimally chosen, then for each M-component class, the new log-likelihood derived as

$$L_{M+1} = \sum_{n=1}^{N} \log f_{M+1}(d_n) = \sum_{n=1}^{N} \log[(1 - \alpha)f_M(d_n) + \alpha\phi(d_n, \theta_{M+1})] \qquad (7)$$

will be maximized, where $\{d_n\}_{n=1}^{N} \in D$. In addition, a notable property of the above maximization problem is that the parameters of old M-component class, $\theta_M$, remain fixed during maximization of $L_{M+1}$.

According to Naive Bayes assumption, in order to utilize Greedy EM algorithm as our use, $\alpha$, $\theta_{M+1}$, $f_M(d_n)$ and $\phi(d_n, \theta_{M+1})$ should be replaced as:

$$\alpha \Rightarrow p(T_{k,|C_k|+1}|C_k) , \qquad (8)$$

$$\theta_{M+1} \Rightarrow p(w_j|C_k, T_{k,|C_k|+1})\forall w_j \in \{w_j\}_{j=1}^{|D|} , \qquad (9)$$

$$f_M(d_n) \Rightarrow \sum_{m=1}^{|C_k|} p(T_{k,m}|C_k)[\prod_{t=1}^{|d_n|} p(w_t|T_{k,m}, C_k)] , \qquad (10)$$

$$\phi(d_n, \theta_{M+1}) \Rightarrow p(d_n|T_{k,|C_k|+1}, C_k) = \prod_{t=1}^{|d_n|} p(w_t|T_{k,|C_k|+1}, C_k) , \qquad (11)$$

for each class $C_k$, where $T_{k,|C_k|+1}$ means the newly added component (keyword) in class, $C_k$, and $|V|$ means the total number of words shown in $D^l$, the *initial* training data.

For each class, $C_k$, utilizing the property that parameters in $\{\theta_m\}_{m=1}^{|C_k|}$ are fixed, we could update the parameters $\theta_{|C_k|+1}$ and $p(T_{k,|C_k|+1}|C_k)$ directly by partial EM:

$$\theta_{|C_k|+1} = \{p(w_j|T_{k,|C_k|+1}, C_k)\}_{j=1}^{|V|}$$

$$= \left\{ \frac{1 + \sum_{n=1}^{|D|} N(w_j, d_n) p(T_{k,|C_k|+1}|d_n, C_k)}{|V| + \sum_{s=1}^{|V|} \sum_{n=1}^{|D|} N(w_s, d_n) p(T_{k,|C_k|+1}|d_n, C_k)} \right\}_{j=1}^{|V|} \quad (12)$$

$$p(T_{k,|C_k|+1}|C_k) = \frac{1 + \sum_{n=1}^{|D|} p(T_{k,|C_k|+1}|d_n)}{(|C_k| + 1) + |D|} , \quad (13)$$

where the numerator and denominator in (13) and (13) are augmented with "pseudo-counts" (one for each word) to assure non-zero frequency of each word in the newly added component, $T_{|C_k|+1}$; where $|V|$ and $|D|$ respectively means the number of vocabularies and the number of documents shown in the *initial* training data.

Since only the parameters of the new components are updated, partial EM steps constitute a simple and fast method for locally searching for the maxima of $L_{M+1}$, without needing to resort to other computationally demanding nonlinear optimization methods.

## 3.4     The Verification Process

This subsection is for step 6 in subsection 3.1. The purpose of the verification process is to further filter out the retrieved documents that are not reliable enough to be included in the set of the augmented training data. For each keyword, $T_{k,m}$, in $C_k$, the distribution $\theta_{T_{k,m}}$ has been initialized with those retrieved "initial" training documents of $C_k$ clustered into $T_{k,m}$; thus the probability of any document $d_i$ assigned to class $C_k$, is defined as:

$$p(C_k|d_i) = \sum_{T_{k,m}} p(C_k|T_{k,m}) p(T_{k,m}|d_i)$$

$$= \sum_{T_{k,m}} p(C_k|T_{k,m}) \frac{p(T_{k,m}) p(d_i|T_{k,m})}{\sum_{T_{k,m}} p(T_{k,m}) p(d_i|T_{k,m})} , \quad (14)$$

where $p(C_k|d_i)$ and $p(T_{k,m}|d_i)$ represent the responsibility of $C_k$ and $T_{k,m}$ to document $d_i$ respectively.

Thus, for any document $d_i$ retrieved via the new keywords $\{T_{k,m}\}_{m=1}^{|C_k|}$ in class $C_k$, if the responsibility of $C_k$ to $d_i$ is lower than the responsibility of $C_{j\neq k}$ to $d_i$, then $d_i$ is dropped out of the collected "augmented" data. This is reasonable, because if $d_i$ is extracted from the keywords belonging to $C_k$, but $C_k$ doesn't get the highest responsibility to $d_i$; then in some sense $d_i$ should be viewed as noises and should be dropped out.

# 4    EM Based Classifier

In this section, we will introduce the text classifier based on EM techniques. Suppose there are $K$ classes and each class has its corresponding model, i.e. word probabilities and mixture weight, as described in Section 2. Given a new document $d_s$ to be classified, its tendency to be assigned into class $C_k$ will be judged from its responsibility in class $C_k$ and the responsibility is defined as:

$$p(C_k|d_s) = \sum_{T_{k,m}} p(C_k|T_{k,m})p(T_{k,m}|d_s) = \sum_{T_{k,m}} p(T_{k,m}|d_s) \quad \forall T_{k,m} \in C_k \ . \ (15)$$

Since no matter with $d_s$ or not, $T_{k,m}$ must belong to $C_k$; thus in (15), $p(C_k|T_{k,m}) \equiv p(C_k|T_{k,m}, d_s) \equiv 1$ and $p(C_k|T_{k,m}, d_s) \equiv 0$ if $j \neq k$ are provided.

## 4.1    Training

It is well known that the result of EM-based algorithm is sensitively depending on the initial models (parameters). A bad initial given modeling will trap the EM algorithm into local extreme value. In section 3, we have provided the initial models for each $T_{k,m}$ conditioned on $C_k$, i.e. $\{p(w_j|T_{k,m}, C_k)\}_{j=1}^{|V|}$.

From (15), it is clear that only $p(T_{k,m}|d_s) \ \forall T_{k,m}$ is needed to classify a new document. Using simple Bayes theorem,

$$p(T_{k,m}|d_s) = \frac{p(T_{k,m})p(d_s|T_{k,m})}{p(d_s)} = \frac{p(T_{k,m})p(d_s|T_{k,m})}{\sum_{T_{k,m}} p(d_s|T_{k,m})p(T_{k,m})} \ , \tag{16}$$

and Naive Bayes assumption, (4), training the classifier is equivalent to training $p(T_{k,m})$ and $p(w_{d_s}|T_{k,m}) \ \forall w_{d_s} \in d_s$. However, compared with (8) and (9), it is shown that if we want to utilize (8) and (9) as the initial modeling of $p(T_{k,m})$ and $p(w_{d_s}|T_{k,m})$, then $C_k$ in (8) and (9) has to be dropped out. Thus, if set $p_0(w_t|T_{k,m})$ and $p_0(T_{k,m})$ as the initial modeling of $p(w_t|T_{k,m})$ and $p(T_{k,m})$ respectively. Using the property that $p(C_j|T_{k,m}) = 1$ if $j = k$, o.w $p(C_j|T_{k,m}) = 0$, $p_0(w_t|T_{k,m})$ and $p_0(T_{k,m})$ could be derivated easily as:

$$p_0(w_t|T_{k,m}) = p(w_t|C_k, T_{k,m}) \ , \tag{17}$$

$$p_0(T_{k,m}) = p(T_{k,m}|C_k)p(C_k) \ . \tag{18}$$

With the initial parameters, the responsibility of $T_{k,m}$ to document $d_s$ could be calculated as

$$p(T_{k,m}|d_s) = \frac{p(T_{k,m})p(d_s|T_{k,m})}{p(d_s)} = \frac{p(T_{k,m})\prod_{j=1}^{|d_s|} p(w_{d_s,j}|T_{k,m})}{\sum_{r=1}^{|C|}\sum_{m=1}^{M_r} p(T_{r,m})\prod_{j=1}^{|d_s|} p(w_{d_s,j}|T_{r,m})} \tag{19}$$

then update $p(T_{k,m})$ and $p(w_t|T_{k,m})$ as

$$p(w_t|T_{k,m}) = \left\{ \frac{1+\sum_{n=1}^{|D|} N(w_t,d_s)p(T_{k,m}|d_s)}{|V|+\sum_{j=1}^{|V|}\sum_{n=1}^{|D|} N(w_j,d_s)p(T_{k,m}|d_s)} \right\}_{t=1}^{|v|} \ , \tag{20}$$

$$p(T_{k,m}) = \frac{1 + \sum_{n=1}^{|D|} p(T_{k,m}|d_s)}{|T| + |D|} \, , \tag{21}$$

where $|T|$ and $|V|$ represents the number of total components (concepts) and the number of total vocabularies respectively, shown in training data which is combined with the initial training set and the "augmented" training set.

There are three points to be noted while training:

a) Those initial training data in $D_k^l \; \forall \; k = 1 \; to \; K$ will always be given responsibility "zero" in all $T_{j,m} \forall j \neq k$.

b) All the newly retrieved "augmented" documents $D^u$ have to pass the verification process.

c) $d_s$ in (19), (20) and (21) belongs to the set of $\bigcup\{D^u, D^l\}$.

## 4.2    Classification

As describing in the start of this section, classifying a new document is by judging its responsibility calculated as (15). The further extension of (15) is:

$$p(C_k|d_s) = \sum_{T_{k,m}} \frac{p(T_{k,m}) \prod_{j=1}^{|d_n|} p(w_{d_n,j}|T_{k,m})}{\sum_{r=1}^{|C|} \sum_{m=1}^{M_r} p(T_{r,m}) \prod_{j=1}^{|d_n|} p(w_{d_n,j}|T_{r,m})} \, . \tag{22}$$

According to (22), the testing document will be assigned to the class with the highest responsibility.

# 5    Experiments

After describing the theoretical part, in this section, we will show the experiment results of embedding keywords generation and Web-based corpora into a text classifier system.

## 5.1    Data Description

Our experiments took the "Computer Science" hierarchy in Yahoo! as the benchmark. There were totally 36 classes in second level in the "Computer Science" hierarchy, 177 classes in the third level and 278 classes in fourth level, all rooted at the class "Computer Science". We used the second-level classes, e.g., "Artificial Intelligence" and "Linguistics" as the target classes and tried to classify text objects into them.

We divided the test text objects into three groups: full articles, which were the Web pages linked from Yahoo!'s Website list under the Computer Science hierarchy, short documents, which were the site description offered by Yahoo!, and text segments, which were the directory names.

## 5.2    Keyword Generation

In section 3.3, the Greedy EM algorithm is treated as the unsupervised learning method to cluster "initial" training data to generate keywords for each class.

**Table 1.** Bi-gram Keywords in 36 second-level classes in Yahoo!'s "Computer Science" Tree

| Class | KeyWords |
|---|---|
| Algorithms | *algorithms {course, project}, source code, {sorting, combinational, discrete, evolutionary, memetic, approximation, collected, graph, parallel} algorithms* |
| Architecture | *{computer, landscape, information, sustainable} architecture, internet resources, world wide, computer science, data structure* |
| Artificial intelligence | *modern approach, intelligence laboratory* |
| Compression | *Mark Nelson, compression {library, software, algorithms, ratio, methods, technologies}, {GIF, video, file, loseless, audio, Huffman, image, data} compression* |
| Computational learning theory | *learning theory* |
| Computational sciences | *Mason university, George Mason* |
| Computer vision | *vision group, computer society* |
| Databases | *{online, image} databases* |
| Distributed computing | *computing project, {java, performance} distributed* |
| DNA based computing | *DNA based, DNA computer* |
| Electronic computer aided design | *aided design, authoring tools, circuit boards* |
| End user programming | *Margaret Burnett, Eisenstadt/Burnett 1000, end-user programming* |
| Finite model theory | *model theory, open problems, abstract model* |
| Formal methods | *critical systems, international school* |
| Graphics | *clip art, clipart images, rights reserved, news front, {country, free, computer} graphics, data structures* |
| Handwriting recognition | *recognition software, press release* |
| Human computer interaction | *ACM sigchi, oclc human-computer, human-computer interaction* |
| Knowledge sciences | *social sciences, pdf/adobe acrobat, computer science, file format* |
| Library and information science | *information science, American library* |
| Linguistics | *applied linguistics* |
| Logic programming | *constraint logic, programming alp* |
| Mobile computing | *computing program, press releases* |
| Modeling | *modeling {group, system, tools, software}, {solid, data, mathematical, protein} modeling* |
| Networks | *{wireless, mobile} networks* |
| Neural networks | *artificial neural, fuzzy logic* |
| Objective oriented programming | *oriented programming, Apple computer, Mac OS* |
| Operating systems | *{alternative, computer} operating* |
| Quantum computing | *gilles brassard, field theory* |
| Real time computing | *real-time computing, computing lab* |
| Robotics | *robotics research, {applied, cognitive} robotics* |
| Security and encryption | *FAQ revision, security HowTo, {Linux, networks, computer, IP, internet, password} security, encryption {technologies, FAQ}* |
| Software engineering | *automated software, engineering institute* |
| Supercomputing and parallel computing | *parallel computing, advanced computing* |
| Symbolic computation | *academic press, common lisp* |
| User interface | *interface {engineering, GUI, software, system, language, programming, library, guidlines}, language xul, interaction design, palm OS* |
| Virtual reality | *exposure therapy, 3d graphics* |

Here comes a problem, too many retrieved documents will cause noises, but too few won't contain enough information about this class. Finally, for each class, we queried 200 short descriptions by sending its own name to Google (http://www.google.com). Table.1 shows the generated bi-gram keywords. The number of keywords in each class was determined automatically by the Greedy EM algorithm.

We observed the bi-gram keywords catch more hidden semantics than the one-gram keywords do. It is encouraging that the algorithm extracted he main idea for some classes. Taking "Modeling" for example, "Modeling" is a rough acronym

**Table 2.** Test result in Yahoo! CS category

| Top-1 | Top-2 | Top-3 | Top-4 | Top-5 |
|-------|-------|-------|-------|-------|
| 40.1% | 46.54% | 51.44% | 55% | 58.1% |

to describe many fields. In the unigram keywords, only "structural" and "agencies" were extracted to represent "Modeling"; however, in the bi-gram keywords, the specific area in "Modeling" were extracted, like "protein modeling", "solid modeling", "mathematical modeling" etc. Again, "handheld" and "challenges" were extracted to represent "Mobile computing" in the unigram keywords; but in the bi-gram keywords "computing program" and "press releases" were extracted. There was a little difference between this two representation ways on "Mobile computing". We can not decide which way is better, thus the following experiments will be tested with the unigram keywords and bi-gram-keywords.

Many of the extracted keywords had represented a certain concept of the class of concern. However, some noises were still extracted like "Eisenstadt/Burnett 1000" in "end user programming". That was arisen from a retrieved document tackling about a $1000 bet during 1985-1995 to predict the scarcity of end user programming. It's interesting but shouldn't be taken to describe the concept of "end user programming". The same situation is shown while "FAQ revision" is extracted in "security and encryption" in bi-gram keywords.

### 5.3    Classification Accuracy

In order to check the classification effect of the proposed approach, the fourth level (450 classes) were used to be the test text objects. For each object it got ten retrieved documents from Google, then transforming the ten documents as a bag of words to represent the class. Thus, if an object got the highest responsibility in its own upper second level, then it was taken as "correctly classified". Table.2 shows the achieved top-1, top-2, top-3, top-4 and top-5 accuracy.

## 6    Conclusions and Future Work

The achieved result is encouraging which demonstrates its great potential that a text classifier can be created via using Web corpora and no labeled training data is required. This, however, will increase the availability of text classification in many real applications., in which to prepare sufficient labeled training data are always be a big problem.

However, there are some challenges which require further investigation:

a) *Temporal factors.* The documents retrieved via Web search engines that might depend on temporal factors. Usually a search term can represent different concepts, like "SARS", but their ranking on search engines are dynamic. Not only means "Severe Acute Respiratory Syndrome", "SARS" is also the abbreviation of "South African Revenue Service".

b) *Word string matching.* Searching with word string matching instead of semantic analysis like Google might cause the incompleteness of concept description. Taking "supercomputing and parallel computing" for example, there is a category "Message Passing Interface (MPI)" in Yahoo!'s "supercomputing and parallel computing" category; however, if sending "supercomputing and parallel computing" to Google, no information about MPI is gained.

c) *Probability framework.* Usually, using probability framework to do text classification performs worse than vector-based framework.

Considering the incompleteness of using a bag of words to represent a document, using N-gram model to represent a document is also what we are on going. In past research, Ngram modeling does provide higher performance than only a bag of words. In addition, combining vector-based modeling and Support Vector Machine will be the next work.

# References

1. Craven, M., DiPasquo, D., Freitag, D., McCallum, A., Mitchell, T., Nigam, K., and Slattery, S. (1998). Learning to extract symbolic knowledge from the World Wide Web. In *Proceedings of the Fifteenth National Conference on Artificaial Intelligence (AAAI-98)*, pp. 509-516.
2. Huang, C. C., Chuang, S. L. and Chien, L.. F. (2004). LiveClassifier: Creating Hierarchical Text Classifiers through Web Corpora, *WWW (2004)*.
3. Nigam, K., McCallum, A., Thrun, S. and Mitchell, T. (2000). Text Classification from Labeled and Unlabeled Documents using EM. In *Machine Learning*, 39 (2/3), pp. 103-134.
4. Shavlik, J. and Eliassi-Rad, T. (1998). Intelligent agents for Web-based tasks: An advice-taking approach. In *AAAI-98 Workshop on Learning for Text Categorization.* Tech. rep. WS-98-05, AAAI Press.
5. Verbeek, J. J., Vlassis, N. and Krose, B. J. A. (2003). Efficient Greedy Learning of Gaussian Mixture Models. In *Neural Computation*, 15 (2), pp.469-485.
6. Vlassis, N. and Likas, A. (2002). A greedy algorithm for Gaussian Mixture Learning. In *Neural Processing Letters (15)*, pp. 77-87.

# Applying CLIR Techniques to Event Tracking

Nianli Ma[1], Yiming Yang[1,2], and Monica Rogati[2]

[1] Language Technologies Institute, Carnegie Mellon University,
[2] Computer Science Department, Carnegie Mellon University,
5000 Forbes Ave., Pittsburgh, PA 15213, U.S.A.
{manianli, yiming, mrogati}@cs.cmu.edu

**Abstract.** Cross-lingual event tracking from a very large number of information sources (thousands of Web sites, for example) is an open challenge. In this paper we investigate effective and scalable solutions for this problem, focusing on the use of cross-lingual information retrieval techniques to translate a small subset of the training documents, as an alternative to the conventional approach of translating all the multilingual test documents. In addition, we present a new variant of weighted pseudo-relevance feedback for adaptive event tracking. This new method simplifies the assumption and the computation in the best-known approach of this kind, yielding a better result than the latter on benchmark datasets in our evaluations.

## 1 Introduction

Information retrieval techniques are quite effective for collecting information given well-defined queries. However, the problem becomes tougher when we need to follow the gradual evolution of events through time. This is the goal of event tracking[1][13]; given an event-driven topic, described implicitly as one or two news stories, we need to recognize which subsequent news stories describe the same evolving and changing event. Notice that, although the task resembles adaptive filtering[8], it is more difficult since the availability of human relevance feedback cannot be assumed. Cross-lingual event tracking (CLET) needs to handle tracking tasks over multiple languages, and is significantly more difficult than monolingual task. The central challenge is finding the most effective way of bridging the language gap between new stories and events/topics.

One popular and effective approach is translating multilingual test documents to a preferred language, and treating the problem as a monolingual task[3]. Whether this is a feasible solution for event tracking depends on the assumptions made regarding the volume of data to be processed.

• If only a small number of pre-selected information sources need to be monitored for event tracking, and if those sources collectively produce a few thousands of multilingual documents daily, then translation of all the documents may be affordable.

• If the information we need is scattered over many sources on the Internet, possibly in many languages, sometimes with restricted access, and typically buried in large

S. H. Myaeng et al. (Eds.): AIRS 2004, LNCS 3411, pp. 24–35, 2005.

volumes of irrelevant documents, then the translate-everything approach is unlikely to scale or be cost effective, given the higher computational cost of machine translation over CLIR methods. These methods require only limited translation of selected training documents, and each document is treated as a bag of words in the translation.

Using the online news on the Web as an example, there are at least 4,500 news sites online (according to Google News indexes), producing hundreds of thousands multi-lingual documents per day, as a rough estimate.    Translating this volume of multilingual documents daily and on the fly is a very demanding proposition for current machine translation, to our knowledge.    Even if this were computationally achievable, whether it is worth the effort is still questionable, given that most of the translated documents are not relevant to the user's interest in event tracking.

Based on the above concerns, we propose a new, more cost-effective approach that preserves the event-specific information by only translating a few sampled training stories per topic.   In addition to the immediate advantage of not having to translate potentially infinite data streams, this approach has two additional advantages. In a realistic Internet scenario, by allowing the translated training stories to be used as queries when downloading data from news sources we can: 1). Limit and focus the input to our system, and 2). Maximize the usefulness of the downloaded stories when a limit is imposed by the news sources. Many query expansion techniques have been successfully applied to improve cross-lingual information retrieval[14][9]. However, the applicability of these methods to CLET has not been studied. In this paper, we are exploring the suitability of these techniques to CLET, and their effects on tracking performance.    In particular, we are focusing on the English-Chinese cross-lingual tracking task, for which benchmark evaluation data and results are available[7].

A unique challenge specific to event tracking is adapting to the evolving event profile in the absence of human relevance feedback.  Using unsupervised learning or pseudo-relevance feedback in an effective fashion is a crucial, as attested to by the use of variable weight adaptation in LIMSI's state of the art system. In the 2001 TDT workshop LIMSI had the best performance on the provided benchmark dataset (denoted by LWAdapt and described in section 2.1 and [4]).   While results obtained using LWAdapt are good, we believe the method has several drawbacks outlined in Section 2.1. The adaptation mechanism also makes it difficult to pinpoint the exact reason or technique responsible for the good results.  Our proposed weighted adaptation technique (denoted by NWAdapt and described in section 2.2) is greatly simplified and yields better results than LWAdapt. In Section 3, we demonstrate the performance advantage NWAdapt has over fixed weight adaptation and LWAdapt.

## 2   Cross-Lingual Event Tracking: Approach

The event-tracking task, as defined in the TDT forum[1] is trying to model a real world setting. The user might be interested in recent events and is willing to provide one or two stories about that event, but not constant online feedback. Example events (named *topics* in the TDT literature) include "Car bomb in Jerusalem" and "Leonid Meteor Shower". After providing the initial stories defining the event, the user is interested in subsequent reports about that event. In other words, the system aims to automatically assign event labels to news stories when they arrive, based on a small number (such as 4) of previously identified past stories that define the event. Note that this task is different from the TREC filtering task. The latter assumes continuous relevance feedback through the entire process of selecting test documents.

Our tracking system is an improved version of those described in[13][11]. We approach the tracking problem as a supervised learning problem, starting with the standard Rocchio formula for text classification:

$$\vec{c}(D,\gamma) = \frac{1}{|R+1|} \sum_{y_i \in R(i)} \vec{y}_i + \gamma \frac{1}{|S_n|} \sum_{y_i \in S_n(i)} \vec{y}_i \qquad (1)$$

where $\vec{c}(D,\gamma)$ is the prototype or centroid of a topic, $D$ is a training set, $\gamma$ is the weight of the negative centroid, $R \in D$ consists of the on-topic documents (positive examples), and $S_n \in D - R$ consists of the $n$ negative instances that are closest to the positive centroid.

An incoming story $x$ can be scored by computing the cosine similarity between it and the centroid:

$$r(\vec{x}, \vec{c}(D,\gamma)) = \cos(\vec{x}, \vec{c}(D,\gamma)) . \qquad (2)$$

A binary decision is obtained by thresholding on $r$.

Unlike text classification where training data is more abundant, for event tracking we have to rely on extremely limited training examples. Naturally, the class prototype trained from the small initial training set is not very accurate. In particular, it cannot capture the different facets of an evolving event. Adaptive learning is useful here because it enables the system to flexibly adapt to the dynamic nature of evolving events by updating the centroid with on topic documents. However, there are two main issues that the adaptation mechanism needs to address:

1. Deciding whether a story is on topic and should be added to the centroid.
2. Choosing a method for adjusting the centroid once a story has been identified as being on topic.

We use pseudo-relevance feedback as a solution to (1): the story ($d$) is added to the centroid ($C$) as long as it has a score $S(d,C)$ that is higher than an adaptation threshold $th_a$. Adapting this threshold is an interesting problem; however, in this paper we are focusing on (2).

To address question (2), we define the new centroid to be:

$$\vec{c}' = \frac{1}{|R+1+\alpha|} \sum_{y_i \in R(i)} (\vec{y}_i + \alpha \cdot \vec{y}_{new}) + \gamma \frac{1}{|S_n|} \sum_{y_i \in S_n(i)} \vec{y}_i \qquad (3)$$

where $\vec{c}'$ is the new centroid after adaptation, $\vec{y}_{new}$ is the vector of the incoming story used to adapt the centroid; $\alpha$ is the weight given to the vector $\vec{y}_{new}$.

One approach is to assign a fixed value to the weight factor $\alpha$ (i.e. fixed weight adaptation). However, intuitively, different stories should have different weights; stories with a higher confidence score $S(d,C)$ should have higher weight. It is not clear, however, what these weights should be. By addressing this problem, LIMSI had the best tracking system on TDT2001 benchmark evaluation. We briefly describe their weighted adaptation method in the next section, followed by our approach.

## 2.1  LWAdapt: LIMSI's Weighted Adaptation

LIMSI developed a novel approach to compute the variable adaptation weight. The similarity between a story and a topic is the normalized log likelihood ratio between the topic model and a general English model[4]. The similarity score $S(d,C)$ is mapped to an adaptation weight $P_r(C,d)$ using a piece-wise linear transformation $P_r(C,d) \approx f(S(d,C))$. This mapping is trained on a retrospective collection of documents (with event labels) by using the ratio of on-topic documents to off-topic documents in that collection. This appears to be a sound approach, and yielded the good performance of the LIMSI's system in TDT 2001. However, it has several problems: 1. A large amount of retrospective data is needed to get a reliable probabilistic mapping function. 2. The stability of the method relies on the *consistency* between events in the retrospective collection and the new events in the collection that the mapping function is applied to.

Recall that news-story events are typically short-lasting, and a retrospective collection may not contain any of the new events in a later stream of news stories. How suitable a mapping function learned for the old events would be for a new set of events is questionable. In the following section, we present a simplified and better performing adaptation mechanism that does not suffer from these drawbacks.

## 2.2  NWAdapt: Normalized Weighted Adaptation

In this section, we outline a new weighted adaptation algorithm (NWAdapt), which is simpler and more effective than LIMSI's approach. The basic idea of our approach is to directly use the cosine similarity scores generated by the tracking system (see formula (2)) as the adaptation weights. The cosine score reflects the similarity between each new and the prototype (centroid) of a topic/event. To ensure that the weights are non-negative, we rescale the cosine scores linearly as follows:

$$P(C,d) = (S(d,C)+1)/2 . \tag{4}$$

This simply maps the [-1,1] cosine range into the usual [0,1] weight range, and makes the weights more intuitive since they now fall in the more familiar probability range. In the adaptation process, the score of each new document is compared to a pre-specified threshold (empirically chosen using a validation set); if and only if the score is higher than the threshold, the new document is used to update the topic prototype (formula (3)).

While this approach may appear to be simplistic, several reasons make it worth investigating: 1. It generalizes LIMSI's approach (i.e., using system-generated confidence scores in PRF adaptation) by examining another kind of confidence scores – the cosine similarity. 2. If it works well, it will provide a strong baseline for future investigations on weighted adaptation methods because cosine similarity is simple, easy to implement, and well-understood in IR research and applications.

In addition, this approach effectively avoids the disadvantages of LIMSI's weighted adaptation: 1. The system does not need any data to train the adaptation weights. 2. The adaptation weights are topic-specific, computed from the similarity scores of documents with respect to each particular topic, not averaged over topics.

As Section 3 shows, the simpler method has the advantage of being slightly more effective, in addition to avoiding LWAdapt's drawbacks.

## 2.3 Cross-Lingual Components

Our cross-lingual event tracking approach involves translating a few sample training documents instead of the large test data set. Therefore, the cross-lingual component is an integral part of our tracking system instead of a preprocessing step. The tracking process can be divided into several steps: 1. Topic Expansion (PRF, optional) 2. Sampling: Choosing training stories to translate. 3. Sample translations: a. using a dictionary (DICT). b. using the CL-PRF technique (below). 4. Segmentation (for Chinese): a. Phrase-based. b. Bigram-based. 5. Adaptation (described in section 2.1 and 2.2).

**Pseudo-relevance Feedback.** *Pseudo-relevance feedback (PRF)* is a mechanism for query (or, in our case, topic) expansion. Originally developed for monolingual retrieval, it uses the initial query to retrieve a few top ranking documents, assumes those documents to be relevant (i.e., "pseudo-relevant"), and then uses them to expand the original query. Let $\vec{q}$ be the original query vector, $\vec{q}'$ be the query after the expansion, $\vec{d}$ be a pseudo-relevant document, and $k$ be the total number of pseudo-relevant documents. The new query is defined as:

$$\vec{q}\,' = \vec{q} + \sum_{i=1}^{k} \vec{d}_i. \tag{5}$$

The adaptation of PRF to cross-lingual retrieval (CL-PRF) is to find the top-ranking documents for a query in the source language, substitute the corresponding documents in a parallel corpus in the target language, and use these documents to form the corresponding query in the target language[12]. Let $\vec{q}'$ be the corresponding query in the target language, $\vec{d}$ be a pseudo-relevant document substituted by target language corresponding document; the updated query/topic is defined to be as follow:

$$\vec{q}\,' = \sum_{i=1}^{k} \vec{d}_i. \tag{6}$$

In our experiments, the positive examples for each topic are the queries.

**Sampling Strategy.** Our sampling strategy is simply using temporal proximity to the 4 stories identified as positive examples as the sole decision factor on whether to translate a story or not. An average of 120 stories per topic are translated, as opposed to a potentially infinite number of stories to translate with the conventional approach. These training stories are the only ones used as training data.

This sampling strategy was chosen due to its mix of convenience, speed and potential effectiveness. The convenience factor is due to the TDT corpus packaging the data using temporal chunks ("files") of about 20 stories. . We then took the chunks containing the on-topic examples as the sample to translate; therefore, the size of the sample can range from 1 to 4 "files". The speed factor is also important: carefully analyzing each story to decide whether to translate it or not can approach the translation cost itself, thereby defeating the purpose of sampling. The *potential* effectiveness comes from the fact that these files can be richer in positive examples and borderline negative examples that mention the recent event of interest in passing. The *actual* effectiveness of the approach when compared to another sampling strategy remains to be proven, since exploring tradeoffs between different sampling strategies is left to future research. In this paper, we are focusing on examining the viability of sample translation itself.

Note that our approach is flexible: the temporal proximity window can be expanded to allow more stories to be translated, if time allows. Tuning this parameter is also left to future sampling strategy research. One interesting challenge is adapting this approach to dealing with many data sources, since the fact that some sources are faster/more prolific than others needs to be taken into account.

**Segmentation.** Segmentation is particularly problematic when translation is involved: BBN's research shows that among the total 25% words which cannot be translated from Chinese into English, 5% result from a segmentation error[3]. In our research, we found that 15% of Chinese tokens cannot be translated into English, due to segmentation errors and unrecognized named entities.

In our experiments, we used both phrase-based segmentation and bigram-based segmentation to separate the terms. For phrase-based segmentation we reconstructed a phrase dictionary from segmented Chinese data provided by the LDC. To segment our Chinese text, we used a longest string matching algorithm. For bigram-based segmentation, we simply used all two consecutive Chinese characters as tokens. Section 3.5.1 compares the tracking effectiveness of these two alternatives.

## 3 Experiments and Results

This section presents our experimental setup and results. Section 3.1 and 3.2 present the data and performance measures used; Section 3.3 discusses previously published results. The following describe our experiments, which can be grouped as follows:

1. Mixed language event tracking: Here, the topics as well as the stories are either in English, or translated into English from Mandarin by SYSTRAN. We present these results in order to demonstrate that our system is comparable to the best teams in recent TDT benchmark evaluations.

2. CLET based on test document translation: This is similar to (1) in that it uses the conventional approach of translating all testing stories, but it does not include the English stories and it establishes a baseline for (3) under the same evaluation conditions.

3. CLET based on translating a few training stories per event: This is the approach we promote in this paper.

### 3.1 Tools and Data

We chose the TDT-3 corpus as our experimental corpus in order to make our results comparable to results published in TDT evaluations[6]. The corpus includes both English and Mandarin news stories. SYSTRAN translations for all Mandarin stories are also provided. Details are as follows:

1. Mixed language event tracking: We used the TE=mul,eng evaluation condition in TDT 2001[7]. This uses the newest publicly available human judgments and allows us to compare our results with benchmark results released by NIST.

2. CLET based on test document translation: We used the TE=man,eng evaluation condition in TDT 1999 (topics are English, news stories are in Mandarin with a SYSTRAN translation provided by NIST). We also provided a dictionary translation using a 111K entries English-Chinese wordlist provided by LDC. In order to compare the results with (3), we use the same experimental conditions.

3. CLET based on translating sampled training stories per event: We used the TE=man,nat evaluation condition in TDT 1999 (topics are in English, documents are Mandarin native stories). In order to model the real test setting, we kept all the Mandarin native data in TDT3 as a test set. For the training phase, we used sampled translated English documents, including the 4 positive examples. There are 59 events. For each event we used around 120 translated stories as the training set. For the query expansion phase we used the first six months of 1998 TDT3 English news stories. For cross-lingual PRF we used the Hong Kong News Parallel Text (18,147 article pairs) provided by LDC. For Chinese phrase segmentation we reconstructed a phrase dictionary from segmented Chinese data provided by the LDC.

## 3.2  Evaluation Measures

To evaluate the performance of our system, we chose the conventional measures for event tracking used in TDT benchmark evaluations[7]. Each story is assigned a label of YES/NO for each of the topics. If the system assigns a YES to a story labeled NO by humans, it commits a *false alarm* error. If the system assigns a NO to a story labeled YES, it commits a *miss* error. The performance measures (*costs*) are defined as:

$$C_{trk} = C_{miss} \cdot P_{miss} \cdot P_{target} + C_{fa} \cdot P_{fa} \cdot P_{non-target} \tag{7}$$

$$(C_{trk})_{norm} = \frac{C_{trk}}{\min(C_{miss} \cdot P_{target}, C_{fa} \cdot P_{non-target})} \tag{8}$$

where $C_{miss}$ and $C_{fa}$ are the costs of a *miss* and a *false alarm*, We use $C_{miss}=1.0$ and $C_{fa}=0.1$, respectively; $P_{miss}$ and $P_{fa}$ are the conditional probabilities of a *miss* and a *false alarm*, respectively; $P_{target}$ and $P_{non-target}$ are the prior target probabilities ($P_{non-target} = 1 - P_{target}$). $P_{target}$ was set to 0.02 following the TDT tradition; $P_{miss}$ is the ratio of the number of *miss* errors to the number of the YES stories in the stream; $P_{fa}$ is the ratio of the number of *false alarm* errors to the total number of NO stories. The *normalized cost* $(C_{trk})_{norm}$ computes the relative cost of the system with respect to the minimum of two trivial systems (Simply assigns "Yes" labels or "No" labels without examining the stories). To compare costs between two tracking approaches, we used the Cost Reduction Ratio ($\delta$):

$$\delta = ((C_{trk})_{norm} - (C'_{trk})_{norm}) / (C_{trk})_{norm} \tag{9}$$

where $(C_{trk})_{norm}$ and $(C'_{trk})_{norm}$ are the normalized costs of two approaches, and $\delta$ is the cost reduction ratio by using approach2 instead of approach1.

We also use the Detection-Error Tradeoff (DET) curve[5] to show how the threshold would affect the trade-off between the miss and false alarm rates.

## 3.3 Mixed Language Event Tracking Results

In the mixed language event tracking task, LIMSI was the best in the benchmark evaluation in TDT2001. The cost using Nt=4 (4 positive instances per topic) is $(C_{trk})_{norm} = 0.1415$, as released by NIST.

In order to compare our new weighted and normalized adaptation (NWAdapt) with LIMSI's weighted adaptation (LWAdapt) and our old adaptation approach (FWAdapt), we implemented LIMSI's approach in our system. We trained LIMSI's confidence transformation and all the parameters on TDT1999 dry-run conditions and applied the adaptation weight and parameters for the TDT 2001 task.

Table 1 shows four results: our system performance without adaptation, our system performance with FWAdapt, with LWAdapt and with NWAdapt. Note that our LWAdapt implementation performed as well as the results reported by NIST. NWAapt reduced the cost more than FWAdapt and LMAdapt when compared to the no adaptation alternative.

**Table 1.** Results of adaptation methods in mixed language event tracking on the TDT-2001 evaluation dataset

| Adaptation Method | Cost | $\delta$ |
|---|---|---|
| Without adaptation | 0.1453 | -- |
| FWAdapt | 0.1448 | 0.3% |
| LMAdapt | 0.1413 | 2.7% |
| NMAdapt | 0.1236 | 14.9% |

## 3.4 CLET Using Test Document Translation

We experimented with the conventional approach to CLET (test document translation) by using both SYSTRAN translations provided by LDC and dictionary translation after topic expansion. For dictionary translation, we used a simple bilingual dictionary and we translate the entire test set word by word. The dictionary translation performs worse than the SYSTRAN translation (0.1336 vs. 0.1745), but this experiment is useful in order to provide a fair comparison with topic translation, which uses the same dictionary. We opted for using a dictionary in our experiments because SYSTRAN is a more costly solution that is also less likely to be available for an arbitrary language. Additionally, while SYSTRAN is better than a dictionary on news stories, this is not necessarily true in a domain with technical vocabulary.

**Table 2.** Parameter Values and Corresponding Labels

| Seg-menta-tion | Ex-pand | Adapta-tion | Label | Seg-menta-tion | Ex-pand | Adapta-tion | Label |
|---|---|---|---|---|---|---|---|
| Phrase | No | No | Phrase | Bigram | No | No | Bigram |
| Phrase | No | FWAdapt | Phrase+FWAdapt | Bigram | No | FWAdapt | Bigram+FWAdapt |
| Phrase | No | LWAdapt | Phrase+LWAdapt | Bigram | No | LWAdapt | Bigram+LWAdapt |
| Phrase | No | NWAdapt | Phrase+NWAdapt | Bigram | No | NWAdapt | Bigram+NWAdapt |
| Phrase | Yes | No | Phrase+TE | Bigram | Yes | No | Bigram+TE |
| Phrase | Yes | FWAdapt | Phrase+TE+FWAdapt | Bigram | Yes | FWAdapt | Bigram+TE+FWAdapt |
| Phrase | Yes | LWAdapt | Phrase+TE+LWAdapt | Bigram | Yes | LWAdapt | Bigram+TE+LWAdapt |
| Phrase | Yes | NWAdapt | Phrase+TE+NWAdapt | Bigram | Yes | NWAdapt | Bigram+TE+NWAdapt |

### 3.5 CLET Using Training Sample Translation

We explored both DICT and CL-PRF as CLIR methods targeted towards bridging the language gap. CL-PRF performed only slightly better than DICT, probably because of the mismatch between the parallel corpus and the news stories. Since DICT is more time efficient and stable with respect to the topics, the experiments described below use DICT instead of CL-PRF. Since there are many factors that affect performance, Table 2 summarizes the different parameter values and their label subsequently used in the result. Refer to Section 2.3 for the detailed description of each factor.

**Topic Expansion and Segmentation.** Table 3 compares the different approaches to topic expansion and Chinese segmentation. As expected, topic expansion does improve the cost significantly ( $\delta$ =41%).

**Table 3.** The Effects of Topic Expansion and Segmentation Method

| Condition: English-Chinese | Cost | $\delta$ |
|---|---|---|
| Phrase(DICT) | 0.5039 | -- |
| Phrase+TE | 0.2974 | 41% |
| Bigram | 0.3848 | 26.3% |
| Bigram+TE | 0.2522 | 50% |

Additionally, using bigrams as linguistic units significantly improves performance over using phrase segmentation. Due to the overlapping nature of bigram segmentation, the segmented text contains more information but also more noise when compared to phrase segmentation. This creates an effect similar to query expansion and is more likely to contain the "true" meaning unit. Our experiments show that, in spite of the added noise, the added information improves the tracking performance. Using both topic expansion and bigram segmentation yields a 50% relative cost reduction.

**Table 4.** The effects of different adaptation approaches

| Condition: English-Chinese | Cost | $\delta$ |
|---|---|---|
| Phrase (no adaptation) | 0.5039 | -- |
| Phrase+FWAdapt | 0.5023 | 0.3% |
| Phrase+LWAdapt | 0.4258 | 15.5% |
| Phrase+NWAdapt | 0.4197 | 16.7% |
| Phrase+TE | 0.2974 | 41% |
| Phrase+TE+LWAdapt | 0.2660 | 47.2% |
| Phrase+TE+NWAdapt | 0.2617 | 48% |
| Bigram+TE | 0.2522 | 50% |
| Bigram+TE+LWAdapt | 0.2467 | 51% |
| Bigram+TE+NWAdapt | 0.2413 | 52.6% |

**Adaptation Approaches for CLET.** In Section 2, we mentioned that adaptation is a useful approach to improve the tracking system performance. Here, we compare the effects of the different adaptation methods, including fixed weight adaptation (FWAdapt), LIMSI weighted adaptation (LWAdapt) and our simplified normalized weighted adaptation (NWAdapt). For fixed adaptation, we chose the best result obtained. We trained LIMSI's mapping approach on the TDT2 data set. Table 4 shows that, while fixed weight adaptation leads to an insignificant improvement over no adaptation, LWAdapt and NWAdapt do significantly better, with our simplified approach being slightly better.

As expected, combining all three favorable approaches yields the best result so far: the cost is 0.2413, with a 52.6% cost reduction with respect to the baseline. Our simplified adaptation performed better than LWAdapt in all parameter combinations.

**Translating Test Documents vs. Sampled Training Documents.** Using SYSTRAN to translate the testing documents and performing the equivalent of monolingual event tracking is the best approach with regard to minimizing tracking cost. However, translating all documents is not practical when dealing with the realities of new streams on the Web. Our goal is to come as close as possible to this upper bound, while avoiding the expense of translating all stories. The approach we have proposed is to translate small training documents per topic, using temporal proximity to positive instances as a sampling criterion.

Figure 1 and Table 5 compare the effectiveness of translating the entire test set with that of translating only small training samples. For comparison purposes, both SYSTRAN and DICT were used when translating the entire test set; this is because the dictionary translation is generally less effective than the best commercial MT system. Topic expansion was also used when translating the test set, to facilitate a fair comparison. The tracking cost is reduced by 27% when the entire test set is translated instead of translating a few training documents per topic. If the data stream is relatively low volume, with few languages represented and comes from a source that does not limit downloads, translating the stories is clearly the best approach. However, if the amount of data, its linguistic variety, or its access limitations makes translating all documents impossible or difficult, the increased cost is an acceptable tradeoff.

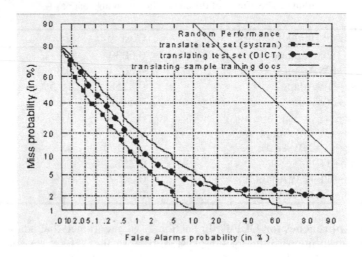

**Fig. 1.** Translating training samples can be a viable alternative

**Table 5.** Translating all test documents vs. Training samples

| Condition | Cost |
|---|---|
| Upper bound: Translating test documents (SYSYTRAN) | 0.5039 |
| Translating test documents (DICT) | 0.2974 |
| Translating training samples (DICT) | 0.2522 |

## 4   Conclusion

In this paper we have proposed a more practical approach to cross-lingual event tracking (CLET): translating a small sample of the training documents, instead of translating the entire test set. The latter approach could be prohibitively expensive when the "test set" is expanded to include the entire Web. In order to implement this approach, we have examined the applicability and performance of several cross-lingual information retrieval techniques to CLET. In addition, we have presented a significantly simplified event tracking adaptation strategy, which is more reliable and better performing than its previously introduced counterpart.

Overall, these strategies (in particular pre-translation topic expansion and bigram segmentation) reduce the cross-lingual tracking cost by more than 50% when compared to simple dictionary translation. This result is not surprising, given that query expansion has been repeatedly shown to improve cross-lingual information retrieval[10], but has not been previously used as a translation aid in true cross-lingual event tracking. We believe these cross-lingual retrieval techniques are an effective way of bridging the language gap in true cross-lingual event tracking.

## 5 Future Work

In this work we focused mostly on expanding topics by using pseudo-relevance feedback and bigram segmentation. In our future work we plan to concentrate on improving the translation accuracy. Potential methods include better machine translation techniques, named entity tracking, and investigating various sampling strategies.

## References

1. Allan, J. (ed.): Topic Detection and Tracking: Event Based Information Retrieval. Kluwer Academic Press (2002)
2. Allan, J., Carbonell, J.G., Doddington, G., Yamron, J., Yang Y.: Topic Detection and Tracking Pilot Study Final Report. In Proc. of the Broadcast News Transcription and Understanding Workshop (1998)
3. Leek, T., Jin, H., Sista, S., Schwartz, R.: The BBN Crosslingual Topic Detection and Tracking System. In Topic Detection and Tracking Workshop (1999)
4. Lo, Y., Gauvain, J.: The LIMSI Topic Tracking System for TDT2001. In Topic Detection and Tracking Workshop (2001)
5. Martin, A., Doddington, G., Kamm, T., Ordowski, M., Przybocki, M.: The Det Curve in Assessment of Detection Task Performance. In Proc. Eurospeech (1997) 1895-1898
6. TDT2001Evaluation. ftp://jaguar.ncsl.nist.gov/tdt/tdt2001/eval/tdt2001_official_results/
7. NIST. The year 2001 Topic Detection and Tracking Task Definition and Evaluation Plan. NIST (2001)
8. Robertson, S., Hull, D.A.: The TREC-9 Filtering Track Final Report. In The Ninth Text REtrieval Conference (2001)
9. Xu, J., Croft, B.: Query expansion using local and global document analysis. In Proc. ACM SIGIR, Zurich (1996) 4-11
10. Xu, J., Weischedel, R.: TREC-9 Cross-lingual Retrieval at BBN. In The Ninth Text Retrieval Conference (2001)
11. Yang, Y., Ault, T., Pierce, T., Lattimer, C.: Improving text categorization methods for event tracking. In Proc. ACM SIGIR (2000) 65-72
12. Yang, Y., Carbonell, J.G., Brown, R., Frederking, R.E.: Translingual Information Retrieval: Learning from Bilingual Corpora. In AIJ special issue: Best of IJCAI-97 (1998) 323-345
13. Yang, Y., Carbonell, J., Brown, R., Pierce, T., Archibald, B., Liu, X.: Learning Approaches for Detecting and Tracking News Events. IEEE Intelligent Systems: Special Issue on Applications of Intelligent Information Retrieval, Vol. 14(4) (1999) 32-43
14. Yang, Y. Ma N.: CMU Cross-lingual Information Retrieval at NTCIR-3. In Proc. of the Third NTCIR Workshop (2002)

# Document Clustering Using Linear Partitioning Hyperplanes and Reallocation

Canasai Kruengkrai, Virach Sornlertlamvanich, and Hitoshi Isahara

Thai Computational Linguistics Laboratory,
National Institute of Information and Communications Technology ,
112 Paholyothin Road, Klong 1, Klong Luang, Pathumthani 12120, Thailand
{canasai, virach}@tcllab.org, isahara@nict.go.jp

**Abstract.** This paper presents a novel algorithm for document clustering based on a combinatorial framework of the Principal Direction Divisive Partitioning (PDDP) algorithm [1] and a simplified version of the EM algorithm called the spherical Gaussian EM (sGEM) algorithm. The idea of the PDDP algorithm is to recursively split data samples into two sub-clusters using the hyperplane normal to the principal direction derived from the covariance matrix. However, the PDDP algorithm can yield poor results, especially when clusters are not well-separated from one another. To improve the quality of the clustering results, we deal with this problem by re-allocating new cluster membership using the sGEM algorithm with different settings. Furthermore, based on the theoretical background of the sGEM algorithm, we can naturally extend the framework to cover the problem of estimating the number of clusters using the Bayesian Information Criterion. Experimental results on two different corpora are given to show the effectiveness of our algorithm.

## 1   Introduction

Unsupervised clustering has been applied to various tasks in the field of Information Retrieval (IR). One of the challenging problems is document clustering that attempts to discover meaningful groups of documents where those within each group are more closely related to one another than documents assigned to different groups. The resulting document clusters can provide a structure for organizing large bodies of text for efficient browsing and searching [15].

A wide variety of unsupervised clustering algorithms has been intensively studied in the document clustering problem. Among these algorithms, the iterative optimization clustering algorithms have demonstrated reasonable performance for document clustering, e.g. the Expectation-Maximization (EM) algorithm and its variants, and the well-known $k$-means algorithm. Actually, the $k$-means algorithm can be considered as a spacial case of the EM algorithm [3] by assuming that each cluster is modeled by a spherical Gaussian, each sample is assigned to a single cluster, and all mixing parameters (or prior probabilities) are equal. The competitive advantage of the EM algorithm is that it is fast, scalable, and easy to implement. However, one major drawback is that it often

S. H. Myaeng et al. (Eds.): AIRS 2004, LNCS 3411, pp. 36–47, 2005.

gets stuck in local optima depending on the initial random partitioning. Several techniques have been proposed for finding good starting clusters (see [3][7]).

Recently, Boley [1] has developed a hierarchal clustering algorithm called the Principal Direction Divisive Partitioning (PDDP) algorithm that performs by recursively splitting data samples into two sub-clusters. The PDDP algorithm has several interesting properties. It applies the concept of the Principal Component Analysis (PCA) but only requiring the principal eigenvector, which is not computationally expensive. It can also generate a hierarchal binary tree that inherently produces a simple taxonomic ontology. Clustering results produced by the PDDP algorithm compare favorably to other document clustering approaches, such as the agglomerative hierarchal algorithm and associative rule hypergraph clustering. However, the PDDP algorithm can yield poor results, especially when clusters are not well-separated from one another. This problem will be described in depth later.

In this paper, we propose a novel algorithm for document clustering based on a combinatorial framework of the PDDP algorithm and a variant of the EM algorithm. As discussed above, each algorithm has its own strengths and weaknesses. We are interested in the idea of the PDDP algorithm that uses the PCA for analyzing the data. More specifically, it splits the data samples into two sub-clusters based on the hyperplane normal to the principal direction derived from the covariance matrix of the data. When the principal direction is not representative, the corresponding hyperplane tends to produce individual clusters with wrongly partitioned contents. One practical way to deal with this problem is to run the EM algorithm on the partitioning results. We present a simplified version of the EM algorithm called the spherical Gaussian EM algorithm for performing such task. Furthermore, based on the theoretical background of the spherical Gaussian EM algorithm, we can naturally extend the framework to cover the problem of estimating the number of clusters using the Bayesian Information Criterion (BIC) [9].

The rest of this paper is organized as follows. Section 2 briefly reviews some important backgrounds of the PDDP algorithm, and addresses the problem causing the incorrect partitioning. Section 3 presents the spherical Gaussian EM algorithm, and describes how to combine it with the PDDP algorithm. Section 4 discusses the idea of applying the BIC to our algorithm. Section 5 explains the data sets and the evaluation method, and shows experimental results. Finally, we conclude in Section 6 with some directions of future work.

## 2    Document Clustering via Linear Partitioning Hyperplanes

Suppose we have a 1-dimensional data set, e.g. real numbers on a line. The question is how to split this data set into two groups. One simple solution may be the following procedures. We first find the mean value of the data set, and then compare each point with the mean value. If the point value is less the mean value, it is assigned to the first group. Otherwise, it is assigned to the second group.

The problem arises when we have a $d$-dimensional data set. Based on the idea of the PDDP algorithm, we can deal with this problem by projecting all the data points onto the principal direction (the principal eigenvector of the covariance matrix of the data set), and then the splitting process can be performed based on this principal direction. In geometric terms, the data points are partitioned into two sub-clusters using the hyperplane normal to the principal direction passing through the mean vector [1]. We refer to this hyperplane as the linear partitioning hyperplane.

Given a matrix $\mathbf{M}$, we obtain the covariance matrix $\mathbf{C} = (\mathbf{M} - \mathbf{m}e^T)(\mathbf{M} - \mathbf{m}e^T)^T = \mathbf{A}\mathbf{A}^T$, where $\mathbf{A} = \mathbf{M} - \mathbf{m}e^T$, $\mathbf{m}$ is the mean vector of $\mathbf{M}$, and $e$ is the vector of ones, $(1, \ldots, 1)^T$. To obtain the principal eigenvector of $\mathbf{C}$, the notion of the Singular Value Decomposition (SVD) is used. The SVD of $\mathbf{A}$ is the factorization $\mathbf{A} = \mathbf{U}\mathbf{\Sigma}\mathbf{V}^T$, where $\mathbf{U}^T\mathbf{U} = \mathbf{V}^T\mathbf{V} = \mathbf{I}$, and $\mathbf{\Sigma}$ is a diagonal matrix, whose non-negative values are sorted in decreasing order. It is known that the left and right singular vectors of $\mathbf{A}$ are equivalent to the eigenvectors of $\mathbf{A}\mathbf{A}^T$ and $\mathbf{A}^T\mathbf{A}$, respectively. If we write $\mathbf{A}$ in terms of the SVD, we get $\mathbf{A}\mathbf{A}^T = \mathbf{U}\mathbf{\Sigma}\mathbf{V}^T\mathbf{V}\mathbf{\Sigma}^T\mathbf{U}^T = \mathbf{U}\mathbf{\Sigma}\mathbf{I}\mathbf{\Sigma}^T\mathbf{U}^T = \mathbf{U}\mathbf{\Sigma}^2\mathbf{U}^T$. Thus, the columns of $\mathbf{U}$, known as the left singular vectors, are the eigenvectors of $\mathbf{A}\mathbf{A}^T$. In our work, we just require the principal eigenvector, which is the first left singular vector denoted by $\mathbf{u}_1$. To efficiently compute a partial SVD of a matrix, the Lanczos algorithm is applied (see [6] for more details).

Let $\mathbf{d}_i$ be a document vector, and $\mathcal{C}$ be a cluster, where $\mathcal{C} = \{\mathbf{d}_1, \ldots, \mathbf{d}_{|\mathcal{C}|}\}$. The principal direction of $\mathcal{C}$ is denoted by $\mathbf{u}_\mathcal{C}$. The mean (or centroid) vector of the cluster can be calculated as follows:

$$\mathbf{m}_\mathcal{C} = \frac{1}{|\mathcal{C}|} \sum_{i=1}^{|\mathcal{C}|} \mathbf{d}_i \ . \tag{1}$$

The linear partitioning hyperplane for splitting $\mathcal{C}$ into the left child $\mathcal{L}$ and right child $\mathcal{R}$ corresponds to the following discriminant function [1]:

$$f_\mathcal{C}(\mathbf{d}_i) = \mathbf{u}_\mathcal{C}^T(\mathbf{d}_i - \mathbf{m}_\mathcal{C}) \ , \tag{2}$$

where

$$\mathbf{d}_i \in \begin{cases} \mathcal{L}, & \text{if } f_\mathcal{C}(\mathbf{d}_i) \leq 0, \\ \mathcal{R}, & \text{if } f_\mathcal{C}(\mathbf{d}_i) > 0. \end{cases} \tag{3}$$

The PDDP algorithm begins with all the document vectors in a large single cluster, and proceeds by recursively splitting the cluster into two sub-clusters using the linear partitioning hyperplane according to the discriminant function in Equations 2 and 3. The algorithm stops splitting based on some heuristic, e.g. a predefined number of clusters. It finally yields a binary tree, whose leaf nodes form the resulting clusters. To keep the binary tree balanced, it selects an un-split cluster to split by using the scatter value, measuring the average distance from the data points in the cluster to their centroid. The basic PDDP algorithm is given in Algorithm 1.

As mentioned earlier, the problem of the PDDP algorithm is that it cannot achieve good results when clusters are not well-separated from one another.

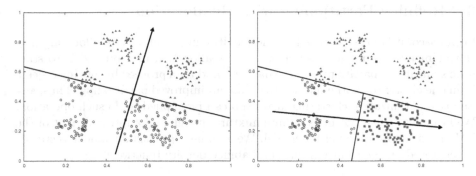

**Fig. 1.** Two partitions after the first iteration (left), and three partitions after the second iteration (right)

---

**Algorithm 1:** The basic PDDP algorithm.

---

    **input**     : A data set representing by a matrix $\mathbf{M} = (\mathbf{d}_1, \mathbf{d}_2, \ldots, \mathbf{d}_n)$, and a desired number of clusters $k_{\max}$.

    **output**  : A binary tree $\mathcal{T}$ with leaf nodes forming a partitioning of the data set.

    **begin**

        Initialize a binary tree $\mathcal{T}$ with a single root node, and $k \leftarrow 0$.

        **while** $k < k_{max}$ **do**

            Select the leaf node $\mathcal{C}$ with the largest scatter value.

            Compute the mean vector $\mathbf{m}_\mathcal{C}$ and the principal direction $\mathbf{u}_\mathcal{C}$.

            For each document $\mathbf{d}_i \in \mathcal{C}$, assign $\mathbf{d}_i$ to $\mathcal{L}$ or $\mathcal{R}$ according to Equations 2 and 3.

            Set $\mathcal{T} \leftarrow \mathcal{T} \cup \{\mathcal{L}, \mathcal{R}\}$, and $k \leftarrow k + 1$.

        **end**

    **end**

---

Let us describe with an empirical example. Figure 1 shows two partitions after the first iteration (left) and three partitions after the second iteration (right) produced by performing the PDDP algorithm on a 2-dimensional data set[1]. The data set consists of 334 points. The actual class labels are not given, but one can observe that it is composed of five compact clusters [8]. Based on the principal direction and the corresponding linear partitioning hyperplane, we can see that the PDDP algorithm starts with significantly wrong partitioning on the middle left-hand cluster. If we further perform the partitioning without making some adjustments, the resulting clusters become worse. This indicates that the basic PDDP algorithm can produce poor solutions in some distributions of the data, which we do not know in advance. Also, we may require some information to suggest whether to split a particular cluster further or keep it as it is.

---

[1] This data set is available at http://www.jihe.net/datasets.htm.

## 3　Refining Partitioning with Reallocation

In practice, it is possible to refine the partitioning results by re-allocating new cluster membership. The basic idea of the reallocation method [12] is to start from some initial partitioning of the data set, and then proceed by moving objects from one cluster to another cluster to obtain an improved partitioning. Thus, any iterative optimization clustering algorithm can be applied to do such operation. We formulate our problem as a finite mixture model, and apply a variant of the EM algorithm for learning the model. We assume that each document vector $\mathbf{d}_i$ is generated independently from a probability density function:

$$p(\mathbf{d}_i|\mathbf{\Theta}) = \sum_{j=1}^{k} p(\mathbf{d}_i|c_j; \boldsymbol{\theta}_j) P(c_j) , \tag{4}$$

where $k$ is the number of components, the conditional probability density function $p(\mathbf{d}_i|c_j; \boldsymbol{\theta}_j)$ is the component density, and the prior probability $P(c_j)$ is the mixing parameter. Thus, we have the model parameters: $\mathbf{\Theta} = \{P(c_1), \ldots, P(c_k); \boldsymbol{\theta}_1, \ldots, \boldsymbol{\theta}_k\}$. To estimate the model parameters, we work with the maximum likelihood (ML) estimation. Given a set of document vectors $\mathcal{D} = \{\mathbf{d}_1, \ldots, \mathbf{d}_n\}$, the incomplete-data log likelihood can be expressed as:

$$\log L(\mathbf{\Theta}) = \log p(\mathcal{D}|\mathbf{\Theta}) = \log \prod_{i=1}^{n} p(\mathbf{d}_i|\mathbf{\Theta}) = \sum_{i=1}^{n} \log \sum_{j=1}^{k} p(\mathbf{d}_i|c_j; \boldsymbol{\theta}_j) P(c_j) . \tag{5}$$

Our objective is to maximize $\log L$. However, Equation 5 is difficult to optimize because it contains the log of sums. Thus, we introduce the variable of the missing data $\mathcal{Z} = \{\mathbf{z}_1, \mathbf{z}_2, \ldots, \mathbf{z}_n\}$, where $\mathbf{z}_i = (z_{i1}, \ldots, z_{ik})^T$ is a vector of binary indicator variables, and $(\mathbf{z}_i)_j = z_{ij} = 1$ if the vector $\mathbf{d}_i$ was generated from the component $c_j$; otherwise $z_{ij} = 0$. Now we can write the complete-data log likelihood as:

$$\log L_c(\mathbf{\Theta}) = \sum_{i=1}^{n} \sum_{j=1}^{k} z_{ij} \log(p(\mathbf{d}_i|c_j; \boldsymbol{\theta}_j) P(c_j)) . \tag{6}$$

Since $\mathcal{Z}$ is unknown, we instead work with its expectation. A local maximum of $\log L_c$ can be found by applying the EM algorithm that performs by iterating two steps:

- **E-step:** $\hat{\mathcal{Z}}^{\langle t+1 \rangle} = E[\mathcal{Z}; \hat{\mathbf{\Theta}}^{\langle t \rangle}]$,
- **M-step:** $\hat{\mathbf{\Theta}}^{\langle t+1 \rangle} = \operatorname{argmax}_{\mathbf{\Theta}} \left( \log L_c(\mathbf{\Theta}; \hat{\mathcal{Z}}^{\langle t+1 \rangle}) \right)$.

So far, the problem is how to estimate the model parameters. Here we assume that the data samples are drawn from the multivariate normal density in $\Re^d$. We also assume that features are statistically independent, and a component $c_j$ generates its members from the spherical Gaussian with the same covariance

---

**Algorithm 2:** The spherical Gaussian EM algorithm.

**begin**

    **Initialization:** Set $(\mathbf{z}_i)_j^{\langle 0 \rangle}$ from a partitioning of the data, and $t \leftarrow 0$.

    **repeat**

        **E-step:** For each $\mathbf{d}_i, 1 \leq i \leq n$ and $c_j, 1 \leq j \leq k$, find its new component index as:

$$(\mathbf{z}_i)_j^{\langle t+1 \rangle} = \begin{cases} 1, & \text{if } j^* = \operatorname{argmax}_j \log(P^{\langle t \rangle}(c_j|\mathbf{d}_i; \boldsymbol{\theta}_j)), \\ 0, & \text{otherwise.} \end{cases}$$

        **M-step:** Re-estimate the model parameters:

$$P(c_j)^{\langle t+1 \rangle} = \frac{1}{n} \sum_{i=1}^{n} (\mathbf{z}_i)_j^{\langle t+1 \rangle}$$

$$\mathbf{m}_j^{\langle t+1 \rangle} = \sum_{i=1}^{n} \mathbf{d}_i (\mathbf{z}_i)_j^{\langle t+1 \rangle} / \sum_{i=1}^{n} (\mathbf{z}_i)_j^{\langle t+1 \rangle}$$

$$\sigma^{2\langle t+1 \rangle} = \frac{1}{n \cdot d} \sum_{i=1}^{n} \sum_{j=1}^{k} \| \mathbf{d}_i - \mathbf{m}_j \|^2 (\mathbf{z}_i)_j^{\langle t+1 \rangle} .$$

    **until** $\Delta \log L_c(\boldsymbol{\Theta}) < \delta$;

**end**

---

matrix [5]. Thus, the parameter for each component $c_j$ becomes $\boldsymbol{\theta}_j = (\mathbf{m}_j, \sigma^2 \mathbf{I}_d)$, and the component density is:

$$p(\mathbf{d}_i|c_j; \boldsymbol{\theta}_j) = \frac{1}{(2\pi)^{d/2} \sigma^d} \exp\left( -\frac{\| \mathbf{d}_i - \mathbf{m}_j \|^2}{2\sigma^2} \right) . \tag{7}$$

Based on Bayes' theorem, the probability that $\mathbf{d}_i$ falls into $c_j$ can be defined as the product of the component density and the mixing parameter:

$$P(c_j|\mathbf{d}_i; \boldsymbol{\theta}_j) = p(\mathbf{d}_i|c_j; \boldsymbol{\theta}_j) P(c_j) . \tag{8}$$

Note that we estimate this probability with $\log(P(c_j|\mathbf{d}_i; \boldsymbol{\theta}_j))$ that corresponds to the log term in Equation 6. Algorithm 2 gives an outline of a simplified version of the EM algorithm. From the assumptions of the model, we refer to this algorithm as the spherical Gaussian EM algorithm (or the sGEM algorithm for short). The algorithm tries to maximize $\log L_c$ at very step, and iterates until convergence. For example, the algorithm terminates when $\Delta \log L_c < \delta$, where $\delta$ is a predefined threshold. In this paper, we set $\delta = 0.001$.

The sGEM algorithm can be combined with the PDDP algorithm in several ways. We can run the sGEM algorithm on the local region after the splitting process by fixing $k = 2$. This idea is closely related to the concept of the bisecting $k$-means algorithm [14]. However, our initial partitions are based on the PDDP algorithm rather than using the random initialization. We can also run the sGEM algorithm at the end of the PDDP algorithm, where $k$ is equal to the number of leaf nodes of the PDDP's binary tree. In other words, we can think that the

---

**Algorithm 3:** The combination of the PDDP and sGEM algorithms using the BIC as the stopping criterion.

---

    **input**    : A data set representing by a matrix $\mathbf{M} = (\mathbf{d}_1, \mathbf{d}_2, \ldots, \mathbf{d}_n)$.

    **output**   : A set of disjoint clusters $\mathcal{C}_1, \mathcal{C}_2, \ldots, \mathcal{C}_k$.

    **begin**

        Initialize a binary tree $\mathcal{T}$ with a single root node, and $k \leftarrow 0$.

        $BIC_{global} \leftarrow BIC(\text{root node})$

        **repeat**

            Select the leaf node $\mathcal{C}$ with the largest scatter value.

            Compute the mean vector $\mathbf{m}_{\mathcal{C}}$ and the principal direction $\mathbf{u}_{\mathcal{C}}$.

            For each document $\mathbf{d}_i \in \mathcal{C}$, assign $\mathbf{d}_i$ to $\mathcal{L}$ or $\mathcal{R}$ according to Equations 2 and 3.

            Run sGEM($\{\mathcal{L}, \mathcal{R}\}$) for the local refinement.

            **if** $BIC(\{\mathcal{L}, \mathcal{R}\}) > BIC(\mathcal{C})$ **then**

                $\mathcal{T}_{tmp} \leftarrow$ all the current leaf nodes except node $\mathcal{C}$.

                $BIC_{tmp} \leftarrow BIC(\mathcal{T}_{tmp} \cup \{\mathcal{L}, \mathcal{R}\})$

                **if** $BIC_{tmp} > BIC_{global}$ **then**

                    $BIC_{global} \leftarrow BIC_{tmp}$

                  Set $\mathcal{T} \leftarrow \mathcal{T} \cup \{\mathcal{L}, \mathcal{R}\}$, and $k \leftarrow k + 1$.

        **until** *the leaf nodes cannot be partitioned*;

        Initialize $\mathcal{C}_1^{\langle 0 \rangle}, \mathcal{C}_2^{\langle 0 \rangle}, \ldots, \mathcal{C}_k^{\langle 0 \rangle}$ with leaf nodes of $\mathcal{T}$, and $t \leftarrow 0$.

        Run sGEM($\{\mathcal{C}_j^{\langle t \rangle}\}_{j=1}^k$) until convergence.

    **end**

---

sGEM algorithm can get trapped in a local optimum due to a bad initialization, so performing the PDDP algorithm as the first round of the sGEM algorithm may help to avoid this problem. The computational time is reasonable. Since the initial partitions generated by the PDDP algorithm are better than random ones, the combined algorithm performs a moderate number of the sGEM iterations.

## 4     Estimating Number of Document Clusters

When we need to apply the clustering algorithm to a new data set having little knowledge about its contents, fixing a predefined number of clusters is too strict and inefficient to discover the latent cluster structures. Based on the finite mixture model described in the previous section, we can naturally extend the combination of the PDDP algorithm and the sGEM algorithm to cover the problem of estimating the number of clusters in the data set. Here we apply a model selection technique called the Bayesian Information Criterion (BIC) [9]. Generally, the problem of model selection is to choose the best one among a set of candidate models. Given a data set $\mathcal{D}$, the BIC of a model $\mathcal{M}_i$ is defined as:

$$BIC(\mathcal{M}_i) = \log L(\mathcal{D}|\mathcal{M}_i) - \frac{p_i}{2} \cdot \log |\mathcal{D}| , \qquad (9)$$

where $\log L(\mathcal{D}|\mathcal{M}_i)$ is the log likelihood of the data according to the model $\mathcal{M}_i$, and $p_i$ is the number of independent parameters to be estimated in the model

$\mathcal{M}_i$. The BIC contains two components, where the first term measures how well the parameterized model predicts the data, and the second term penalizes the complexity of the model [4]. Thus, we select the model having the largest value of the BIC, $\mathcal{M}^* = \text{argmax}_i BIC(\mathcal{M}_i)$.

In the context of document clustering, the log likelihood of the data can be derived from Equation 6. As a result, we directly obtain the value of the first term of the BIC from running the sGEM algorithm. However, we can also compute it from the data according to the partitioning. The number of parameters is the sum of $k-1$ component probabilities, $k \cdot d$ centroid coordinates, and 1 variance.

Boley's subsequent work [2] also suggests a dynamic threshold called the centroid scatter value (CSV) for estimating the number of clusters. This criterion is based on the distribution of the data. Since the PDDP algorithm is a kind of the divisive hierarchical clustering algorithm, it gradually produces a new cluster by splitting the existing clusters. As the PDDP algorithm proceeds, the clusters get smaller. Thus, the maximum scatter value in any individual cluster also gets smaller. The idea of the CSV is to compute the overall scatter value of the data by treating the collection of centroids as individual data vectors. This stopping test terminates the algorithm when the CSV exceeds the maximum cluster scatter value at any particular point.

We can think of the CSV as a value that captures the overall improvement, whereas the BIC can be used to measure the improvement in both the local and global structure. As mentioned earlier, in the splitting process, we need some information to make the decision whether to split a cluster into two sub-clusters or keep its current structure. We first calculate the BIC locally when the algorithm performs the splitting test in the cluster. The BIC is calculated globally to measure the overall structure improvement. If both the local and global BIC scores improve, we then split the cluster into two children clusters. Algorithm 3 describes the unified algorithm based on the PDDP and sGEM algorithms using the BIC as the stopping criterion.

## 5   Experiments

### 5.1   Data Sets and Experimental Setup

The Yahoo News (K1a) data set[2] consists of 2340 news articles obtained from the Web in October 1997. Each document corresponds to a web page listed in the subject hierarchy of YAHOO! This data set is divided into 20 categories. Note that the number of documents in each category varies considerably, ranging from 9 to 494.

The 20 Newsgroups data set consists of 20000 articles evenly divided among 20 different discussion groups [10]. This data set is collected from UseNet postings over a period of several months in 1993. Many categories fall into confusable clusters. For example, five of them are computer discussion groups, and three

---

[2] The     21839×2340     term-document     matrix     is     publicly     available     at
http://www-users.cs.umn.edu/~karypis/cluto/download.html.

of them discuss religion. We used the Bow toolkit [11] to construct the term-document matrix (sparse format). We removed the UseNet headers, and also eliminated the stopwords and low-frequency words (occurring less than 2 times). We finally obtained the 59965×19950 term-document matrix for this data set.

We also applied the well-known tf·idf term weighting technique. Let $\mathbf{d}_i = (w_{i1}, w_{i2}, \ldots, w_{im})^T$, where $m$ is the total number of the unique terms. The tf·idf score of each $w_{ik}$ can be computed by the following formula [13]:

$$w_{ik} = tf_{ik} \cdot \log\left(\frac{n}{df_k}\right) , \qquad (10)$$

where $tf_{ik}$ is the term frequency of $w_k$ in $\mathbf{d}_i$, $n$ is the total number of documents in the corpus, and $df_k$ is the number of documents that $w_k$ occurs. Finally, we normalized each document vector using the $L_2$ norm.

For the purpose of comparison, we chose the basic PDDP algorithm as the baseline. We first tested three different settings of Algorithm 3, including sGEM-PDDP-L, sGEM-PDDP-G, and sGEM-PDDP-LG. sGEM-PDDP-L refers to running the sGEM algorithm locally on each PDDP splitting process, whereas sGEM-PDDP-G refers to running the sGEM algorithm globally after the PDDP algorithm converged. sGEM-PDDP-LG refers to running the sGEM algorithm with the PDDP algorithm both locally and globally. The number of clusters $k$ is varied in the range $[2, 2k]$, and no stopping criterion was used. Then we applied both the CSV and the BIC to the above settings in order to test the estimation of the number of clusters.

## 5.2    Evaluation Method

Since all the documents are already categorized, we can perform evaluation by comparing clustering results with the true class labels. In our experiments, we used the normalized mutual information ($NMI$) [16]. In the context of document clustering, mutual information can be used as a symmetric measure for quantifying the degree of relatedness between the generated clusters and the actual categories. Particularly, when the number of clusters differs from the actual number of categories, mutual information is very useful without a bias towards smaller clusters. By normalizing this criterion to take values between 0 and 1, the $NMI$ can be calculated as follows:

$$NMI = \frac{\sum_{h,l} n_{h,l} \log(n \cdot n_{h,l}/n_h n_l)}{\sqrt{(\sum_h n_h \log(n_h/n))(\sum_l n_l \log(n_l/n))}} , \qquad (11)$$

where $n_h$ is the number of documents in the category $h$, $n_l$ is the number of documents in the cluster $l$, and $n_{h,l}$ is the number of documents in the category $h$ as well as in the cluster $l$. We can interpret that the $NMI$ value is 1 when clustering results exactly match the true class labels, and close to 0 for a random partitioning [17].

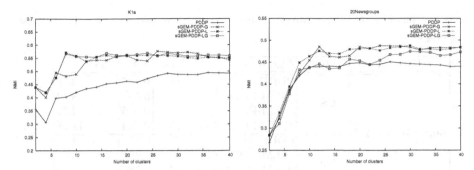

**Fig. 2.** $NMI$ results on Yahoo News (K1a) (left), and 20 Newsgroups (right)

## 5.3    Experimental Results

In Figure 2, we show clustering results on the Yahoo News (K1a) data set (left), and the 20 Newsgroups data set (right). The horizontal axis indicates the number of clusters, while the vertical axis shows the $NMI$ values. On the Yahoo News (K1a) data set, we can see that our combined algorithms significantly outperform the basic PDDP algorithm over the entire range. Furthermore, we can observe the correlation between PDDP and sGEM-PDDP-G. This indicates that performing the global refinement with the sGEM algorithm can considerably improve the quality of the clustering results in this data set. We can also observe the correlation between sGEM-PDDP-G and sGEM-PDDP-LG. From the curves, performing the global refinement after the local refinement seems to slightly affect the quality of the clustering results. As the number of clusters increases, the results of all the combined algorithms tend to converge to similar values.

On the 20 Newsgroups data set, we can see that our combined algorithms perform relatively better than the basic PDDP algorithm, particularly with sGEM-PDDP-G and sGEM-PDDP-L. However, performing the global refinement after the local refinement as in sGEM-PDDP-LG degrades the quality of the clustering results. The global refinement with the sGEM algorithm leads to more decisions to move each document from its cluster to other candidate clusters. As described earlier, since some categories in this data set have similar or related topics, it has a chance that the document is reassigned to the wrong cluster.

Table 1 summarizes clustering results by varying the stopping criteria. Note that both the data sets have the actual number of categories at 20. We can see that both criteria can yield reasonable results in terms of the number of clusters and $NMI$ values. For the Yahoo News (K1a) data set, applying the CSV to all the combined algorithms as well as the basic PDDP algorithm gives the same result at 15 clusters. With the BIC, PDDP and sGEM-PDDP-G underestimate the number of clusters at 5, while sGEM-PDDP-L and sGEM-PDDP-LG find 12 clusters. Since the BIC is calculated locally on each splitting test and measures how well the partitioning can model a given data set, it can be affected by the local refinement more than the CSV. For the 20 Newsgroups data set, with the CSV, PDDP and sGEM-PDDP-G find 34 clusters, while sGEM-PDDP-L and sGEM-PDDP-LG find 26 clusters. When we tested with the BIC, the results are

**Table 1.** Clustering results by varying stoping criteria on `Yahoo News` (K1a) and 20 `Newsgroups`

| Data set | Criterion | Algorithm | $k$ found | $NMI$ | Time (sec.) |
|---|---|---|---|---|---|
| Yahoo News (K1a) | CSV | PDDP | 15 | 0.447 | 2.79 |
| | | sGEM-PDDP-G | 15 | 0.538 | 8.14 |
| | | sGEM-PDDP-L | 15 | 0.555 | 4.77 |
| | | sGEM-PDDP-LG | 15 | 0.564 | 11.98 |
| | BIC | PDDP | 5 | 0.478 | 2.01 |
| | | sGEM-PDDP-G | 5 | 0.496 | 3.64 |
| | | sGEM-PDDP-L | 12 | 0.589 | 4.78 |
| | | sGEM-PDDP-LG | 12 | 0.571 | 6.34 |
| 20 Newsgroups | CSV | PDDP | 34 | 0.443 | 15.838 |
| | | sGEM-PDDP-G | 34 | 0.482 | 105.39 |
| | | sGEM-PDDP-L | 26 | 0.478 | 50.40 |
| | | sGEM-PDDP-LG | 26 | 0.466 | 219.33 |
| | BIC | PDDP | 25 | 0.426 | 14.70 |
| | | sGEM-PDDP-G | 25 | 0.463 | 78.45 |
| | | sGEM-PDDP-L | 28 | 0.473 | 50.58 |
| | | sGEM-PDDP-LG | 28 | 0.476 | 164.74 |

slightly different. PDDP and sGEM-PDDP-G find 25 clusters, and sGEM-PDDP-L and sGEM-PDDP-LG find 28 clusters.

## 6   Conclusion and Future Work

We have presented several strategies for improving the basic PDDP algorithm. When the principal direction is not representative, the corresponding hyperplane tends to produce individual clusters with wrongly partitioned contents. By formulating the problem with the finite mixture model, we describe the sGEM algorithm that can be combined with the PDDP algorithm in several ways for refining the partitioning results. Preliminarily experimental results on two different document sets are very encouraging.

In future work, we intend to investigate other model selection techniques for approximating the number of underlying clusters. Recently, work by [7] has demonstrated that estimating the number of clusters in the $k$-means algorithm using the Anderson-Darling test yields very promising results, and seems to outperform the BIC. We think that this statistical measure can also be applied to our proposed algorithm.

## References

1. Boley, D.: Principal direction divisive partitioning. *Data Mining and Knowledge Discovery*, 2(4):325–344. (1998)
2. Boley, D., and Borst, V.: Unsupervised clustering: A fast scalable method for large datasets. *CSE Report TR-99-029, University of Minnesota.* (1999)

3. Bradley, P. S., and Fayyad, U. M.: Refining initial points for k-means clustering. In *Proceedings of the Fifteenth International Conference on Machine Learning*, pages 91–99. (1998)
4. Chickering, D., Heckerman, D., and Meek, C.: A bayesian approach to learning bayesian networks with local structure. In *Proceedings of the thirteenth Conference on Uncertainty in Artificial Intelligence*, pages 80–89. Morgan Kaufmann. (1997)
5. Dasgupta, S., and Schulman, L. J.: A two-round variant of em for gaussian mixtures. *Sixteenth Conference on Uncertainty in Artificial Intelligence (UAI)*. (2000)
6. Golub, G., and Loan, C. V.: *Matrix Computations*. The Johns Hopkins University Press, Baltimore. (1989)
7. Hamerly, G., and Elkan, C.: Learning the $k$ in $k$-means. In *Proceedings of the seventeenth annual conference on neural information processing systems (NIPS)*. (2003)
8. He, J., Tan, A.-H. , Tan, C.-L., and Sung, S.-Y.: *On Quantitative Evaluation of Clustering Systems*. In W. Wu and H. Xiong, editors, Information Retrieval and Clustering. Kluwer Academic Publishers. (2003)
9. Kass, R. E., and Raftery, A. E.: Bayes factors. *Journal of the American Statistical Association*, 90:773–795. (1995)
10. Lang, K.: Newsweeder: Learning to filter netnews. In *Proceedings of the Twelfth International Conference on Machine Learning*, pages 331–339. (1995)
11. McCallum, A. K.: Bow: A toolkit for statistical language modeling, text retrieval, classification and clustering. `http://www.cs.cmu.edu/~mccallum/bow`.
12. Rasmussen, E.: *Clustering algorithms*. In W. Frakes and R. Baeza-Yates, editors,Information retrieval: data structures and algorithms. Prentice-Hall. (1992)
13. Salton, G., and Buckley, C.: Term-weighting approaches in automatic text retrieval. *Information Processing and Management: an International Journal*, 24(5):513–523. (1988)
14. Steinbach, M., Karypis, G., and Kumar, V.: A comparison of document clustering techniques. *KDD Workshop on Text Mining*. (1999)
15. Strehl, A., Ghosh, J., and Mooney, R. J.: Impact of similarity measures on webpage clustering. In *Proceedings of AAAI Workshop on AI for Web Search*, pages 58–64. (2000)
16. Strehl, A., and Ghosh, J.: Cluster ensembles - a knowledge reuse framework for combining multiple partitions. *Journal on Machine Learning Research*, 3:583–617. (2002)
17. Zhong, S., and Ghosh, J.: A comparative study of generative models for document clustering. *SDM Workshop on Clustering High Dimensional Data and Its Applications*. (2003)

# Summary Generation
# Centered on Important Words

Dongli Han[1], Takashi Noguchi[2], Tomokazu Yago[2], and Minoru Harada[1]

[1] Department of Integrated Information Technology, Aoyama Gakuin University,
5-10-1 Fuchinobe, Sagamihara-shi, Kanagawa, Japan
{Kan, Harada}@it.aoyama.ac.jp
http://www-haradalb.it.aoyama.ac.jp/index.html
[2] Department of Industrial and Systems Engineering, Aoyama Gakuin University,
5-10-1 Fuchinobe, Sagamihara-shi, Kanagawa, Japan

**Abstract.** We developed a summarizing system ABISYS based on the output of semantic analysis system SAGE. ABISYS extracts important words from an article and generates summary sentences according to the word meanings and the deep cases among the words in the output from SAGE. In this paper, we define five kinds of scores to evaluate the importance of a word respectively on repetition information, context information, position information, opinion word information and topic-focus information. We first calculate the above scores for each substantive and reflect them in a five-dimensional space. Then the probability of each substantive to be important is calculated using a pan-distance of Mahalanobis. Finally, we complement the indispensable cases for verbs and the Sahen nouns that have been selected as important words, and use them as the summary element words to generate easy-to-read Japanese sentences. We carried out a subjectivity evaluation for our system output by referring to the summaries made by human. In comparison with the subjectivity evaluations made for other summarizing systems, we found that the point of readability was on a par with other systems, while the point of content covering was much better. And 95% of the summary sentences generated by ABISYS were acknowledged as correct Japanese sentences.

## 1 Introduction

Automated summarization has attracted more and more attention for years as a field of text mining [15,16]. A major method used in earlier days for Japanese summarizing tasks is based on the extraction of important sentences. Recent years, a methodological change has been seen in some research where the extraction of important words, the deletion of unnecessary words, or a hybrid model of the former two is used. However, neither seems to have made semantic analysis. As a result, some important words are neglected, while many unnecessary words remain in the summary.

For instance, Hatayama etc. [4] extracted the important words using surface information and structural information, and selected the words for generating sentences around the main verb only. Ueda etc. made a proposal to summarize by listing the word pairs that are considered important [18]. Oguro etc. made a natural summary

S. H. Myaeng et al. (Eds.): AIRS 2004, LNCS 3411, pp. 48 – 60, 2005.
© Springer-Verlag Berlin Heidelberg 2005

relying on the structural information among the phrases in a sentence [13]. Ishizako etc. [6] and Ohtake etc. [14] put their focus on the deletion of words and the extraction of importance sentences.

On the other hand, there are some summarizing systems for non-Japanese documents that have employed semantic information. For instance, the work of Hahn and Reimer [1], and that of Hovy and Lin [5].

As denoted above, except several summarizing systems for non-Japanese documents, no summarization for Japanese has taken account of semantic information up to now. In this paper we describe an automatic summarizing system ABISYS developed on the basis of a Japanese semantic analysis. We aimed to extract semantically important words and generate correct and natural Japanese sentences.

## 2  Outline of Abisys

Our idea for summarizing a Japanese document is similar to those used in various summarizing tasks. We all try to extract important words and generate refined sentences. However, the difference is that we use word meanings and the deep cases[1] among the words in the output from a semantic analysis system, rather than the widely used surface information or structural information in other works.

### 2.1  System Input

We take the output case-frames from the semantic analysis system SAGE Harada etc. have built [2,3] as the input to ABISYS. SAGE determines the word meanings and

```
frame(1,'全','ゼン','JN5','JT4','全','接頭詞-数接続','接頭辞-名詞接頭辞','none','none','0fa5ca',[],[1,1.1]).
frame(2,'17','イチナナ','JN3','JN3','17','名詞-数','名詞-数詞','none','none','000111',[],[1,1.2]).
frame(3,' 日 ',' ニ チ ','JN6','JN6',' 日 ',' 名 詞 - 接 尾 - 助 数 詞 ',' 接 尾 辞 - 名 詞 性 名 詞 助 数 辞 ','none','none','1022cc',[[head,3]],[1,1.3]).
frame(4,' 全  17  日 ','none','JNP','*',' 全  17  日 ','none','*','none',' の ','000000',[[consist,1],[consist,2],[consist,3]],[1,1]).
frame(5,' 日 程 ',' ニ ッ テ イ ','JN1','JN1',' 日 程 ',' 名 詞 - 一 般 ',' 名 詞 - 普 通 名 詞 ','none',' を ','102332',[[modifier,3]],[1,2]).
frame(6,'終え','オエ','JVE','*','終える','動詞-自立','*','連用形','て','1f23fc',[[implement,5]],[1,3]).
frame(7,'シドニー','シドニー','JN2','JN2','シドニー','名詞-固有名詞-地域-一般','名詞-地名','none','none','3c5fb7',[],[1,4.1]).
frame(8,' 五 輪 ',' ゴ リ ン ','JN1','JN1',' 五 輪 ',' 名 詞 - 一 般 ',' 名 詞 - 普 通 名 詞 ','none','none','0f2a51',[[head,8]],[1,4.2]).
frame(9,' シ ド ニ ー 五 輪 ','none','JNP','*',' シ ド ニ ー 五 輪 ','none','*','none',' は ','000000',[[consist,7],[consist,8]],[1,4]).
frame(10,' 終 了 ',' シ ュ ウ リ ョ ウ ','JSA','JSA',' 終 了 ',' 名 詞 - サ 変 接 続 ',' 名 詞 - サ 変 名 詞 ','none','none','3cea4c',[[head,10],[sequence,6],[object,8],[main,10]],[1,5.1]).
frame(11,' し ',' シ ','JVE','*','せ','動詞-自立','*','連用形','none','3d079a',[],[1,5.2]).
frame(12,'た','タ','JJD','*','た','助動詞','*','基本形','none','2621c8',[],[1,5.3]).
frame(13,'   終   了   し   た   ','none','JPR','*',' 終 了 し た ','none','*','none','none','000000',[[consist,10],[consist,11],[consist,12]],[1,5]).
```

**Fig. 1.** Output case-frames from SAGE for the sentence "全17日の日程を終えてシドニー五輪は終了した。"(The Sydney Olympic Games ended over 17 days.)

---

[1] Deep case means the semantic relationship among words.

the deep cases among words according to the definition in the EDR Dictionary[2]. Each case-frame corresponds to a morpheme in the sentence. And as shown in Fig. 1, a case-frame is composed of 14 elements including 'number', 'denotation', 'reading', 'EDR-POS[3] from Chasen-POS', 'EDR-POS from Juman-POS', 'dictionary entry denotation', 'Chasen-POS', 'Juman-POS', 'inflection', 'particle', 'concept ID' (meaning), 'deep case information', 'sentence number' and 'phrase number'.

The summarizing process starts when a file containing a group of case-frames is provided and both the document style (1: a newspaper article / 2: a leading article) and the summarizing rate (e.g., 30%) are given.

---

(1)The original->
全17日の日程を終えてシドニー五輪は終了した。
(The Sydney Olympic Games ended over 17 days.)
移民の国オーストラリアで開催する五輪が掲げたテーマは民族の融和だったのだと言われる。
(It is said that the theme of the Olympic Games held in the immigrant nation, Australia, has been set up to be ethnic reconciliation.)
異文化の相互理解はアボリジニ先住民族と白人の少女が手をつないでいた開会式の場面に象徴されるが、盛り上がりを見せたこの五輪をアボリジニはどのように感じたのか。
(Although mutual understanding among different cultures was symbolized by the scene that aboriginal people held hands with the white girls in the opening ceremony, how did the aboriginal people feel about this Olympic Games?)
アボリジニ市民評議会のハーブシムズ幹事に話を聞くと相互のコミュニケーションの重要性を訴えている。
(The secretary of the Aboriginal citizen congress, Mr. Herb Simms, made an appeal about the significance of mutual communication.)
(2)The deleted words indicated by [ ] ->
全17日の日程を終えてシドニー五輪は終了した。
[移民の：immigrant] [国：nation]オーストラリアで開催する五輪が[掲げた：set up][テーマは：theme]民族の融和だったのだと言われる。
異文化の 相互理解は [アボリジニ：aborigine] [先住民族と：aboriginal people] [白人の：the white race] [少女が：girl] [手を：hand] [つないでいた：held hands with] [開会式の：opening ceremony] [場面に：scene] 象徴されるが [盛り上がりを：upsurge] [見せた：showed] [この：this] 五輪を アボリジニは どのように 感じた のか?
[アボリジニ市民評議会の：Aboriginal citizen congress] [ハーブシムズ幹事に：Secretary Herb Simms] [話を：talk] [聞くと：ask] [相互の：mutual] [コミュニケーションの：communication] [重要性を：significance] [訴えている：appeal] 。
(3)The summary->
全17日の日程を終えてシドニー五輪は終了した。
(The Sydney Olympic Games ended over 17 days.)
オーストラリアで開催する五輪が民族の融和だったのだと言われる。
(It is said that the Olympic Games held in Australia represented ethnic reconciliation.)
異文化の相互理解は象徴されるが五輪をアボリジニはどのように感じたのか?
(Although mutual understanding among different culture was symbolized, how did the aboriginal people feel about this Olympic Games?)

---

**Fig. 2.** Output example of ABISYS

## 2.2 System Output

The output of ABISYS includes 3 parts: the original document, the rewritten document with the deleted words shown with "[" and "]", and the summary as shown in Fig. 2. For the readers' better understanding, we give their English translation as well.

---

[2] http://www.iijnet.or.jp/edr/index.html
[3] POS in this paper represents part of speech.

## 2.3  Summarizing Process

ABISYS summarizes a document in 4 steps. First, the important words are extracted. Then the element words to compose a summary are selected according to the important words extracted in the first step. After that, the unnecessary verbose words are deleted from the element word list. Finally, using the case-frames of the remaining element words, correct and natural sentences are generated.

We define five  kinds of scores to evaluate the importance of a word. In the next section, we give detailed description for each of them.

# 3  Scores for Importance Evaluation

Generally, indeclinable words[4] tend to show the key point of a document. Along this line in this paper, we take indeclinable words as the candidates for selecting important words. We define five kinds of scores to evaluate the importance of a word respectively on repetition information, context information, position information, opinion information and topic-focus information. The probability of an indeclinable word to be important is calculated by consolidating the five scores. All indeclinable words are ranked in a list by their probabilities and important words are extracted from the list in a descending sequence until the summarizing rate is reached.

## 3.1  Repetition Information Score

It is said that a repetition word tends to show the entire content of a document and to remain in the summary [17]. Here, a repetition word indicates a word or its synonym that appears twice or more times in different sentences in a document. Referring to this heuristics, we assign scores to repetition words. We first find out all repetition words from the case-frame set, then calculate their frequencies of appearance. Here in the calculating process, we count the numbers of the case-frames that have the same 'dictionary entry denotation', and the numbers of the case-frames whose 'dictionary entry denotation' are close to each other in the EDR Concept Dictionary (in fact we limit the distance to be 0 or 1). Finally, a score of 20, 30, 40, or 50 points is given to the repetition word if its frequency of appearance is 2-3, 4-5, 6-7, or 8 and over.

```
frame(2,'茶',‥ ;'JN1', ‥·,'茶', ···,'odc56a', ‥).

frame(76,'茶',‥ ;'JN1', ‥·,'茶', ·····).

frame(163,'飲み物', ···,'JN1', ···,'1032b4', ···).
```

**Fig. 3.** An example of repetition word "茶"

Here is an example. Suppose we have a document about "茶" (tea), and its case-frame set as shown in Fig. 3. Here, "茶" is a repetition word as it appeared in two

---

[4] An indeclinable word in Japanese is a substantive word.

case-frames with the 'number' of 2 and 76. Further more, we know that the word "飲み物" (liquid refreshments) is close to the word "茶" in the thesaurus defined by EDR Concept Dictionary (distance = 1 as shown in Fig. 4). That is to say, we would take into account the case-frame of "飲み物" as well when assigning scores to the repetition word "茶".

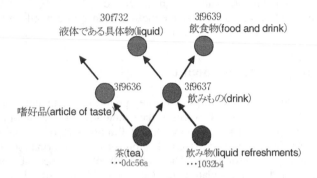

**Fig. 4.** Concept distance

## 3.2  Context Information Score

In most cases, semantic relation holds between two adjacent sentences. Kawabata etc. have built a system InSeRA, which determines the semantic relation between two

**Table 1.** Scores assigned according to the semantic relation between sentences

| Semantic relation | Meaning | Score |
|---|---|---|
| inter-reason(S1,S2) | reason | S2+20 |
| inter-cause(S1,S2) | cause | S2+20 |
| inter-enable(S1,S2) | possibility | |
| inter-purpose(S1,S2) | purpose | |
| inter-change(S1,S2) | change of state | |
| inter-detail(S1,S2) | detail | S2-30 |
| inter-example(S1,S2) | example | S1+30,S2-30 |
| inter-explanation(S1,S2) | explanation | S2-30 |
| inter-answer(S1,S2) | question and answer | |
| inter-contrast(S1,S2) | contrast | |
| inter-contradiction(S1,S2) | contradiction | S1+30,S2+20 |
| inter-or(S1,S2) | choice | |
| inter-logicalCondition(S1,S2) | logical condition | S2+20 |
| inter-timingCondition(S1,S2) | time condition | |
| inter-subjunctive(S1,S2) | subjunctive | |
| inter-sequence(S1,S2) | sequence in time | S2+5 |
| inter-cooccurrence(S1,S2) | duplication in time | |
| inter-synchronous(S1,S2) | concurrent start | |
| inter-equivalent(S1,S2) | coordination | S2+10 |
| inter-conversion(S1,S2) | conversion | S2+10 |
| inter-parallel(S1,S2) | parallel | |

sentences [7] using the 21 relations defined in Table 1. We made an investigation and it turned out that the former or the latter sentence tends or not to remain in the summary for some certain semantic relation between them. Along this line, we assign the indeclinable words in the former or the latter sentence some adequate scores as shown in Table 1 according to the particular semantic relation between the two sentences.

### 3.3  Position Information Score

Usually, the beginning sentence is crucial in the content of a newspaper article, and the final sentence talks mostly about the future of the current theme. The same thing can be said about a leading article too, where both the lead sentence and the last sentence tend to act as a summary. Many summarizing studies adopt this simple but effective position information. In this paper, we also take account of the position of a sentence. We assign the indeclinable words in the first sentence 30 points each, the ones in the second sentence 10 points each, and the ones in the last sentence 20 points each.

### 3.4  Opinion Information Score

For a leading article, the argument, suggestion, or wish of the author is regarded most important, and ought to be left in the final summary. Ohtake collected 55 patterns for the argument or suggestive expression in Japanese [14]. In this paper, we make an induction from Ohtake's efforts. We consolidate all the expressions Ohtake collected into 3 opinion concepts 考える('consider', 30f878), 思考する('think', 444dda), and 表現する('express', 30f86f). Any word that takes either of the 3 opinion concepts as its upper concept in EDR Concept Dictionary is called an opinion word. A word holding deep case with an opinion word will be assigned an opinion information score (10 points).

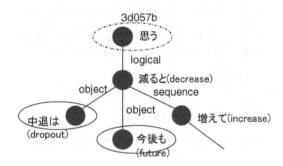

**Fig. 5.** An example to assign opinion information score

For instance, 思考する(444dda) is the upper concept of the word 思う in Fig. 5 implying that 思う is an opinion word. As a result, the two indeclinable words 中退 and 今後 are assigned 10 points as they are linked to 思う in deep case.

### 3.5  Topic-Focus Information Score

The topic and the focus of a document generally remain in the summary. However, it is not easy to identify neither the topic nor the focus strictly. Here we take the following steps to find the topic and the focus of a document, and assign the corresponding words the topic-focus information scores.

Step1: Search the expressions like

(noun)A+ (particle)は/も/が (topic) or

(noun)A+ (particle)を/に/で (focus).

Step2: If A represents an agent-, object-, or goal-case of another word, assign A 20 points.

## 4  Consolidation of the Scores

The method for consolidating a number of scores so far is to multiple each score with its weight, and then to add all the products together to obtain general score. However, the weights used are generally given empirically, and hence not statistically reliable. We propose here a method to calculate the general score using a pan-distance of Mahalanobis. We first select 1000 sample words from documents of the same catalogue, and divide them into two sets: valuable ($v$) and non-valuable ($n$) by handcraft according to the content of documents. Then we set up a five-dimensional space with the repetition information, context information, position information, opinion information and topic-focus information as the five dimensions. Finally, we reflect the sample words of the two sets in the five-dimensional space according to the coordinate of each sample word.

The general score of an important word candidate is calculated using the distance between the candidate word and each of the 2 sets in the five-dimensional space. The procedure of calculation is described below.

Let $x_1, x_2, x_3, x_4, x_5$ denote the repetition information, context information, position information, opinion information and topic-focus information, and get the distribution-covariance matrix for the valuable set and the non-valuable set respectively.

Let the distribution-covariance matrix and its inverse matrix be

$$V_v = \begin{pmatrix} s_{11} & \cdots & s_{15} \\ \vdots & \ddots & \vdots \\ s_{51} & \cdots & s_{55} \end{pmatrix}, \quad V_v V_v^{-1} = I \tag{1}$$

and the average vector of the valuable set be $u_v$ and the score vector of the important word candidate $X$ be

$$u_v = (x_{v1}, \cdots, x_{v5}), \quad X = (x_1, \cdots, x_5). \tag{2}$$

Then the Mahalanobis distance $D_v$ from the valuable set to the important word candidate will fulfill the following equation,

$$D_v^2 = (X - u_v) V_v^{-1} (X - u_v). \tag{3}$$

The Mahalanobis distance from the non-valuable set to the important word candidate $D_n$ is calculated in the same way. Finally, we calculate $f_v$ and $f_n$ as shown below,

$$f_v = e^{-\frac{1}{2}D_v^2}, \quad f_n = e^{-\frac{1}{2}D_n^2} \tag{4}$$

and get $P_v$, the probability of the important word candidate to be a valuable word. This is the final score we have been seeking so far.

$$P_v = 100 \times \frac{f_v}{f_v + f_n} \tag{5}$$

Here, we switch the number of dimensions to 4 when processing a newspaper article for the reason that the opinion information score is employed only in the summarizing process of leading articles. We observed from the real distribution-covariance matrix obtained in the experiment for summarizing leading articles that repetition information score is the most crucial one among the 5 variables used, and the importance of the left 4 are in a descending sequence of position information, topic-focus information, context information, and opinion information.

## 5   Generation of Sentence

Sentences are generated using words from the valuable word list that has been achieved in section 4. We employ two strategies to guarantee the correctness and the naturalness of the generated sentences.

### 5.1   The Use of Indispensable Case

Indispensable case is a necessary element in order to generate a meaningful and correct Japanese sentence whose focus is on the declinable word. A simple method to identify such an indispensable case is to find the postpositional particle が or を around the verb and take the word in front of が or を as the indispensable case. However, an indispensable case does not always appear with a postpositional particle. Sometimes, an indispensable case does exist for a verb in a sentence while no surface information exposes this fact.

In a study to summarize newspaper articles, Hatayama etc. used a case-frame dictionary to compensate indispensable cases for a verb [4]. As their method compensates for the primary verb only, in case that multiple verbs appear in a sentence, indispensable cases will not be compensated properly except for the primary verb. In another study by Minami etc., in order to compensate indispensable cases for verbs, threshold values are assigned to the agent-case and the object-case with 0.08 and 0.15 respectively [11]. Threshold values used here are empirically set and accordingly untrustworthy.

In this paper we propose a method to determine the threshold values for each deep case statistically. Specifically, we first select 1000 declinable words from Asahi Shimbun[5], then search the EDR Co-occurrence Dictionary for the co-occurrence of each declinable word and each of its deep cases, $x$, finally calculate the average ap-

---

[5] Asahi Shimbun is a daily newspaper in Japan.

pearing ratio of each deep case and take it as the threshold value of the deep case[6], called $m_x$. When $\theta_x$, the appearing ratio of a deep case $x$ for a declinable word $v$ is larger than $m_x$, we assign $x$ to $v$ as one of its indispensable cases.

$$\theta_{v,x} = \frac{frequency of \; x \; for \; v}{frequency of \; all \; deep \; cases \; for \; v} \tag{6}$$

### 5.2 Extraction of Summary Element Words

After compensating indispensable cases for the declinable words, we select the summary element words in accordance with the valuable words extracted in section 4. This process is composed of 4 steps. Every sentence here is represented as a dependency tree holding the primary predicate verb as the root.

> Step1 Select all words on the path starting from the valuable word and ending at the primary predicate verb. Also select the indispensable cases of a declinable word or a Sahen noun[7], if any exists along the path.
> Step2 Select the declinable word $v_1$ that is linked to a valuable word in deep case, then select the declinable word $v_2$ that is linked to $v_1$ with the deep case of reason, cause, or sequence, and also select the indispensable cases of $v_1$ and $v_2$.
> Step3 Select the word that is adjacent to the valuable word downwards in the dependency tree.
> Step4 Select the which- and that-case of the previously selected summary element words containing the characters こと, の, or という, as they tend to associate with some other words to mean something important.

By Step1, we get the proper expression to show how the valuable word is mentioned. Then in Step2 and Step3, we compensate the words that hold important relations with the valuable words. Finally, we make up for the words that can hardly express anything meaningful individually in Step4. In this way, we attempt to extract the necessary summary element words to generate the sentence as correct as possible.

## 6   Shortening of the Sentence

The final process of the summarizing task is to generate sentences using the summary element words extracted in Section 5. However, even among the element words, there exist some duplicated ones or the ones which are not that valuable as they have been assessed in Section 5. Here, we are to find them and delete them in order to generate shorter and briefer sentences. Specifically, we focus on the linguistic phenomena, quoting verb.

A quoting verb is a verb leading a quotation, and often appears in a leading article. We follow the steps shown below to delete the quoting verbs. Fig. 6 and Fig. 7 are an

---

[6] The reason we take the average as the threshold value is on the purpose to be able to generate correct sentence by compensating as more indispensable cases as possible.

[7] Sahen noun is a special type of noun in Japanese that holds the function of verbs.

example of deleting the quoting verb in the expression とんでもないと思うのだが (I think it is unbelievable).

Step1  Find the declinable word A among the summary element words whose upper concept in EDR Concept Dictionary is 考える(30f878) or 思考する(444dda).

Step2  Find the word B in the dependency tree with which A holds a logical-, time-, or purpose-case.

Step3  If B is a declinable word and the postpositional particle following B is と, ように, or とか, estimate A to be a quoting verb and delete it.

Step4  Delete all deep cases of A in the dependency tree except B, the part that has been considered to be a quotation.

---

frame(192,'とんでも','···','fffff',[[head,192]],[16,11.1]).
frame(193,'ない','···','2621be',[],[16,11.2]).
frame(194,'とんでもない','···','JPR','と','000000',[[consist,192],[consist,193]],[16,11]).
· · · · · · · · · · · · · · · · · · · · · · · · · · · · ·
frame(208,'思う','···','JVE','···','3cfa55',[[logical,192],[nil,209]],[16,12]).
frame(209,'の','···','1033a2',[[which,208],[head,209],[main,209]],[16,13,1]).
frame(210,'だ','···','2621c6',[],[16,13,2]).
frame(211,'のだ','···','000000',[[consist,209],[consist,210]],[16,13]).

---

**Fig. 6.** Case-frames of the example

With the procedure in mind, let us see the sentence とんでもないと思うのだが. First we get that the upper concept of 思う is思考する(444dda), and that 思う holds a logical-case with とんでも. Then, we find that とんでも is a constituent of the compound word とんでもない whose POS is JPR implying a declinable word. Finally, we see that とんでもない is followed by the postpositional particle と. In other words, we make a conclusion that 思う is a quoting verb, and accordingly delete all deep cases of it except the quotation as shown in Fig. 7.

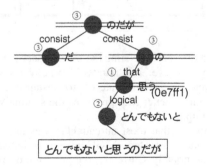

**Fig. 7.** Deletion of the quoting verb

## 7  Evaluation

It has been very difficult to evaluate the performance of a text summarizing system objectively. To make the issues clear, the first text summarization challenge was offered in the NTCIR Workshop[8] 2 (NTCIR-2) held in 2001 for Japanese documents. In this paper, we evaluate ABISYS using the subjective evaluation method designed in NTCIR-2. In addition, we evaluate the output of our system for its correctness. The former evaluation was made by 5 graduate students and the latter was by the authors.

### 7.1  Subjective Evaluation

Human judges evaluate and rank 4 types of summaries in 1 to 4 scale (1 is for the best and 4 is for the worst) in terms of two points of views, the covering rate in content of each summary and the readability of it [12]. The first two summaries are human-produced summaries. The third is the one produced by the baseline method of sentence extraction, and the fourth is the system output.

We use the data prepared in NTCIR-2 for the subjective evaluation. In Table 2, we compared the evaluating results between ABISYS and the systems that participated in NTCIR-2 employing 20% as the summarizing rate. It is clear that the point of readability of ABISYS is on a par with other systems, while the point of content covering is much better than others.

**Table 2.** Results of subjective evaluation

| System | Content covering | Readability |
|--------|------------------|-------------|
| I | 3.4 | 3.1 |
| II | 3.0 | 2.5 |
| III | 3.5 | 3.8 |
| IV | 3.4 | 3.4 |
| V | 3.2 | 2.7 |
| VI | 3.3 | 3.3 |
| VII | 3.1 | 3.0 |
| Lead | 3.3 | 3.2 |
| Ave. | 3.25 | 3.11 |
| ABISYS | 2.65 | 3.11 |

The summary produced by merely deleting the unnecessary parts tends to be highly evaluated in its readability. However, it is impossible to cover the total content of the original text with the given summarizing rate of, for instance, 20% or 30%. In that regard, ABISYS behaves much better as it produces the summary by generating sentences from semantically important words.

Further examination reveals that most unreadable cases are due to the incorrect output from the semantic analysis system SAGE, where the indispensable cases have not been compensated appropriately. Currently, the precision of morphological analy-

---

[8] http://research.nii.ac.jp/ntcir/index-en.html

sis and dependency analysis have reached about 90% [8,10], and SAGE also shows a precision of about 90% for the analysis of both word meaning and deep case [9]. We believe that the readability of the summary produced by ABISYS will improve further if the mistakes occurred in these preceding analyses could have been removed.

### 7.2   Evaluation of Linguistic Correctness

It is necessary to examine the linguistic correctness of the generated sentences in a sentence-generation based summarizing work. Here a linguistically correct sentence means a sentence holding all the necessary and correct deep cases, and being easy-to-read as Japanese. We randomly select 10 newspaper articles and 10 leading articles and produce summaries for them using ABISYS. The result of the evaluation shows that about 95% of the summarized sentences have been acknowledged as correct Japanese sentences.

## 8   Conclusion

We have developed a summarizing system ABISYS. Specifically, using the case-frames output from the semantic analysis, we assign scores to corresponding words in accordance with repetition information, context information, position information, opinion information and topic-focus information. Then we determine the valuable words by consolidating the five scores statistically. Finally, we attempt to generate correct and natural summary by compensating the indispensable deep cases and extracting the necessary summary element words.

The comparison with the subjective evaluations made for other summarizing systems in NTCIR-2 indicates that our system is on a par with other systems in regard to the readability, while the point of content covering is much better. And about 95% of the summary sentences generated by ABISYS are acknowledged as correct Japanese.

## Acknowledgements

This work has been supported in part by the Ministry of Education, Culture, Sports, Science and Technology, under Grants-in-Aid for Scientific Research (C) 13680461.

## References

1. Hahn, U., and Reimer, U. Topic essentials. In *Proceedings of the 11th COLING* (Bonn, Germany, 1986). 497-503.
2. Harada, M., and Mizuno, T. Japanese semantic analysis system SAGE using EDR. *Transactions of the Japanese Society of Artificial Intelligence, 16,* 1(2001), 85-93. (in Japanese)
3. Harada, M., Tabuchi, K., and Oono, H. Improvement of speed and accuracy of Japanese semantic analysis system SAGE and its accuracy evaluation by comparison with EDR corpus. *Transactions of Information Processing Society of Japan, 43,* 9(2002), 2894-2902. (in Japanese)

4. Hatayama, M., Matsuo, Y., and Shirai, S. Summarizing newspaper articles using extracted informative and functional words. *Journal of Natural Language Processing, 9*, 4 (2002), 55-70. (in Japanese)

5. Hovy, E. H., and Lin, C. Automated text summarization in SUMMARIST. In *Proceedings of the ACL97/EACL97 Workshop on Intelligent Scalable Text Summarization.* (Madrid, Spain, 1997).

6. Ishizako, Y., Kataoka, A., Masuyama, S., and Nakagawa, S. Summarization by reducing overlaps and its application to TV news texts. *SIG Notes, NL*-133 (1999), 45-52. (in Japanese)

7. Kawabata, T., and Harada, M. Development research on system InSeRA which analyzes the semantic relations between Japanese sentences. *SIG Notes, NL*-142 (2001), 105-112. (in Japanese)

8. Kurohasi, S., and Nagao, M. *Nihongo Koubun Kaiseki System KNP manual (version 2.0b6).* http://www.kc.t.u-tokyo.ac.jp/nl-resource/index.html. 1998.

9. Maezawa, T., Menrai, M., Ueno, M., Han, D., and Harada, M. Improvement of the precision of the semantic analysis system SAGE, and generation of conceptual graph. In *Proceedings of the 66th National Conference of Information Processing Society of Japan*, 2 (2004), 177-178. (in Japanese)

10. Matsumoto etc. *Morphological analysis system chasen version 2.2.8 manual.* http://chasen.aist-nara.ac.jp/hiki/ chasen. 1999.

11. Minami, A., and Harada, M. Development of anaphoric analysis system which uses similarity of vocabulary. In *Proceedings of the 64th National Conference of Information Processing Society of Japan*, 2 (2002), 53-54. (in Japanese)

12. Nanba, H., and Okumura, M. Analysis of the results and evaluation methods of text summarization challenge (TSC), a subtask of NTCIR Workshop 2. *Technical Report of IEICE, NLC2001*-28 (2001), 46-52. (in Japanese)

13. Oguro, R., Ozeki, K., Zhang, Y., and Takagi, K. A Japanese sentence compaction algorithm based on phrase significance and inter-phrase dependency. *Journal of Natural Language Processing, 8*, 3 (2001), 3-18. (in Japanese)

14. Ohtake, K., Okamoto, D., Kodama, M., and Masuyama, S. A summarization system YELLOW for Japanese newspaper articles. *Transactions of Information Processing Society of Japan, 43*, SIG2 (2002), 37-47. (in Japanese)

15. Okumura, M., and Nanba, H. Automated text summarization: a survey. *Journal of Natural Language Processing, 6*, 6 (1999), 1-26. (in Japanese)

16. Okumura, M., and Nanba, H. New topics on automated text summarization. *Journal of Natural Language Processing, 9*, 4 (2002), 97-116. (in Japanese)

17. Sakuma, M. *Bunshou Kouzou To Youyakubun No Showou (Version 3)* Kurosio Publisher, 2000. (in Japanese)

18. Ueda, Y., Oka, M., Koyama, T., and Miyauchi, T. Development and evaluation of a summarization system based on phrase-representation summarization method, *Journal of Natural Language Processing, 9*, 4. (2002), 75-96. (in Japanese)

# Sentence Compression Learned by News Headline for Displaying in Small Device

Kong Joo Lee[1] and Jae-Hoon Kim[2]

[1] School of Computer & Information Technology,
KyungIn Women's College, Gyeyang-gu, Incheon 407-740 Korea
kjoolee@kic.ac.kr
[2] Dept. of Computer Engineering,
Korea Maritime University,
Yeongdo-gu, Busan 606-791 Korea
jhoon@mail.hhu.ac.kr

**Abstract.** An automatic document summarization is one of the essential techniques to display on small devices such as mobile phones and other handheld devices. Most researches in automatic document summarization have focused on extraction of sentences. Sentences extracted as a summary are so long that even a summary is not easy to be displayed in a small device. Therefore, compressing sentences is practically helpful for displaying in a small device. In this paper, we present a pilot system that can automatically compress a Korean sentence using the knowledge extracted from news articles and their headlines. A compressed sentence generated by our system resembles a headline of news articles, so it can be one of the briefest forms preserving the core meaning of an original sentence. Our compressing system has shown to be promising through a preliminary experiment.

## 1 Introduction

In recent years, wireless access to the WWW through mobile phones and other handheld devices has been growing significantly. Also new models of such mobile and handheld devices have been developed rapidly, however, there are many shortcomings associated with these devices, such as small screen, low bandwidth, and memory capacity. As the limited resolution and small screen restrict the amount of information to be displayed, it is very difficult to load and view a large document on the devices. Summarizing a document is currently becoming one of the most crucial solutions for dealing with this problem on these devices. Most automatic summarization systems extract important sentences from a source document and concatenate them together according to several criteria [5]. Sentences extracted as a summary tend to be so long and complex that even a summary is not easy to be displayed in a small device [15]. Therefore, compressing a sentence into an abstraction is practically helpful to display in a small device.

Let us pay attention to a news article. In general, a news article is composed of a headline and a body. A headline of a news article is the most concise summary

S. H. Myaeng et al. (Eds.): AIRS 2004, LNCS 3411, pp. 61–70, 2005.

that represents a body. Also, Wasson [12] reported that a body usually begins with a good summarizing leading text. Generally speaking, a headline is shorter than a leading text in most cases. Therefore, our work is based on the premise that a headline can be considered as a compressed form of a leading text in news articles. We will refer a first sentence of a first paragraph in a news article as the *leading sentence*, and the compressed form for a leading sentence as *compressed sentence* in this work. Let us take a look at the following example.

| ① | ② | ③ | ④ | ⑤ | ⑥ | ⑦ | ⑧ | ⑨ |

(1a) *대구시/는* 빠른 시일에 *경북도와* *시행정구역에* 대한 공식적인 *협의를* 갖기로_했다.

Daegu City  within a few days  with KB province  on administraticve region  official  discussion  will have
(SUBJ)                                                                                      (OBJ)

*meaning*: Daegu City will have an official discussion on administrative region with KB province within a few days.

(1b) 대구시   경북도와   시행정구역   협의
Daegu City  with KB province  administraticve region  discussion

The sentence (1a) is the leading sentence extracted from a news article and the sentence (1b) is the headline corresponding to (1a). Imagine that we make a headline for the given sentence (1a). As you can see, the relatively unimportant words are dropped from the leading sentence (1a), and then the remaining words $1^{st}$, $4^{th}$, $5^{th}$ and $8^{th}$ of (1a) build up the headline (1b). When generating a headline, even some postpositions that are attached to the tail of words are omitted from the words in the leading sentence. The postpositions (in grayed boxes) in (1a) are deleted from the words $1^{st}$, $5^{th}$ and $8^{th}$, respectively.

In an examination of a large number of news articles, we found three basic ideas how to convert a leading sentence into a headline: The first is to remove inessential words such as additional adjunct phrases from leading sentences. The second is to replace longer and complex expressions with shorter and simpler ones. The third is to omit functional words such as postpositions in Korean if the meaning of a sentence remains same even without the functional words.

In this paper, we present a model that can automatically learn how to generate a compressed sentence from the collection of pairs of leading sentences and their headlines. A compressed sentence generated automatically resembles a headline of news articles, so it can be one of the briefest forms preserving the core meaning of an original sentence. Also, this approach can be easily applicable to other languages only if a parallel corpus for the language is available online.

This paper is organized as follows: Section 2 introduces the related works briefly. Section 3 and 4 describe a model and a system for sentence compression respectively. Section 5 discusses results and an evaluation of our system. Finally we draw conclusions in Section 6.

## 2  Previous Approaches to Sentence Compression

Many researches on automated text summarization have been performed since late 1950s and progressed very briskly while the amount of online text on the Internet keeps growing and growing [9]. Recently the researches have focused on extractive summarization, which extracts the most important sentences in texts and re-arrange them to be more readable and coherent. In this paper, we use leading sentences as summaries based on lead-based approach [12]. In case that the length of the leading sentences is very short, the leading sentences are enough as summaries for mobile devices and other handheld devices with limited screen size. To lighten this problem on limited screen size, a few of researchers have started to address the problem of generating compressed sentences.

The system proposed by Knight and Marcu [6] generates grammatical and coherent abstracts that capture the most important pieces of information in the original documents. To do this, it uses a noisy-channel and a decision-tree approach. Their basic idea is originated from sentence-to-sentence translation based on a noisy-channel model, that is, they rewrite documents as source sentences into abstracts as target sentences. This works had been improved by their colleague, Lin [8] by using the concept on re-ranking using an oracle.

Yang and Wang [14] developed fractal summarization based on the fractal theory. At the first step of the fractal summarization, a brief summary is generated, and then a detail summary is produced on demands of users interactively. In this work, the important information is captured from the source text by exploring the hierarchical structure and salient features of the document. The compressed document is produced iteratively using the contractive transformation in the fractal theory.

The basic concept of Balazer [1] is very similar to that of Knight and Marcu [6]. This work is to learn how to decide what words or constituents can be dropped using a neural network model of Weka machine learning tool [13] and generates grammatically compressed sentences along with the number of words in the original sentence, the number of words in the constituent of a rewriting rule, and the depth of the constituent (distance to the root).

Vandeghinste and Tjong Kim Sang [11] described a method to automatically generate subtitles for the deaf from transcripts of a television program, and used a hybrid method combining a statistic approach and a rule-based approach. The statistic approach was used for getting a ranking in the generated sentence alternatives and the rule-based approach for filtering out ungrammatical sentences that are generated.

Carroll et al. [2] proposed a method for text simplification to assist aphasic readers. This method is composed of two components: one is a sentence analyzer which provides a syntactic analysis and disambiguation of the newspaper text, and the other is a simplifier which adapts the output of the analyzers to aid readability of aphasic people. The analyzer is language-dependent because it can treat the special syntax like passive voice and complex sentences. Also this system has some difficulties in learning rules on the text simplification.

# 3   Modeling for Sentence Compression

## 3.1   Training Corpus

Our training corpus is the collection of pairs of a leading sentence and its headline of news article, which is one of the data sources that we can easily gather. For the sake of simplifying a model, we collect only the pairs that satisfy the following 4 conditions. Assume a leading sentence $S_l = (w_1, w_2, ..., w_N)$ and its headline $S_c = (x_1, x_2, ..., x_M)$;

C1: $N \geq M$

C2: exists a simple function $g : S_c \rightarrow S_l$; it indicates that every word in $S_c$ is mapped into one of the words in $S_l$, but not vice versa.

C3: $position(g(x_i)) < position(g(x_j))$ in a leading sentence, for all $position(x_i) <$ $position(x_j)$ in a headline; where $position(word)$ returns the position of a $word$ in a sentence; it means that the words in both $S_l$ and $S_c$ occur in the same order.

C4: $g(x) = x$ or $stem(g(x)) = stem(x)$, where $stem(x)$ returns a stem of a word $x$; it says that the common words to $S_l$ and $S_c$ have the same stems.

The pair of the sentences (1a) and (1b) satisfies the above 4 conditions completely. We collect 1,304 pairs automatically from the news article corpus [10], which satisfy all the above conditions.

## 3.2   Marked Tree

A marked tree for a sentence is a syntactic dependency tree with marked nodes. Its nodes are labeled over a word, its part-of-speech, and a mark symbol (0 or 1) and its edges are also labeled over syntactic categories between nodes. A leading sentence is syntactically analyzed into a marked tree. The nodes for the words common to a leading sentence and its headline are marked with 1 in a marked tree. The nodes for the words not included in a headline are marked with 0. Fig. 1 as an example shows the marked tree for the sentence (1a). In Fig. 1, the mark symbol on the node is actually represented by a background color of white (0) or gray (1). The nodes corresponding the words $1^{st}$, $4^{th}$, $5^{th}$ and $8^{th}$ are marked with 1 because they are also included in the headline (1b).

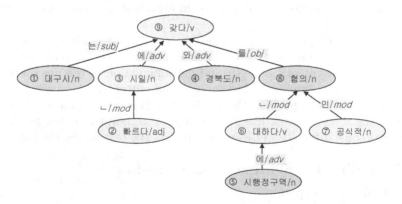

**Fig. 1.** A marked tree for the leading sentence (1a)

## 3.3    Modeling for Sentence Compression

Our main goal is to generate compressed sentences resembling headlines of news articles. This is compression problem which is to produce a shorter one from an original longer sentence. In this work, a compressed sentence is generated from the marked nodes (henceforth a node marked with 1 is simply referred as 'marked node') in a marked tree, so it is very important to determine which nodes are marked with '1' out of all nodes in a syntactic dependency tree. To achieve this goal, we can compute the score of a marked tree for a given sentence as in [3] and generate a compressed sentence with marked nodes in a marked tree.

Consider a syntactic dependency tree to be modeled. Let $n_i$ be a node which is composed of the $i$-th word $w_i$, its POS tag $t_i$, and its mark $m_i$, that is $n_i = (w_i, t_i, m_i)$, and let $h(i)$ be a position of the head of the node $n_i$ in a sentence, and let $r_i$ be a relation between the node $n_i$ and its head $n_{h(i)}$ and $e_i$ be an edge connecting $n_i$ and $n_{h(i)}$ with $r_i$, that is $e_i = (n_i, r_i, n_{h(i)})$. Then let $T_m$ be a marked tree represented as $T_m = (e_1, e_2, ..., e_N)$ and the score of the marked tree $T_m$ can be calculated by Equation (1):

$$Score(T_m) = \prod_{i=1}^{N} Pr(m_i = 1|T_m) \qquad (1)$$

where $N$ is the number of the nodes in a marked tree. Our goal is to find the marked tree that maximizes Equation (1). It is not easy to estimate $Pr(m_i = 1|T_m)$, therefore it can be approximated as Equation (2).

$$Pr(m_i = 1|T_m) = Pr(m_i = 1|e_1, e_2, ..., e_N)$$
$$\cong Pr(m_i = 1|e_i) = Pr(m_i = 1|n_i, r_i, n_{h(i)}) . \qquad (2)$$

It can be estimated using the maximum likelihood estimation. In an experiment described later, we will smooth the score of Equation (2) using Equation (3) to avoid the data sparseness problem.

$$Pr(m_i = 1|n_i, r_i, n_{h(i)}) \cong Pr(m_i = 1|t_i, r_i, t_{h(i)}, m_{h(i)})$$
$$= \frac{count(m_i=1, t_i, r_i, t_{h(i)}, m_{h(i)})}{count(t_i, r_i, t_{h(i)}, m_{h(i)})} \qquad (3)$$

where $count(x)$ is a frequency of $x$. In order to simplify the calculation of Equation (1), we assume that the number of dependency trees for a given sentence is only one. Suppose that a dependency tree has $N$ nodes. Then we should calculate the scores for the possible $2^N$ marked trees to find the marked tree that maximizes Equation (1). To compute the calculation more efficiently, we use a greedy method which is not optimal, but near-optimal. The selection policy of the greedy method first selects the highest significance node described immediately below. The higher the significance of the node is, the higher the possibility that the node over the other nodes is marked with 1 is. The significance of a node is propositional to how many times the node is included in a compressed sentence, how many times the node given its parent is marked is included in a compressed sentence, and the inverse of a node depth. The significance $M(n_i)$ of a node can be defined as Equation (4).

$$M(n_i) = Pr(m_i = 1|t_i, r_i) \times Pr(m_i = 1|m_{h(i)} = 1, t_i, r_i, t_{h(i)}) \times \frac{1}{d(n_i) + \alpha} \quad (4)$$

where $d(n_i)$ is the depth of the node $n_i$ in a dependency tree, and $\alpha$ is a constant. The first and the second of the right-hand side of Equation (4) will be estimated as the same manner as Equation (3).

## 4      Sentence Compression

Our system for sentence compression puts three steps together as shown in Fig. 2. The first step is syntactic parsing which analyzes an input sentence and produce a syntactic dependency tree. The second step is node-marking which decides if each node on the dependency syntactic tree is marked or not. The final step is surface realization which selects a lexical string and then generates a compressed sentence as a final output.

### 4.1      Syntactic Parsing

Using a Korean probabilistic parser [7], we parse an input sentence. The original parser produces restricted phrase structures of a sentence although dependency structures are required in this work. The two structures are convertible [4] and it is not difficult to convert each other. In this work, we modify the original parser in order to provide dependency structures, which consist of nodes and relations as you can see in Fig. 1.

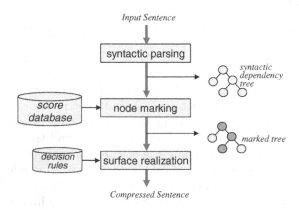

**Fig. 2.** The process for compression sentences

---

node_marking($T, N$)
**begin**
    $score_{max} = 0$;
    $tm_{max} = $ NULL;
    calculate $M(n_i)$ for all nodes $n_i$ in $T$ using Equation (4);
    **for** $k = 1$ **to** $N$ **do**
        choose the most proper marked tree $T_m$
            with the maximum $Score(T_m)$ of Equation (1) for $k$ marked nodes;
        **if** ( $\sqrt[k]{Score(T_m)} > score_{max}$ ) **then**
            $score_{max} = \sqrt[k]{Score(T_m)}$;
            $tm_{max} = T_m$;
        **end if**
    **end for**
    **return** $tm_{max}$;
**end**

---

**Fig. 3.** The algorithm to find the marked tree with the best score

## 4.2 Node Marking

Node marking is to determine whether each node in a syntactic dependency tree is included in a compressed sentence, or not. As previously mentioned, we use a greedy method to find a maximum marked tree with $K$ marked nodes. The significance of each node can be calculated by Equation (4). We set the constant $\alpha$ in Equation (4) as 2.0 in this version. The simplified algorithm to determine a marked tree with the best score is shown in Fig. 3. The score of a marked tree in Equation (1) is an absolute score, which can not reflect the difference of the number of marked nodes. Therefore, the score cannot be directly compared with each other in case that there is a difference in the number of marked nodes. For the sake of compensation of the difference in the number of marked nodes, the score is normalized by the average of products, $\sqrt[k]{Score(T_m)}$ as you can see in Fig. 3.

## 4.3 Surface Realization

Surface realization takes a marked tree and produces a surface form as a compressed sentence by solving some morphology and syntactic problems. As our system is designed to generate a compressed sentence of which word order is same as that of an original sentence, the word order in surface realization is beyond question. In this work, there are basically three ways to realize surface strings for a compressed sentence: *as-is*, *omission*, and *replacement*. The way of *as-is* produces an original word exactly as it is without any changes. The way of *omission* generates a reduced form of an original word which omits functional morphemes like postpositions. The way of *replacement* brings forth a totally different form, which preserves the meaning of an original word. The *replacement* is kind of paraphrase in short form such as abbreviation, Korean sino-word[1] and so on. How to realize a surface form in a compressed sentence is surprisingly important as it can affect the fluency and understandability of a compressed sentence. In the following sentences, (1c) is the list of the marked nodes in the marked tree; (1b) is the headline for the sentence (1a). An input for the surface realization is (1c), and an output should be a similar one to (1b).

(1b)  대구시      경북도와      시행정구역      협의
      (Daegu City)   (with KB province)   (administraticve region)   (discussion)

(1c)  대구시는      경북도와      시행정구역 에      협의를
      (Daegu City-*SUBJ*)   (with KB province)   (*on* administraticve region)   (discussion-*OBJ*)

In the current version of the system, we do not implement *replacement* for surface realization. The case of *replacement* is considered as *as-is*. Using C4.5, we train the surface realization how to generate surface forms for a list of marked nodes. A part of the training data we used in C4.5 are shown in Table 1[2]. The position in a sentence, the part-of-speech of a content part and a functional part of a word are used as the features in training data. The fourth column of Table 1 can be easily inferred by the difference between two words in (1b) and in (1c). Using decision rules that C4.5 induces, we can finally generate a compressed sentence similar to (1b) from the string (1c).

**Table 1.** The training data extracted from (1b) and (1c) for C4.5

| position | Surface form in leading sentence (1c) | | How to realize |
|---|---|---|---|
| | POS of content word | POS of functional word | |
| 1 | proper_noun(대구시) *(Daegu City)* | jxt (는) *(SUBJ)* | omission |
| 2 | proper_noun(경북도) *(KB Province)* | jca(와) *(with)* | as–is |
| 3 | noun (시행정구역) *(administrative region)* | jca(에) *(on)* | omission |
| 4 | vnoun (협의) *(discussion)* | jco(를) *(OBJ)* | omission |

## 5    Experimental Results

### 5.1    Training and Testing Corpus

The 1,304 pairs of the leading sentences and the headlines of news articles are used as a training/testing corpus in this paper. The average number of words in the leading sentences in the corpus is 12.67, while that in the headlines is 5.77. We divide the corpus into the 10 sub-parts for 10-fold cross validation.

---

[1] Korean sino-word is Chinese-derived, and even a single sino-word is full of meaning. Therefore, a phrase consisting of a few words in Korean sometimes can be paraphrased into a single sino-word with the same meaning.

[2] 'vnoun' denotes active-predicative noun, 'jxt' auxiliary particle, 'jca' adverbial case particle, and 'jco' objective case particle.

## 5.2    Experimental Results

The measures of the precision and the recall are used for the evaluation. The precision is calculated as the number of correctly marked nodes divided by the total number of nodes marked by the system. The recall is calculated as the number of correctly marked nodes divided by the number of marked nodes in the corpus. The average number of words in the compressed sentences created by our system is 6.85, which is longer than that in the headlines in the corpus. It results in far higher recall than precision as shown in Table 2. The interpretation of 77.3% precision is that one or two words among 6.85 words are incorrectly chosen in the compressed sentences.

**Table 2.** The precision and the recall of the system

|              | precision | recall  | correctness of surface realization |
|--------------|-----------|---------|------------------------------------|
| Testing data | 77.3%     | 91.95%  | 94.5%                              |

The third column in Table 2 shows the correctness of the surface strings created by the surface realization. As the surface realization is simply implemented in this version, we can get 94.5% accuracy in the compressed sentences.

As the usefulness of the compressed sentences cannot be judged by the precision and the recall, we estimate the acceptability of the compressed sentences by humans. We present the original leading sentences and the generated compressed sentences to judge, and he/she is asked to assign 'O' on the compressed sentences if they are acceptable as a compressed sentence, and assign 'X' if they are not. The 74.51% of the total sentences are allowable as the compressed, as shown in Table 3.

**Table 3.** The acceptability of the compressed sentences

|                 | Human Judge |
|-----------------|-------------|
| Acceptable      | 74.51%      |
| Non-accepatable | 25.49%      |

## 6    Conclusion

In this paper, we propose the system that can automatically learn how to generate a compressed sentence from a collection of the pairs of leading sentences and their headlines. A compressed sentence generated automatically resembles a headline of news articles, so it can be one of the briefest forms preserving the core meaning of an original sentence. Also, this approach can be easily applicable to other languages only if a parallel corpus in the languages is available online. The surface realization should be improved to enable to paraphrase a longer expression into a shorter one in order to compress an input sentence more efficiently and more fluently.

# References

1. Balazer, J.: Sentence Compression Using a Machine Learning Technique. http://www.eecs.umich.edu/ balazer/sc/. (2004)
2. Carroll, J., Minnen, G., Canning, Y., Devlin, S., Tait, J.: Practical simplification of English newspaper text to assist aphasic readers. Proceedings of the AAAI-98 Workshop on Integrating AI and Assistive Technology (1998)
3. Chung, H., Rim, H.-C.: A new probabilistic dependency parsing model for head-final free word-order languages. IEICE Trans. on Information and Systems. Vol. E86-D, No. 11 (2003)
4. Collins, M.: Head-Driven Statistical Models for Natural Language Parsing. Ph.D. Thesis. Department of Computer and Information Science, University of Pennsylvania (1999)
5. Hovy, E., Lin, C.-Y.: Automated text summarization in SUMMARIST system. Eds. I. Mani and M. T. Maybury Advances in Automatic Text Summarization. MIT Press. (1999) 81–94
6. Knight, K., Marcu, D.: Summarization beyond sentence extraction: A probabilistic approach to sentence compression. Artificial Intelligence **139** (2002) 91–107
7. Lee, K. J., Kim, J.-H., Han, Y. S., G. C. Kim.: Restricted representation of phrase structure grammar for building a tree annotated corpus of Korean. Natural Language Engineering **3-2&3** (1997) 215–230
8. Lin, Chin-Yew.: Improving summarization performance by sentence compression – A pilot study. Proceedings of IRAL 2003 (2003)
9. Mani, I., Maybury, M. T.: Advances in Automatic Text Summarization. MIT Press (1999)
10. Myaeng, S. H., Jang, D., Song, S., Kim, J., Lee, S., Lee, J., Lee, E., Seo, J.: Construction of an information retrieval test collection and its validation. Proceedings of the Conference on Hangul and Korean Language Information Processing (in Korean) (1999) 20–27
11. Vandeghinste, V., Tjong Kim Sang, E.: Using a parallel transcript/subtitle corpus for sentence compression. Proceedings of LREC2004 ELRA Paris (2004)
12. Wasson, M.: Using leading text for news summaries: Evaluation results and implications for commercial summarization applications. Proceedings of COLING-ACL98 (1998) 1364–1368
13. Witten, I. H., Frank, E.: Data Mining: Practical Machine Learning Tools and Techniques with Java Implementations. Morgan Kaufmann (1999)
14. Yang, C. C., Wang, F. L.: Fractal summarization: summarization based on fractal theory. Proceedings of SIGIR (2003) 391–392
15. Yoshihiro, U., Mamiko, O., Takahiro, K., Tadanobu, M.: Toward the at-a-glance summary: Phrase-representation summarization method. Proceedings of the International Conference on Computational Linguistics (2000) 878–884

# Automatic Text Summarization Using Two-Step Sentence Extraction

Wooncheol Jung[1], Youngjoong Ko[2], and Jungyun Seo[1]

[1] Department of Computer Science and Program of Integrated Biotechnology,
Sogang University, Sinsu-dong 1, Mapo-gu
Seoul, 121-742, Korea
wcjung@nlprep.sogang.ac.kr, seojy@sogang.ac.kr
[2] Division of Electronics and Computer Engineering, Dong-A University,
840 Hadan 2-dong, Saha-gu, Busan, 604-714, Korea
yjko@dau.ac.kr

**Abstract.** Automatic text summarization sets the goal at reducing the size of a document while preserving its content. Our summarization system is based on Two-step Sentence Extraction. As it combines statistical methods and reduces noise data through two steps efficiently, it can achieve high performance. In our experiments for 30% compression and 10% compression, our method is compared with Title, Location, Aggregation Similarity, and DOCUSUM methods. As a result, our method showed higher performance than other methods.

## 1 Introduction

The goal of text summarization is to take an information source, extract content, and present the most important content to a user in a condensed form and in a manner sensitive to the user's or application's needs [10]. To achieve this goal, text summarization system should identify the most salient information in a document and convey it in less space than original text. The most widespread summarization strategy is still sentence extraction. Traditional text summarization methods have used linguistic approaches and statistical approaches to extract salient sentences. But some problems have occurred in both methods for text summarization. Despite high performance, linguistic approaches have some difficulties in requiring to use high quality linguistic analysis tools such as discourse parser and linguistic resources such as WordNet, Lexical Chain, and Context Vector Space [1][7][12]; they are very useful resources for summarization systems but a weak point of them is to take much time and high cost to construct. On the other side, statistical approaches are easy to understand and implement, but generally they show low accuracy.

In this paper, we propose a new high performance summarization method to efficiently combine statistical approaches. By combining several statistical methods in two steps, our method can obtain higher accuracy than other statistical methods or the linguistic method (DOCUSUM) [7]. Moreover, our method can also a have low cost and robust system architecture because it does not require any linguistic resources.

S. H. Myaeng et al. (Eds.): AIRS 2004, LNCS 3411, pp. 71–81, 2005.

In the first step of our system, our method creates bi-gram pseudo sentences by combining two adjacent sentences for solving feature sparseness problem; it occurs in text summarization because of using only one sentence as the linguistic unit. And then our method applies the combined statistical method (Title and Location) to the bi-gram pseudo sentences for calculating the importance of them. The goal of the first step is not to extract salient sentences but to remove noisy sentences. Thereafter, we can get more useful pseudo sentences through removing noisy pseudo sentences. Then, in the second step, our method separates the bi-gram pseudo sentences into each original single sentence and it performs second sentence extraction by adding Aggregation Similarity method [5] to the linear combination of the first step. Because the Aggregation Similarity method estimates the importance of sentences by calculating the similarities of all other sentences in a document, it can be more efficient in our system after removing the noisy sentences. Since our system carries out a summarization task without any linguistic resources such as WordNet and discourse parser, it could be low cost and robust. As shown in experimental results, our system showed higher performance than other statistical methods and DOCUSUM as a linguistic method.

The rest of this paper is organized as follows. Section 2 describes related works in existing summarization systems. In Section 3, we present the methods used in each step in detail. Section 4 is devoted to evaluating experimental results. In the last section, we draw conclusions and present future works.

## 2   Related Works

The summarization system has two main categories: Linguistic approaches and Statistical approaches. The former uses semantic relations between words, phrase, and clause structural information in a sentence by using the linguistic resources while the latter uses title, frequency, location, and cue words and so on.

### 2.1  Linguistic Approaches

Bazilay and Elhadad constructed Lexical Chain by calculating semantic distance between words using WordNet [1]. Strong Lexical Chains are selected and the sentences related to these strong chains are chosen as a summary. The methods which use semantic relations between words depend heavily on manually constructed resources such as WordNet [12]. WordNet is not available in several languages such as Korean and this kind of linguistic resources are hard to maintain.

To overcome the limitation of this problem, Ko, Kim, and Seo constructed Lexical Clusters [7]. Each Lexical Cluster has different semantic categories while they are more loosely connected than Lexical Chains. They call their system DOCUSUM.

Marcu used discourse structural information [11]. This is based on contextual structure through analyzing sentence relations and sentence semantics.

These Linguistic approaches are producing high quality summary but, in case of time and expansion manner, they leave much room for improvement.

## 2.2  Statistical Approaches

The pioneering work studied that most frequent words represent the most important concepts of the text [8]. This representation abstracts the source text into a frequency table. Therefore, this method ignores the semantic content of words and their potential membership in multi-word phrases.

In other early summarization system, Edmundson studied that first paragraph or first sentences of each paragraph contain topic information [2]. Also he studied that the presence of words such as significant, hardly, impossible signals topic sentences.

A query-based summarization makes a summary by extracting relevant sentences from a document [3]. The criterion for extraction is given as a query. The probability of being included in a summary increases according to the number of words co-occurred in the query and a sentence.

# 3  Two-Step Sentence Extraction

## 3.1  Overall Architecture

Our summarization system is based on statistical approaches. Generally, they have several weak points: feature sparseness and low performance. The former problem is caused by extracting features from only one sentence and the latter is caused by depending on the particular format and the style of writing.

In order to successfully deal with the feature sparseness problems, we made an assumption. The salient sentences are grouped at a definite position without regarding to location of subjective sentences. To apply it to our system, we combine two adjacent sentences into bi-gram pseudo sentence. This extension of semantic unit can resolve the feature sparseness problem in part.

In order to improve the performance of statistical methods regardless of the particular formats, we make an efficient combination of them by means of estimating various statistical methods.

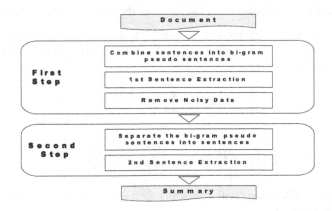

**Fig. 1.** Illustration of our summarization system

In the first step, the system estimates the importance score of bi-gram pseudo sentences by the combination of Title and Location methods. And then it removes invaluable bi-gram pseudo sentences which are called by noisy data. In the second step, it extracts salient sentences for summary from original single sentences which are separated from remaining bi-gram pseudo sentences. Here, we add Aggregation Similarity method to our combination method to achieve better performance.

The Fig. 1 shows the architecture of our summarization system.

## 3.2 General Statistical Methods

Our system uses several statistical methods. We here present several typical statistical methods.

**Title Method.** The score of sentences is calculated as how many words are commonly used between the sentence and title. This calculation is acquired by a query from title in Boolean weighted vector space model. The inner product method is exploited for similarity between a sentence and a query.

$$sim(S_i, Q) = \sum_{k=1}^{n} w_{ik} w_{qk} .$$ (1)

$$Score(S_i) = sim(S_i, Q) .$$ (2)

where $S_i$ is an $i$-th sentence and $Q$ is a title query. And $w_{ik}$ is the weight of $k$-th word in $i$-th sentence and $w_{qk}$ is the weight of $k$-th word in the query.

**Location Method.** It has been said that the leading several sentences of an article are important and a good summary [15]. Therefore, the leading sentences in compression rate are extracted as a summary by the location method.

$$Score(S_i) = 1 - \frac{i-1}{N} .$$ (3)

where $S_i$ is an $i$-th sentence and $N$ is the total number of sentences in the text.

**Aggregation Similarity Method.** The score of a sentence is calculated as the sum of similarity with other all sentence vectors in document vector space model. Each score is computed as follows [5].

$$sim(S_i, S_j) = \sum_{k=1}^{n} w_{ik} w_{jk} .$$ (4)

$$Score(S_i) = asim(S_i) = \sum_{j=1, j \neq i}^{m} sim(S_i, S_j) .$$ (5)

Equation 4 calculates the similarity with other sentence and $w_{ik}$ is the weight of $k$-th word in $i$-th sentence.

**Frequency Method.** The frequency of term occurrences within a document has often been used for calculating the importance of sentences [14]. In this method, the score of a sentence can be calculated as the sum of the score of words in the sentence. The score of important score $w_i$ of word $i$ can be calculated by the traditional $tf.idf$ method as follows [13].

$$w_i = tf_i \times \log \frac{N}{df_i} .$$ (6)

where $tf_i$ is the term frequency of word $i$ in the document, $N$ is the total number of texts, and $df_i$ is the document frequency of word $i$ in the whole data set.

### 3.3  The TF-Based Query Method

As described above, the title has usually been used for a query and the Title method has shown higher performance than other approaches in general. However, in special cases, it can be hard to extract a title from documents or any style of documents has no-title. For these cases, we propose a method to extract topic words for a query. The TF-based query method uses a query which consists of words with the highest term frequency in a document. This method considers words with high frequency as important concepts such as [8].

Like the Title method, the inner product metric is used as the similarity measure between a sentence and a TF-based query. To represent sentences, only proper and common nouns are used after eliminating stop words. For sentence vectorization, the Boolean weighting is used as follows:

$$S_i = (w_{i1}, w_{i2}, w_{i3}, ..., w_{in})$$
$$w_{ik} = \begin{cases} 1 & \text{if } tf_{ik} > 0 \\ 0 & \text{otherwise} \end{cases} .$$ (7)

where $tf_{ik}$ is the term frequencies of $k$-th word in sentence $i$ and $S_i$ is sentence vector. In general, $tf.idf$ representation has shown better performance in information retrieval. However, binary representation has generally showed better performance in summarization [4].

By the following Equation 7, we calculate the similarity between sentences and the TF-based query.

$$sim(S_i, TfQ) = \sum_{k=1}^{n} w_{ik} w_{TFQk} .$$ (8)

where $n$ is the number of words which is included in a document. $w_{ik}$ is the weight of $k$-th word in $i$-th sentence and $w_{TFQk}$ is the weight of $k$-th word in the TF-based query.

**Table 1.** Performances according to the number of word

|                | 30%   | 10%   |
| -------------- | ----- | ----- |
| 1 Topic Word   | 0.500 | 0.526 |
| 2 Topic Words  | 0.490 | 0.434 |
| 3 Topic Words  | 0.478 | 0.367 |
| 4 Topic Words  | 0.455 | 0.330 |
| 5 Topic Words  | 0.450 | 0.313 |

To verify our TF-based query method, we did experiments according to the number of topic words as shown in Table 1. As a result, we achieved the best performance when using one topic word, *the most frequent word*.

### 3.4   The Combination of Statistical Methods in Two Steps

In this section, we describe a new combinational method for statistical approaches. Before introducing our method in detail, we observed the performance of each statistical method to choose statistical methods used in our combinational method. The performance of general statistical methods are shown in Table 2.

**Table 2.** Experimental results of statistical methods

| Method | 30% | 10% |
|---|---|---|
| Title | 0.488 | 0.435 |
| Location | 0.494 | 0.466 |
| TF based query | 0.456 | 0.465 |
| Aggregation  Similarity | 0.406 | 0.239 |
| Frequency | 0.352 | 0.130 |

Our combinational method exploits Title, Location, Aggregation Similarity methods as considering their performances and characteristics. For a case of no-title documents, the TF-based method is used instead of the Title method.

### 3.4.1   Removing the Noisy Sentences in the First Step

In the first step, our goal is to reduce noisy sentences. First of all, our system generates bi-gram pseudo sentences to solve the feature sparseness problem; they simply are made by combining two adjacent sentences by sliding window technique [9]. Then, since Title and Location methods show high performances, our system linearly combines these methods in the first step as follows:

$$Score(S_i) = sim(S_i, Q) + (1 - \frac{i-1}{N}) . \tag{9}$$

where notations in this Equation follow those of Equation (1) and (3). After all the bi-gram pseudo sentences are scored by Equation (9), about 50% of them are removed because they are regarded as noisy sentences.

### 3.4.2   Extracting Summary in the Second Step

After the first step, the system can get more salient bi-gram pseudo sentences. Thereafter, it separates the bi-gram pseudo sentences remained from the first step into original single sentences. We here add Aggregation Similarity method to linear combination method of the first step (Equation (9)). Since noisy sentences are eliminated in the first step, the score of sentences could be improved as adding the Aggregation Similarity method; it considers the sum of similarities with other all sentences as the important score of a sentence. The final Equation is as follows:

$$Score(S_i) = sim(S_i, Q) + (1 - \frac{i-1}{N}) + w_a asim(S_i) . \tag{10}$$

where $w_a$ is a weight value reflecting the importance of Aggregation Similarity method.

If documents have no title, TF-based query method is used instead of the Title method as shown in the following Equations (11) and (12).

$$Score(S_i) = sim(S_i, TfQ) + (1 - \frac{i-1}{N}) . \tag{11}$$

$$Score(S_i) = sim(S_i, TfQ) + (1 - \frac{i-1}{N}) + w_a asim(S_i) . \tag{12}$$

## 4 Empirical Evaluation

### 4.1 Data Sets and Experimental Settings

In our experiments, we used the summarization test set of KOrea Research and Development Information Center (KORDIC). This data is composed of news articles. The articles consist of several genres such as politics, culture, economics, and sports. Each test document has title, content, 30 % summary, and 10 % summary. The 30 % and 10 % summaries of the test document are made by manually extracting sentences from content. In our experiments, we used 841 document-summary pairs after eliminating duplicate articles and inadequate summary pairs although the size of summarization test set was reported as 1,000 documents [6].

We have two parameters to be determined; one of them is to adjust how many pseudo sentences are removed and the other is a weight parameter of Aggregation Similarity method in Equation (10), (12). For this parameter setting, we used 280 documents as a validation set which are selected at random. Hence, the test set in our experiments is composed of the rest of data set (561 documents).

To measure the performance of the method, F1 measure is used as the following Equation (13).

$$F_1 = \frac{2(P \times R)}{P + R} . \tag{13}$$

where $P$ is precision and $R$ is recall.

### 4.2 Experimental Results

#### 4.2.1 Comparing with One-Step Combination

In order to verify the effectiveness of Two-step combination method, we compared the performance of Two-step combination method with that of One-step combination method. At first, our Two-step combination method without Aggregation Similarity method (*2 step T&L*: Two-step Title and Location combination) reported higher performance than One-step combination method (*T&L*: One-step Title and Location combination) as shown in Fig. 2. Next, to verify the addition of Aggregation Similarity method in the second step, we compared *2 step T&L* and *2 step T&L&A* (Two-step Title, Location, and Aggregation Similarity combination). We also achieved the improved performance in *2 step T&L&A*. Finally, we obtained 3.9% advanced score on the 10% summary and 1.5% advanced score on the 30% summary.

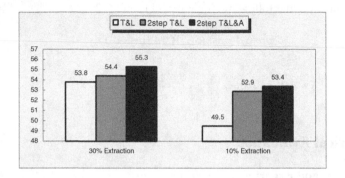

**Fig. 2.** Comparison between One-step and Two-step methods in case of using title

**Table 3.** Comparison between One-step and Two-step methods in case of using title

|       | One-step | Two-step | Improvement |
|-------|----------|----------|-------------|
| 10%   | 49.5     | 53.4     | **+3.9**    |
| 30%   | 53.8     | 55.3     | **+1.5**    |

In case of no-title, we used TF-based query method in place of Title method. In this cases, 2step *TfQ&L&A* (Two-step TF-based query, Location and Aggregation Similarity combination) showed 1.1% and 2.3 % advanced scores higher than *TfQ&L* (One-step TF-based query and Location combination) in 10% and 30% summary respectively as shown in Table 4.

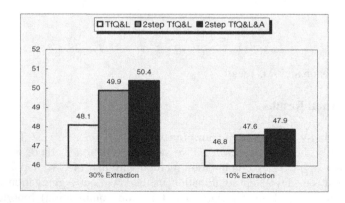

**Fig. 3.** Comparison between One-step and Two-step methods in case of using no-title

**Table 4.** Comparison between One-step and Two-step methods in case of using no-title

|       | One-step | Two-step | Improvement |
|-------|----------|----------|-------------|
| 10%   | 46.8     | 47.9     | **+1.1**    |
| 30%   | 48.1     | 50.4     | **+2.35**   |

Note that we chose *T&L* method for One-step because it showed the best perform-ance in above both two cases (especially better than *T&L&A*).

### 4.2.2 Comparing with Other Summarization Methods

In this section, we make a comparison between our Two-step method and other sys-tems such as Title, Location, and DOCUSUM. Especially, DOCUSUM is used for comparison with linguistic approaches.

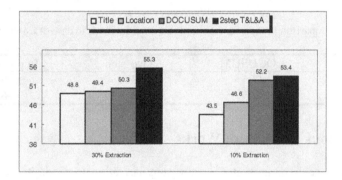

**Fig. 4.** Comparison between our method and other methods in case of using title

**Table 5.** Comparison between our method and other methods in case of using title

|       | DOCUSUM | Two-step | Improvement |
|-------|---------|----------|-------------|
| 10%   | 52.2    | 53.4     | **+1.2**    |
| 30%   | 50.3    | 55.3     | **+5.0**    |

As shown in Fig. 4 and Table 5, our system showed much better performance than title, location and even DOCUSUM. Even though DOCUSUM used knowledge resource such as context vector space for lexical clustering, it rather showed 1.2% and 5% lower performance than our method in 10% summary and 30% summary respec-tively.

Moreover, we conducted experiments in no-title case. DOCUSUM* used only a topic keywords query without a title query extracted by its lexical clustering method. But the results in Fig. 5 also showed that our system performed much better than any other method (even DOCUSUM).

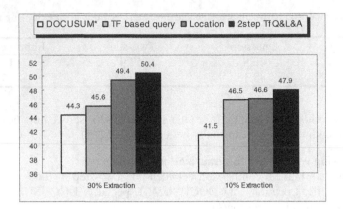

**Fig. 5.** Comparison between our method and other methods in case of using no-title

**Table 6.** Comparison between our method and other methods in case of using no-title

|      | DOCUSUM | Two-step | Improvement |
|------|---------|----------|-------------|
| 10%  | 41.5    | 47.9     | **+6.4**    |
| 30%  | 44.3    | 50.4     | **+6.1**    |

## 5   Conclusions and Future Works

In this paper, we have presented a new summarization method. It used bi-gram pseudo sentences to solve feature sparseness problem and Two-step combination method to improve the performance. As a result, we achieved higher performance than other statistical methods and DOCUSUM. Even though our system does not use any knowledge resource, it reported much better performance than DOCUSUM. In this paper, we implemented the high performance summarization system only to combine simple statistical methods on two steps. Our system has not only high performance but also the strong point to easily implement because it uses only simple statistical methods.

We plan to further researches on the multi-document summarization to apply our Two-step sentence extraction method to multi-document summarization. We will study the differentiation between single and multi-document summarization and implement new multi-document summarization system by using Two-step sentence extraction method.

## Acknowledgements

This work was supported by grant No. R01-2003-000-11588-0 from the basic Research Program of the KOSEF and special research grant of Sogang University.

# References

1. Barzilay, R., and Elhadad, M., Using Lexical Chains for Text Summarization, Advances in Automatic Summarization, The MIT Press (1999) 111-121.
2. Edmundson, H. P., New Methods in Automatic Extraction, Journal of the ACM 16(2) (1968) 264-285.
3. Goldstein, J., Kantrowitz, M., Mittal, V., and Carbonell, J., Summarizing Text Documents: Sentence Selection and Evaluation Metrics, In Proceedings of ACM-SIGIR'99 (1999) 121-128.
4. Han, K., Baek, D., and Rim, H., Automatic Text Summarization Based on Relevance Feedback with Query Splitting, In Proceedings of the Fifth International Workshop on Information Retrieval with Asian Languages (2000) 201-202.
5. Kim, J. H., and Kim, J. H., Korean indicative summarization using aggregation similarity, In Proceedings of the 12th Conference on Hangul and Korean Language Information Processing, 238-244.
6. Kim, T. H., Park, H. R., Shin, J. H., "Research on Text Understanding Model for IR/Summarization/Filtering", The Third Workshop on Software Science, Seoul, Korea. (1999).
7. Ko, Y., Kim, K., and Seo, J., Topic Keyword Identification for Text Summarization Using Lexical Clustering. IEICE transaction on information and system, Vol. E86-D, No. 9. (2003) 1695-1701.
8. Luhn, H. P., The Automatic Creation of Literature Abstracts, IBM Journal of Research and Development (1959) 159-165.
9. Maarek, Y., Berry, D., and Kaiser, G., 1991, An Information Retrieval Approach for Automatically Construction Software Libraries, IEEE Transaction on Software Engineering, Vol. 17, No. 8 (1991) 800-813.
10. Mani, I., Automatic Summarization, John Benjamins Publishing Co. (2001) 1-22.
11. Marcu, D., "Building up Rhetorical Structure Tree", In Proceedings of the 13th National Conference on Artificial Intelligence, Vol. 2 (1996) 1069-1074.
12. Miller, G., Beckwith, R., Fellbaum, C., Gross, D., and Miller, K., Introduction to WordNet: An on-line lexical database. International Journal of Lexicography (special issue) 3(4) (1990)234-245.
13. Salton, G., Automatic Text Processing: The Transformation, Analysis, and Retrieval of Information by Computer, Addison-Wesley Publishing Company (1989).
14. Wasson, M., Using leading text for news summaries: Evaluation results and implications for commercial summarization applications, In Proceedings of the 17th International Conference on Computational Linguistics and 36th Annual Meeting of the ACL, (1998) 1364-1368.
15. Zechner, K., Fast generation of abstracts from general domain text corpora by extracting relevant sentences, In Proceedings of the 16th international Conference on Computational Linguistics (1997) 986-989.

# Sentence Extraction Using Time Features in Multi-document Summarization

Jung-Min Lim, In-Su Kang, Jae-Hak J. Bae[+], and Jong-Hyeok Lee

Division of Electrical and Computer Engineering Pohang University of Science and
Technology (POSTECH), Advanced Information Technology Research (AITrc)
San 31 Hyoja-dong, Nam-gu, Pohang 790-784, Republic of Korea
[+]School of Computer Engineering and information Technology, University of Ulsan
{beuuett, dbaisk, jhlee}@postech.ac.kr, [+]jhjbae@ulsan.ac.kr

**Abstract.** In multi-document summarization (MDS), especially for time-
dependent documents, humans tend to select sentences in time sequence. Based
on this insight, we use time features to separate documents and assign scores to
sentences to determine the most important sentences. We implemented and
compared two different systems, one using time features and the other not. In
the evaluation of 29 news article document sets, our test method using time
features turned out to be more effective and precise than the control system.

## 1 Introduction

In MDS, target documents mostly give explicit time information as additional
information, especially in newspaper articles. The time feature usually consists of a
year, date, and time, and most documents for the input of MDS contain a year and a
date. In the typical three stage of summarization, they are used to order sentences in a
realization (generation) phase, and are used as a weighing scheme in the content
identification (sentence extraction) phase. However, we use time information to
separate documents in order to obtain the main contents according to the time
sequence in the content identification phase. As we sort documents by time
information, we identify the time-slot that has the most documents. Although a set of
documents is constructed not considering which time slot is more important than the
other, we assume that a time slot having more documents have a high probability for
containing important content like term frequency in information retrieval.

Separating documents by time information, we also have the benefit of detecting
redundant content in documents. As the input of MDS is a set of topically related
documents, similar content is inevitable. To find similar content, we check the
similarity between each sentence that is extracted. In time dependent documents such
as a newspaper article, they are published according to events that occur in sequence,
we assume that similar contents are detected in documents that were published at an
adjacent time. We thus do not need to compare all extracted sentences, but only detect
redundant content among documents published at an adjacent time. Most of
techniques for extracting sentences are not new, but we add time features to improve
extracting sentences.

S. H. Myaeng et al. (Eds.): AIRS 2004, LNCS 3411, pp. 82–93, 2005.
© Springer-Verlag Berlin Heidelberg 2005

In this paper, we propose a method of extracting sentences and detecting redundant sentences, using time features based on the above two assumptions. In experiments, we attempt to find how separating documents and giving weights to sentences by time features affect sentence extraction in time-dependent multiple documents.

## 2  Related Works

To identify important content in multiple documents, Mani and Bloedern built activation networks of related lexical items to extract text spans from a document set [7], and used similarities and differences among related news articles for multi-document summarization [8]. SUMMONS [15] is a knowledge-based MDS system that extracts information, using specialized domain specific knowledge sources, and generates summaries from the templates. Centroid-based summarization [16] uses as input the centroids of the clusters, to identify which sentences are central to the topic of the cluster, rather than individual articles. Maximal Marginal Relevance-Multi-Document [3] is a query-based method to minimize redundancy and maximize both relevance and diversity. The Columbia university system [2, 11] identifies similar sections and produces natural language summary using machine learning and statistical techniques. NeATS [5] is an extracted-based system using term clustering, to improve topic coverage and readability. It uses time features for ordering sentences and resolving time ambiguity, not for assigning weights to sentences. DEMS [21] relies on concepts instead of individual nouns and verbs. Like SDS using rhetorical structure theory (RST), Radev [17] introduced CST, a theory of cross-documents structure, which tried to find a relation between documents, such as subsumption, update, and elaboration, to cross-document conceptual links in MDS [18, 19]. However, Radev does not formalize these relationships [10].

Time-based summarization techniques, evolving summaries of new documents related to an ongoing event were proposed by Radev [20]. Mani and Wilson [9] have focused on how to extract temporal expressions from a text, looking for and normalizing references to dates and times. Allan [1] produced a summary that indicates only what has changed over time. It shares the idea of finding all of the events within a topic over time with our method. However, using time features, we not only classify documents according to time, but also identify important content from a document set.

## 3  Preliminary Experiments

To find important sentences that can be the candidate of a final summary, four graduate students selected sentences from multiple documents. Each student read all the articles in the document set, as well as the nine selected sentences that must be included in a summary. The five sets of topically-related documents consisted of newspaper articles from the 1998 edition of Hankook Ilbo (Korean newspaper). To obtain reasonable results, we counted the number of selected sentences from each result, and determined the final sentences using the frequency of sentences are chosen. If the number of documents was less than 13, we chose six sentences for a result; otherwise, nine sentences were selected. The distribution of human-selected sentences is shown in Table 1.

**Table 1.** Group of documents by data and distribution of human-selected sentences

| Document Set | | Time slot (By date) | | | | | | | | | |
|---|---|---|---|---|---|---|---|---|---|---|---|
| | | 1 | 2 | 3 | 4 | 5 | 6 | 7 | 8 | 9 | 10 |
| A | No. of documents (18) | 1 | 1 | 1 | 1 | 5 | 2 | 1 | 1 | 4 | 1 |
| | No. of sentences (175) | 7 | 13 | 4 | 9 | 35 | 21 | 9 | 8 | 26 | 10 |
| | No. of human-selected (9) | 1 | | 1 | 1 | 3 | | | | 2 | 1 |
| B | No. of documents (12) | 1 | 2 | 3 | 1 | 1 | 1 | 2 | 1 | | |
| | No. of sentences (78) | 5 | 17 | 18 | 3 | 5 | 7 | 14 | 9 | | |
| | No. of human-selected (6) | 1 | 1 | 1 | | | | 1 | 2 | | |
| C | No. of documents (13) | 2 | 6 | 4 | 1 | | | | | | |
| | No. of sentences (104) | 7 | 53 | 39 | 5 | | | | | | |
| | No. of human-selected (6) | | 3 | | 3 | | | | | | |
| D | No. of documents (11) | 8 | 1 | 1 | 1 | | | | | | |
| | No. of sentences (98) | 75 | 8 | 6 | 9 | | | | | | |
| | No. of human-selected (6) | 4 | 1 | 1 | | | | | | | |
| E | No. of documents (13) | 8 | 2 | 3 | | | | | | | |
| | No. of sentences (115) | 71 | 18 | 26 | | | | | | | |
| | No. of human-selected (6) | 3 | 1 | 2 | | | | | | | |

We found that extracted sentences for a final summary will be located along a time line, and a time slot that containing more documents has a high probability to including important sentences. Four document sets (A,B,C,E) among five sets have important sentences in recent time slots. Our preliminary experiment has confirmed Goldstein [3] who gave a weight to recent documents. We also found that the first time slot contained important sentences in four sets (A,B,D,E). The number of important sentences in the first time slot was different according to sets. We assumed that the difference depended on the characteristic of each document set, and assigned little weight to the first time slot. For summarizing time-dependent documents, we used these assumptions to design our method.

## 4  Sentence Extraction Using Time Features

### 4.1  System Overview

First, we classified target documents by their time features. The number of time slot was determined, according to the time feature, and each time slot contained documents. We extracted sentences from each document using the method of SDS. Next, among extracted sentences, we detect redundant sentences, and build the term cluster of a time slot from remaining sentences, to find a topic sentence. Using the term cluster, we assign a mark to sentences, and choose a sentence with highest score as a topic sentence in a time slot. A topic sentence receives more marks as a representative of a time slot. From each time slot, we obtain a term cluster, and construct the global term cluster. Using these global terms, we give a mark to all

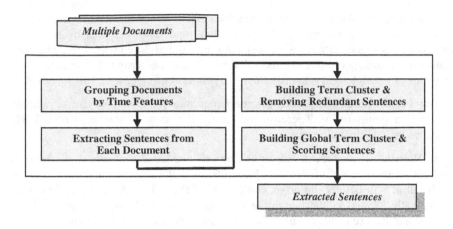

**Fig. 1.** System Architecture

sentences again. Finally, we rearrange all sentences ordered by their scores, and extract sentences.

In the following sections, we will explain each phase of the system illustrated in Figure 1.

### 4.2  Grouping Documents by Time Features

The time information given by news articles is usually a combination of date and time. However some articles do not contain a time feature. In these cases, we group the documents by their publication date.

### 4.3  Extracting Sentences

To extract important sentences in a single document, we used traditional methods: sentence position, length, stigma word, and keywords in a headline. As we dealt with newspaper articles, we mainly relied on sentence position. The score of the $i$-th sentence $S_i$ is calculated as follows:

$$S_{ext}(S_i) = w_{position} \times S_{pos}(S_i) - Pen(S_i) - w_{stigma} \times | S_i \cap P | + w_{lead} \times | S_i \cap L | \qquad (1)$$

Where,  - $S_{pos}(S_i)$ : A position score of $S_i$.     - $Pen(S_i)$: A length penalty of $S_i$.
   - $L$: A set of lead words.       - $P$: A set of stigma terms.

In our system, we selected 50% of the sentences that had a higher score than others in each document, using the combination of each scoring function. Selected sentences are used for the input of the next phase. The following subsections describe each weighing scheme.

**Sentence Position.** Sentence position has been used to find important content since the late 60s. We gave a mark to sentences according to their position. A sentence located at the beginning of the document obtained a higher score.

**Sentence Length.** We gave a length penalty Pen(Si) to a sentence Si which was too short or too long. In Korean, the length of important sentences that are selected by humans in newspaper articles is usually between 10 and 30 eojeols[1] [22]. The length of an eojeol is usually five to six syllables. Thus, we assigned a penalty 0.5 to sentences of which length is shorter than 50 syllables or longer than 180 syllables.

**Stigma Terms.** When some sentences contain quotation marks, they can be redundant content in a summary [5]. Therefore we reduce the score of sentences by 0.4 when sentences include quotation marks.

**Lead Words.** A headline is the most simple and concise method of delivering information about news articles. The basic idea is that a sentence that contains words in a headline will deliver important content about the article. Also, the main contents are typically located at the beginning of a news article. Thus, we use words in a headline and lead sentences to identify important sentences in a document.

### 4.4 Building Term Cluster of a Time Slot and Removing Redundant Sentences

**Building Term Cluster of a Time Slot.** After extracting sentences, we built the term cluster of a time slot to find a topic sentence. We chose sentences that had a higher score from each document. Identifying terms from these sentences, we calculated the frequency of the terms, and constructed a term cluster with terms with a higher frequency than a threshold $T_{ts}$. Each term cluster of a time slot could be used to identify a main sentence which could represent important content in that time slot.

**Removing Redundant Sentences.** After extracting the important sentences in multiple documents, we grouped them by their similarity. We added a module that prevents similar contents from inclusion in the final result. Our system is based on sentence-extraction.

Thus, we checked redundant content on the sentence-level, and estimated the redundancy between sentences. The redundancy value is used to construct a cluster of semantically redundant sentences. Our system basically calculated the Dice coefficient as a similarity measure based on the number of words. We developed an improved term-weighing method that assigns weights to words, using syntactic information [6].

When measuring redundancy between sentences, we did not rely on term frequency (TF), and inverse document frequency (IDF), because they cannot distinguish words that are more syntactically important from others. When we compared two sentences, we expected that syntactically important words would obtain a higher score than others. Basically, main clauses will deliver more important information than sub clauses. In addition, we believe that subjects, objects, and verbs are syntactically important, compared to others in a sentence. Therefore, we gave weights to each word according to its syntactic role and the type of sentences that it locates.

When comparing words, we used not only the surface form of a word, but also the concept code of it. In topically related document sets, sentences that contain the same

---

[1] An *eojeol* is a one or more morphemes and is identified with a preceding and following space. It is similar to the notion of a *word* in English.

contents can be represented with the combination of different words, and the surface form of a word in news article can be varied. We assumed that verbs have a high probability to be represented in different way. Thus we used the concept code only for predicates. For conceptual generalization, we used the concept codes of the Kadokawa thesaurus, which has a 4-level hierarchy of 1,110 semantic classes [14]. Concept nodes in level L1, L2 and L3 are further divided into 10 subclasses and nodes in level L4 have 3-digit code between 000 and 999. Formally, the redundancy between two sentences S1 and S2 is calculated as follows:

$$Sim \ (S_1, S_2) = \frac{\sum\limits_{(t_i, t_j) \in S_1 \cap S_2}^{*} W \ (t_i, S_1) + W \ (t_j, S_2)}{\sum\limits_{t \in S_1} W \ (t, S_1) + \sum\limits_{t \in S_2} W \ (t, S_2)} \tag{2}$$

Where, $-S = \{t_1, \cdots, t_n\}$, $-\cap = \cap \cdot \cap$, $-W \ (t, S) = W_{st} \ (t, S) + W_{gr} \ (t, S)$

- $W_{st} \ (t, S)$: A weight function for a term $t$ located which type of a clause.
- $W_{gr} \ (t, S)$: A weight function for a term $t$ by its grammatical role.
- $S_1 \overset{L}{\cap} S_2 = \{(t_i, t_j) \mid t_i \in S_1, t_j \in S_2, t_i = t_j\}$
- $S_1 \overset{C}{\cap} S_2 = \{(t_i, t_j) \mid t_i \in S_1, t_j \in S_2, C(t_i) \cap C(t_j) \neq \phi\}$
- $C(t)$: The semantic code set of term $t$.

We used a single-link clustering algorithm. Sentence pairs having similarity value higher than the threshold value $T_r$ are regarded as similar, and are included in a redundant cluster. The threshold value $T_r$ is set to 0.5. After clustering, each cluster is expected to have several redundant sentences. From these clusters we chose only one sentence that had the highest score.

We compared our redundancy-measuring (System1) to a method which does not use syntactic information (System2). Test data is fifteen sets of redundant sentences from four topically-related newspaper articles (the 1998 edition of Hankook Ilbo). The average number of sentences in test sets is 2.9. Thus, if there were two or more redundant sentences in both the test sets and the sets generated by systems, we consider that the systems correctly detected redundant sets. The threshold value $T_r$ is set 0.5.

**Table 2.** Result of identifying redundant sets

| System ID | No. of redundant sets | No. of correct redundant sets |
|---|---|---|
| 1 | 10 | 7 |
| 2 | 13 | 6 |

**Table 3.** Result of performance

| System ID | Precision | Recall |
|---|---|---|
| 1 | 83.3% | 47.7% |
| 2 | 53.5% | 39.3% |
| Improvement | 56% | 21% |

Table 3 shows that syntactic information improves detecting redundant sentences. Thus, we used a Korean dependency parser to obtain syntactic information. The cluster of redundant sentences was also used as a feature to give a weight to sentences. If several documents contain the same information, the repeated information can be the topic among documents and deliver important content [20]. Thus, we add it to a scoring formula in this phase.

**Find a Topic Sentence in a Time Slot.** Using the term cluster, the number of redundant sentences, and a previous score of sentences, we assigned a score to sentences, and regarded a sentence with the highest score among all extracted sentences to be a topic sentence in a time slot. A topic sentence receives more marks as a representative of a time slot in the next phase. We also consider the syntactic information of terms in sentences. If the syntactic role of terms is a subject, object, verb, or terms are identified as an unknown word, we give more marks to them. A sentence S is calculated as follows:

$$S_{time\_slot}(S) = w_{ext} \times S_{ext}(S) + w_{term} \times \sum_{t_k \in S \cap C} t_k \cdot f_k + w_{redendant} \times R(S) \qquad (3)$$

$$+ w_{syn} \times \sum_{t_k \in S \cap C} W_{gr}(t_k, S)$$

Where, - $S_{ext}(S)$: A score of a extraction-phase.   - $C$: A term cluster of a time slot.
  - $f_k$: A frequency of a term $k$ in $C_i$.
  - $R(S)$: The number of sentences in a redundant cluster that contains $S$.
  - $W_{gr}(t,S)$: A weight function for a term $t$ by it's grammatical role.

## 4.5  Building Global Term Cluster and Scoring Sentences

To extract sentences from multiple documents, we constructed a global term cluster using a term cluster from each time slot. Like constructing the term cluster of time slot, we identified terms and select terms that had a higher frequency than the threshold $T_g$. Sentences selected as a topic sentence from each time slot obtained a higher score. According to the results of the preliminary experiment, we gave a high probability to a time slot that contained relatively more documents than others, and to a time slot in the beginning and ending of time sequence. We used an implicit time feature to assign weight to the time slot. Some sentences contained time feature that indicated a specific time slot. Although we do not know what happens at this reference time slot, we assume that some event related to a topic occurred at that time. If the time slot is referenced by a sentence that is in a different time slot, we assigned weights to the referenced time slot. Also, we considered a sentence's score produced in the previous phase, and the syntactic information of terms mapped with global terms. A sentence $S$ is finally calculated as follows:

$$S_{final}(S) = w_{time} \times S_{time\_slot}(S) + w_{\#doc} \times \frac{|D(S)|}{N} + w_{slot} \times TS(S) \qquad (4)$$

$$+ w_{topic\_sen} \times Tsen(S) + w_{G\_term} \times \sum_{t_k \in S \cap G} t_k \cdot f_k + w_{syn} \times \sum_{t_k \in S \cap G} W_{gr}(t_k, S)$$

Where, - $S_{time\_slot}(S)$: A score of a previous phase.  - $N$: No. of whole documents.

- $|D(S)|$: No. of documents in a time slot which contains $S$.
- $Tsen(S)$: A weight for topic sentences.   - $TS(S)$: A weight of a time slot.
- $G$: A global term cluster.            - $f_k$: A frequency of a term $k$ in $G_i$.
- $W_{gr}(t,S)$ : A weight function for a term t by it's grammatical role.

# 5  Documents for Experiments

We used 29 document sets which are concerned with a certain topic. These document sets were used for the formal run evaluation of NTCIR TSC3 (NII-NACSIS Test Collection for IR Systems, Text Summarization Challenge 3). In NTCIR TSC3, documents consisted of newspaper articles from the 1998, 1999 edition of the Mainichi and Yomiuri (Japanese newspaper). NTCIR also provided a scoring tool to evaluate precision and coverage for sentence extraction. We used this tool for evaluating our experiments. Because the document sets were written in Japanese, we first needed to translate the sets into Korean to using a Korean parser and keyword extractor. Using a Japanese-to-Korean machine translator, we translated all the document sets. The documents for experiments no doubt contained translation errors. However, there are no available Korean documents sets and tool for evaluating multi-document summarization systems. Thus, it was necessary to use translated documents in this experiment.

# 6  Evaluation for Sentence Extraction

We briefly describe the evaluation method from Hirao and Okumura's overview paper of NTCIR TSC3 [4].

## 6.1  Precision

Precision is the ratio of how many sentences in the system output are included in the set of the corresponding sentences. It is defined by the following equation:

$$precision \quad = \frac{m}{h} \tag{5}$$

Where $h$ is the least number of sentences needed to produce the abstract, and $m$ is the number of correct sentences in the system output.

**Table 4.** Important Sentences

| Sentence ID of Abstract | Set of Corresponding Sentences |
|:---:|:---:|
| 1 | $\{s_1\}\_\{s_{10}, s_{11}\}$ |
| 2 | $\{s_3,s_5,s_6\}$ |
| 3 | $\{s_{20},s_{21},s_{23}\}\_\{s_1,s_{30},s_{60}\}$ |

$\{s_i, s_j\}$ : Intersection,  $s_i \_ s_j$ : Union

## 6.2 Coverage

Coverage is an evaluation metric for measuring how close the system output is to the abstract taking into account the redundancy found in the set of sentences in the output. The set of sentences in the original documents that corresponds correctly to the $i$-th sentence of the human-produced abstract is denoted here as $A_{i,1}$, $A_{i,2}$,....., $A_{i,m}$. In this case, we have $m$ sets of corresponding sentences. Here, $A_{i,j}$ indicates a set of elements each of which corresponds to the sentence number in the original documents, denoted as $A_{i,j} = \{G_{i,j,1}, G_{i,j,2} ...., G_{i,j,k}\}$. In TSC3, they assume that there are correspondences between sentences in original documents and their abstract as in Table 4. For example, from Table 4, $A_{1,2} = G_{1,2,1}$, $G_{1,2,2}$ and $G_{1,2,1,} = s_{10}$, $G_{1,2,2} = s_{11.}$ Then, we define the evaluation score $e(i)$ for the $i$-th sentence in the abstract:

$$e(i) = \max_{i \leq j \leq m} \left( \frac{\sum_{k=1}^{|A_{i,j}|} v(Gi,j.k)}{|A_{i,j}|} \right) \qquad (6)$$

Where $v(\acute{a})$ is one(1), if the system output contains $\acute{a}$, otherwise, $v(\acute{a})$ is zero(0). Function e returns one(1) when any $Ai,j$ is outputted completely. Otherwise, it returns a partial score according to the number of sentences $|Ai,j|$. Given function e and the number of sentences in the abstract n, Coverage is defined as follows:

$$Coverage = \frac{\sum_{i=1}^{n} e(i)}{n} \qquad (7)$$

## 7 Experiments and Discussion

Our experiments focused on how separating documents and assigning weights to sentences with the time feature which affects sentence extraction in MDS. To identify the effect of time feature, we implemented two systems. The first system (System1) used the time feature to separate documents and to give a mark to sentences. The other system (System2) did not use the time feature. According to our proposed method, we implemented System1 to detect the redundancy only in the same time slot. As the time feature, we use the date information (yyyymmdd). We attempted to use the referenced time feature in sentences, but we did not yet implement a detecting module for the time feature in our system. For System2, we extracted sentences from each document, and built the global term cluster. We assigned a score to sentences, checked redundancy between all sentence pairs, and rearranged them by their score. For 29 document sets, we generated two types of sentence extraction: short and long. The number of extracted sentences is already defined for each document set. We calculated precision and coverage for the results of two systems. The evaluation of the two systems is shown in Table 5.

**Table 5.** Evaluation results of two systems

| Type | Short | | Long | | Average | |
|---|---|---|---|---|---|---|
| Evaluation | Prec. | Cov. | Prec. | Cov. | Prec. | Cov. |
| System 1 | 0.553 | 0.319 | 0.563 | 0.330 | 0.558 | 0.325 |
| System 2 | 0.465 | 0.321 | 0.519 | 0.337 | 0.492 | 0.329 |

System1 using the time feature had high precision values in both short and long type extraction. However, coverage values are slightly lower than System2 in both short and long. The result of experiments shows that time feature can be used as a feature to improve sentence extraction from time-dependent multiple documents. Coverage is an evaluation metric for removing redundant contents. Although the result of coverage depends on the algorithm of measuring redundancy between sentences and several parameters, redundant contents were found in different time slots in our results. We assumed that "a time slot having more documents might have a high probability for containing important content". Although we used this assumption in weighing schemes, we didn't consider it for producing a final summary. It possibly resulted in lower coverage in System1.

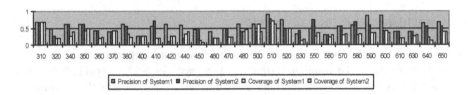

**Fig. 2.** Evaluation of short-sentence extraction

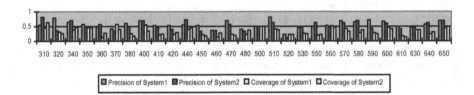

**Fig. 3.** Evaluation of long-sentence extraction

**Table 6.** Comparisons to others in NTCIR TSC 3

| ID | Short | | Long | |
|---|---|---|---|---|
| | Cov. | Prec. | Cov. | Prec. |
| F0301(a) | 0.315 | 0.494 | 0.355 | 0.554 |
| F0301(b) | 0.372 | 0.591 | 0.363 | 0.587 |
| F0303(a) | 0.222 | 0.314 | 0.313 | 0.432 |
| F0303(b) | 0.293 | 0.378 | 0.295 | 0.416 |
| F0304 | 0.328 | 0.496 | 0.327 | 0.535 |
| F0306 | 0.283 | 0.406 | 0.341 | 0.528 |
| F0307 | 0.329 | 0.567 | 0.391 | 0.680 |
| F0309 | 0.308 | 0.505 | 0.339 | 0.585 |
| F0310 | 0.181 | 0.275 | 0.218 | 0.421 |
| F0311 | 0.251 | 0.476 | 0.247 | 0.547 |
| Our System | 0.319 | 0.553 | 0.330 | 0.563 |
| Ranking | 4/11 | 3/11 | 6/11 | 4/11 |

Figure 2 and 3 show that the effectiveness of time features strongly depends on the characteristic of document sets. Some document sets (topic id: 310, 440, 550) consisted of time-independent documents, and have a tendency not to follow time-dependent events. Therefore, these document sets revealed high precision in System2. However, topic ID 440 showed high precision in Long type. Although the document set of a topic id 410 consisted of time-dependent documents, System2 obtained higher precision in a short type extraction. For increasing the performance of our system, we needed to analyze document sets. We also compared our results to other systems that participated in NTCIR TSC3. The results are shown in Table 6. Their results were produced from Japanese documents. Although we used translated documents, our method performed well.

# 8   Conclusion and Future Work

We propose a method of extracting sentences and detecting redundant sentences, using time features. To improve extracting sentences in MDS, especially in time-dependent documents, the time feature can be used effectively. Using the time feature, separating documents and assigning weight to sentences can improve the output of a system that is based on a sentence-extraction method. Our system needs to refine several additional parameters for increased efficiency, and therefore further experimentation is necessary.

In the future, we will try to scrutinize document sets to increase the performance of our system, and discover how to apply the time feature to documents that have time-independent tendency. Also we will apply category information [12, 13] for distinguishing between time-dependent documents and time-independent ones. We believe that this will greatly enhance our research.

**Acknowledgements.** This work was supported by the Korea Science and Engineering Foundation, through the Advanced Information Technology Research Center (AITrc) and the BK 21 project.

# References

1. Allan, J., Gupta, R., and Khandelwal, V. *Temporal Summaries of News Topics*, SIGIR, 2001.
2. Barzilay, R., McKeown, K. R., and Elhadad, M. *Information Fusion in the Context of Multi-Document Summarization*. In Proceedings of the 38th Annual Meeting of the ACL, 1999.
3. Goldstein, J., Mittal, V., Carbonell, J., and Kantrowitz, M. *Multi-Document Summarization by Sentence Extraction*. ANLP/NAACL Workshop, 2000.
4. Hirao, T., and Okumura, M. *Text Summarization Challenge 3-Text summarization evaluation at NTCIR Workshop4*. NTCIR Workshop 4, 2004.
5. Lin, C. Y., and Hovy, E. *Form Single to Multi-document Summarization: A Prototype System and its Evaluation*. In Proceedings of the 40th Annual Meeting of the ACL, 2002.
6. Lim, J. M., Kang, I. S., Bae, J. H., and Lee, J. H. *Measuring Improvement of Sentence-Redundancy in Multi-Document Summarization*. In Proceedings of the 30th KISS fall 2003

7. Mani, I., and Bloedorn, E. *Multi-document summarization by graph search and merging.* In Proceedings of AAI-1997, pages 622-628.
8. Mani., I., and Bloedorn, E. *Summarizing similarities and differences among related documents.* Information Retrieval, 1:35-67, 1999.
9. Mani, I., and Wilson, G. *Robust Temporal Processing of News,* In Proceedings of the 38th Annual Meeting of the ACL, 2000.
10. Mani, I. *Automatic Summarization.* John Benjamins Publishing Company, 2001.
11. McKewon, K. R., Klavans, J. L., Hatzivassiloglou, V., Bazilay, R., and Eskin, E. *Towards Multi-document Summarization by Reformulation: Progress and Prospects.* In Proceedings of the AAAI, 1999.
12. McKeown, K. R., Barzilay, R., Evans, D., Hatzivassilogou, V., Kan, M. Y., Schiffman, B., and Teufel. S. *Columbia Multi-Document Summarization: Approach and Evaluation.* In Online Proceedings of DUC, 2001.
13. Nobata, C., Sekine, S., Uchimoto, K., and Isahara, H. *A Summarization system with categorization of document sets.* NTCIR Workshop 3 meeting TSC2, 2002.
14. Ohno, S., and Hamanishi, M. *New synonyms Dictioinary, Kadokawa Shoten.* Tokyo, 1981.
15. Radev, D. R., and McKeown, K. R. *Generating natural language summaries from multiple on-line sources.* Computational Linguistics, 24 (3), pages 469-500, 1998.
16. Radev, D. R., Jing, H., and Budzikowska, M. *Centroid-Based Summarization of Multiple Documents.* ANLP/NAACL Workshop, 2000.
17. Radev, D. R., *A Common Theory of Information Fusion from Multiple Text Sources Step One: Cross-Document Structure.* ACL SIGDIAL Workshop, 2000.
18. Radev, D. R., Blair-Goldensohn, S., and Zhang, Z. *Experiments in Single and Multi-Document Summarization Using MEAD.* In Online Proceedings of DUC, 2001.
19. Radev, D. R., Otterbacher, S. J., Qi, H., and Tam, D. *MEAD ReDUCs: Michigan at DUC 2003.* In Online Proceedings of DUC, 2003.
20. Radev, D. R., *Topic Shift Detection-Finding new information in threaded news.* Technical Report CUCS-026-99, Columbia University Department of Computer Science. 1999.
21. Schiffman, B., Nenkova, A., and McKeown, K. Experiments in Multidocument Summarization. HLT conference, 2002.
22. Yoon, J. M. Automatic summarization of newspaper articles using activation degree of 5W 1H. Master's thesis,s POSTECH, 2002.

# Extracting Paraphrases of Japanese Action Word of Sentence Ending Part from Web and Mobile News Articles

Hiroshi Nakagawa[1] and Hidetaka Masuda[2]

[1] Information Technology Center,
The University of Tokyo, 7-3-1 Hongou,
Bunkyo, Tokyo 113-0033, Japan
nakagawa@dl.itc.u-tokyo.ac.jp
[2] Tokyo Denki University 2-2 Kandanishikicho,
Chiyoda, Tokyo 101-8457, Japan
masuda@im.dendai.ac.jp

**Abstract.** In this research, we extract paraphrases from Japanese Web news articles that are long and aimed at displaying on personal computer screens and mobile news articles that are short and compact and aimed at mobile terminals' small screens. We have collected them for more than two years, and aligned them at article level and then at sentence level. As the result, we got more than 88,000 pairs of aligned sentences. Next, we extract paraphrases of the final part of sentences from this aligned corpus. The paraphrases that we try to extract are the sentence final nouns of mobile article sentences and their counterpart expressions of Web article sentences. We extract character strings and word sequences for paraphrases based on branching factor, frequency and length of string. The precision is 90% for highest ranked candidate and 83% to 59% for each top three candidates of 100 most frequently used action nouns.

## 1 Introduction

Paraphrase extraction became one of the main research topics of computational linguistics these days. Enormous amount of research results have been published through many workshops as well as conferences [8, 4] and so on. In paraphrase acquisition, extraction of candidates from an entire corpus is the first and tough task. This difficulty can be reduced to some extent by using parallel corpus. [1] is one of the most successful research using parallel corpus. Their idea is that they use contexts to extract paraphrases from aligned sentences that are translation of the same sentence of the other language. As for paraphrase extraction from monolingual corpus, [6] proposed an unsupervised method using contexts. [3] improves these works by employing syntactic structures.

If we find new useful language resources of monolingual corpus, however, it is desirable we have to utilize them for paraphrase extraction. In fact, we are witnessing the rapid growth of mobile terminals such as mobile phones or PDAs

S. H. Myaeng et al. (Eds.): AIRS 2004, LNCS 3411, pp. 94–105, 2005.

that are used by ordinary people every day. These kinds of mobile devices have a small and low resolution screen. On the other hand, an ordinary personal computer has a big and high resolution screen. In this circumstance, two types of Web pages, one for mobile devices and the other for personal computers, are developed separately even though they describe the same topic or contents written in the same language. These two types of Web pages are, from the viewpoint of computational linguistics, regarded as comparable corpora because the topics are same but the surface texts are not. This is why we propose to use these two types of Web pages for the purpose of paraphrases extraction. As expected, Web news articles for mobile devices are much shorter and more compact than Web news articles for ordinary personal computers. Due to this, a paraphrase we can extract from these two types of Web pages is a pair of ordinary expression and compressed form of it, like "The bomb has exploded" and "bomb explosion." They are, obviously, useful to compress a formal sentence into a shorter phrase. In addition, this kind of short sentences are used many places, for instance, short messages displayed on low resolution screens in a train, on advertisement screens on building wall or inside of shops, etc. In this situation, our research topic is important for a new but widely used short and compact expression of language.

In order to extract paraphrases, we align sentences of two types of Web pages at first. We can easily find the place where candidates of paraphrases exist in the texts if we focus on the aligned sentences. On the contrary, if we use non-aligned corpus, detecting candidates is hard and has a high computational cost. We detour this problem by using the two types of Web pages collected from the Web. Then the problem is reduced to ranking of many candidates of paraphrases extracted from them.

Below, we describe the corpus we collected from the Web in Section 2, and alignment in Section 3. Section 4 is for paraphrase extraction, and Section 5 is the conclusion.

## 2　Web Articles and Mobile Articles

### 2.1　Characteristics

Henceforth, we call the Internet newspaper articles aimed at personal computers "Web articles" and those aimed at mobile phones "mobile articles." More than a hundred Web newspaper articles written in Japanese are distributed on the Web every day by Mainichi newspaper company (http://www.mainich.co.jp/). Their lengths are a few hundreds to five hundreds characters and the average length is about 250 characters. An Web article consists of several key words, a title and a body of text. These articles consist of ordinary and formal written sentences.

About 70 mobile newspaper articles also written in Japanese are distributed by Mainichi newspaper company on the Web. The average length of one mobile article is around 50 characters for the old types of mobile phones. A mobile article consists only of the body text.

Since they are available on the Web only for a few days, we have routinely downloaded them on a day-to-day basis. Actually we have collected 48,075 pairs of Web articles and mobile articles of Mainichi newspaper from April 26th 2001 to March 30th 2003. Since one mobile article often consists of more than one sentences, the total number of sentences of a mobile article is 88,333.

## 2.2   Final Parts of Mobile Articles' Sentence

Mobile articles are short and compact. We find this compactness especially appearing at a final part of mobile article's sentences. Ordinary formal Japanese sentences, which are obviously used in Web articles, almost always end with a verb or an auxiliary verb of present or past tense because Japanese is a head final language. On the contrary, sentences of mobile articles end variety of POSs as shown in Table 1, where the ratio is the total of each case against the above described 88,333 mobile sentences. We got a POS tag of each word by Japanese morphological analyzer: Chasen [7].

**Table 1.** Patterns of sentence final parts of mobile articles

| POS | Ratio (%) |
|---|---|
| action noun | 38.8 |
| other noun | 18.0 |
| post positional particle | 16.4 |
| verb | 18.3 |
| auxiliary verb | 7.6 |
| others | 0.9 |

In this table, action noun which we call "SA-HEN MEISI", is a kind of noun mainly expressing an action, etc. Its English counterpart is a noun appearing in the pattern of light verb + noun, i.e. "tennis" in "do tennis." All of SA-HEN MEISIs do not necessarily mean an action. Some are for mental state change or whatever. Nevertheless we, henceforth, call SA-HEN MEISI *action noun* in the remainder of this paper for simplicity.

## 3   Alignment

### 3.1   Article to Article Alignment

As stated previously, the number of Web articles is larger than the number of mobile articles. Since mobile news articles of a day are the excerpts of the whole Web news articles of the day, every mobile article finds its counterpart in the same day's Web articles. Thus the search space for a counterpart Web article of the mobile article is significantly narrowed down to the same day's articles. In this circumstance, the first thing to do is to find the Web article

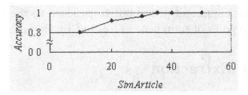

**Fig. 1.** Accuracy of article alignment

which corresponds to each mobile article. For this we use the similarity score: $SimArticle(W, M)$ where $W$ means a Web article and $M$ means a mobile article defined below,

$$SimArticle(W, M) = a \times K + b \times T + NN \ . \qquad (1)$$

where $K$ is the number of $W$'s key words which also appear in $M$, $T$ is the number of nouns in the title of $W$ which also appear in $M$, and $NN$ is the number of nouns that appear in both of $W$'s body and $M$. The parameters $a$ and $b$ are weights of the first and second factors respectively and both are chosen to be 3.0 experimentally. Figure 1 shows the relation between $SimArticle$ and accuracy of article alignment for randomly selected 605 mobile articles, where machine output alignments are checked by hand. As seen from Figure 1, sentence pairs whose $SimArticles$ are more than 35 are correct pairs and resulting in 481 correct pairs. Thus we apply this threshold of 35 to all of Web articles and mobile articles described in Section 2.

### 3.2 Sentence to Sentence Alignment

Next, we extract sentence pairs from these aligned pairs of articles. Since newspaper articles always put the most important information in the first few sentences, we only focus on the first paragraph of Web articles. Practically, sentences of Web article aligned to the sentence of mobile article are identified by the following method where $Ws$ means a sentence in the Web article's first paragraph, $\{Ws\}$ means the set of them, $Ms$ means a sentence of mobile article, and $Ws(Ms)$ is a Web article sentence aligned to $Ms$,

$$\text{foreach}(Ms)$$
$$\{Ws(Ms) = \text{a sentence in } \{Ws\} \text{ having}$$
$$\text{the highest similarity with } Ms\} \ . \qquad (2)$$

where the similarity is defined as a number of nouns appearing in both of $Ws$ and $Ms$. We extract 88,333 aligned pairs of sentences by this method. We choose 500 pairs randomly from these pairs and checked them by hand and found that 92.8% of them were correctly aligned. This figure might not be sufficiently high for alignment task itself. The main objective of our research, however, is extraction of paraphrases by means of some statistical method. Therefore we decided not to pay more effort for alignment per se but to proceed to the task of paraphrases extraction using the pairs of sentences extracted by this method.

# 4 Paraphrase Extraction

## 4.1 Background

Paraphrases would be used for many purposes including text simplification [5]. Our target, which is a little bit similar with their work, is to extract expressions of the same meaning in Web sentences and mobile sentences. The latter is a more compact and simplified form of the former. Then what we want to extract is paraphrases by which we simplify Web sentences into sentences that can be used as mobile sentences.

Paraphrases we focus on here are the last clause of each of Web sentences and the last noun phrases of mobile sentences. A last part of ordinary Japanese sentence is usually a verb, on the contrary, a last part of Japanese sentence of compact text like mobile articles is often an action noun as already seen in Table 1. For instance, Japanese verb phrase "akira-ka ni na-tta"('be discovered') is sometimes paraphrased with a noun "han-mei" which can be translated into "be known", "proved to be", etc. As shown in this example, "XXX"(YYY) means that XXX is a part of Japanese sentences and YYY is its English gloss.

Now we selected distinct expressions at the end of mobile sentences by the following way. Firstly, post positional particles (PPP) are function words like prepositions of English, and are one or two characters length. Then we combine a word right adjacent of PPP to make one distinct expression. Secondly, since our paraphrase extraction algorithm is based on frequencies as described later, we select expressions that occur more than two times. Then we have had 4566 distinct expressions, which contain 1085 distinct action nouns.

Looking at 30 most frequent expressions within these 4566 expressions, 15 of them are action noun like "han-mei." From the viewpoint of paraphrasing, this is a compression of a phrase which contains a pattern of action noun + light verb. Considering these factors, we focus on this type of expressions as our target of paraphrase extraction in this paper. The following is an example where a character string separated by a space or hyphen corresponds to one Japanese character. Note that a character connected by hyphens is one morpheme. The italic parts of these two sentences are the paraphrases we try to extract by the system we will propose in the remaining of this paper.

Mobile sentence ending part: "zi-ko gen-in *han-mei*" (the cause of the accident identified).
Web sentence ending part: "zi-ko gen-in *ga akira-ka ni na-tta*" (the cause of the accident is identified).

## 4.2 Extraction Framework

What we want to extract is a set of expressions appearing in Web sentences that are the paraphrases of the action noun appearing at the end of mobile sentences.

> **Step 1:** Gather mobile article sentences: $Ms$ having the same $AN$ at the last part of mobile sentences to make $\{Ms(AN)\}$.
> **Step 2:** Gather Web sentences paired with each of $\{Ms(AN)\}$ to make $\{Ws(AN)\}$.
> **Step 3:** Extract candidates of paraphrase of $AN$ from $\{Ws(AN)\}$.

**Fig. 2.** The framework of paraphrase extraction

Then the first thing to do is to extract action nouns from mobile sentences. We have already done it and showed the result in Table 1. Here, a set of mobile sentences having an action noun:$AN$ is denoted as "$\{Ms(AN)\}$." The next thing to do is to gather a set of Web sentences which may have the paraphrase of the given $AN$. Since we have already had is the big amount of aligned pairs of Web sentence and mobile sentence as stated in Section 3, it is easily accomplished by simply gathering Web sentences being aligned with each mobile sentence in $\{Ms(AN)\}$. We denote it as "$\{Ws(AN)\}$" henceforth. Once we get $\{Ws(AN)\}$, the remaining problem is to extract a candidate of paraphrase of $AN$ from $\{Ws(AN)\}$. This process is formally described as follows.

Owing to the aligned sentences described in Section 3.2, it becomes much easier to extract sentences which may contain paraphrases than paraphrase extraction researches that do not use this kind of aligned sentences. Even though we use aligned sentences to extract paraphrases, we still have many possible types of paraphrases like a noun, a noun phrase, verb, verb phrases, and so on. Thus we make the procedure at Step 3 one step easier by focusing on the last parts of sentences in $\{Ws(AN)\}$. By this narrowing down, we can identify where paraphrases exist. The remaining problem is how to implement Step 3 of Figure 2.

### 4.3    Character Based Extraction with Branching Factor and Frequency

Since $AN$s are located at the end of mobile sentences, it is reasonable to expect many paraphrases of $AN$ are also located at the last part of sentences in $\{Ws(AN)\}$. Of course, some paraphrases might be located not at the end of sentence. The method we use is, however, based on statistics of distribution of target expressions. Thus we only focus on the last part of sentences.

The next thing to note is that we adopt character based extraction as well as word based extraction. This is because 1) we are free from error of morphological analyzer whose accuracy is still around 95% for Japanese, 2) the proposed method is independent on language and can be applied any language, and 3) the proposed method can extract paraphrases exhaustively including character string not yet recognized as a word. One example of 3) is an abbreviation like "X-ing" meaning "crossing." Of course we loose a sophisticated linguistic information a morphological analyzer gives us, however, we take the merits 1) 2) and 3) more in our character based extraction.

Then the problem of extracting paraphrases of $AN$ is how many characters from the end of sentence of $\{Ws(AN)\}$ we should extract. To solve this problem, we first cut out some character strings which are probably paraphrases of $AN$ from $Ws(AN)$. For this we introduce an important notion that we call "branching factor." A similar idea, "accessor variety", has been proposed in [2], however their aim is word segmentation which is completely different application.

**Branching Factor.** Here, we firstly introduce a forward branching factor of a character string $Cs$ in a set of sentences which makes it easy to understand the notion of branch. If the length of $Cs$ is necessary to express explicitly, we explicitly write down $Cs(n)$ where $n$ is the length in character, henceforth.

Forward branching factor: $FB(Cs)$ is defined as the number of distinct characters which are right adjacent to $Cs$ in a set of sentences. Let $Cs$ be a character "n." Then $FB(\text{"n"})$ is big because we may have many kinds of characters after "n" of the first character of words, like "na", "ne", etc. After "na", "m" of "name", "t" of "nature", etc. may come, and still $FB(\text{"na"})$ is high. But after "natu", very few kinds of character can come like "r" of "nature." Thus $FB$ decreases as we proceed right within a word. Obviously, once a word ends, $FB$ suddenly increases. Thus we would extract linguistically meaningful expression by cutting out character strings at the point where a $FB$ increases. The notion of $FB$ is proven to be useful to segment out the meaningful expression [9].

However, we would like to cut out expression from the last part of sentence. Thus, what we need is a branching factor of the other direction, namely a backward branching factor: $BB$. $BB(Cs)$ is defined as the number of distinct characters that are left adjacent to $Cs$ in a set of sentences. We expect that if we scan character strings backwards from the end of sentence, the same situation as described in a forward branching case is expected to happen. We depict the situation with more concrete example. Consider, for instance, a Japanese sentence.

$$\text{"ke-nen wo hyou-mei si ta"} \qquad \text{(The suspicion was expressed)} \ . \qquad (3)$$

In (3) a sequence of character separated by a space or hyphen such as "si" and "ta" indicates one Japanese character. If we take a set of Web sentences $\{Ws(AN)|AN = \text{"hyou-mei"}\}$ ( $= \{Ws(\text{"hyou - mei"})\}$), the following figure depicts the situation of backward branching.

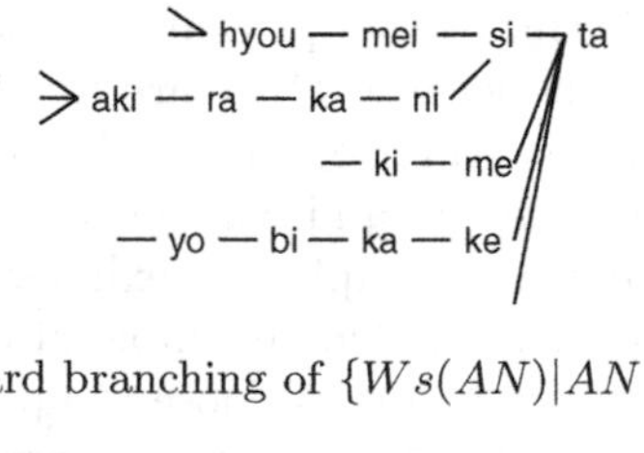

**Fig. 3.** Backward branching of $\{Ws(AN)|AN = \text{"hyou-mei"}\}$

A Japanese character taken from the end of sentence is "ta" which is an independent morpheme indicating past tense. Therefore very many kinds of character come to the left of "ta" as shown in Figure 3. Thus $BB(\text{"ta"})$ is very large. $BB($ "si-ta"$)$, where "si-ta" means "did" in English, is rather low because there can be several possible action nouns whose meaning is same or similar with "hyou-mei" (express). Then longer the string is, the smaller $BB$ becomes. However, any word can come to the left of "hyou-mei si-ta", that means $BB$ turns to increase at this point. Here "increase" of $BB(Cs(n))$ means:

$$BB(Cs(n)) > BB(Cs(n-1)) \ . \tag{4}$$

We show an example of $BB$ for a string "kanga-e wo aki-ra-ka ni si-ta" calculated using a set of sentences $\{Ws($ "hyou-mei"$)\}$ in Figure 4. Clearly, character strings whose $BB$ is increasing coincide with coherent expressions like "aki-ra-ka ni si-ta"(disclosed) and "kanga-e wo aki-ra-ka ni si-ta"(disclosed his/her opinion).

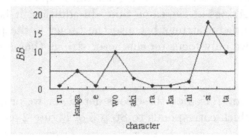

**Fig. 4.** $BB($ "hyou-mei"$)$ vs. each character from the end of sentence

This situation generally holds because a sentence is always a sequence of fixed expression like words or phrases. Therefore, $BB$ is useful to extract character strings of fixed and/or coherent expression. Concretely, if we cut out strings: $Cs(n)$ where $BB(Cs(n))$ increases, that $Cs(n)$ is a linguistically coherent and fixed expression which turns out to be a good candidate of paraphrase.

**Frequency, Length and Their Combination.** As all of the $Ws(AN)$s contain paraphrases of $AN$, we expect that many of them share the same expression which has the same meaning $AN$ has. The character string which has a high frequency within $\{Ws(AN)\}$ probably is a paraphrase of $AN$. Therefore a frequency of character string is a good indicator of how likely the character string is a paraphrase.

The second factor we have to consider is a length of the character string, namely $n$ of $Cs(n)$. If we take character strings only based on their frequencies, we probably encounter the following cases. One extreme case is short strings like "si-ta"(did). Since every action noun can come just left of "si-ta", $BB($ "si-ta"$)$ is high and possibly increasing, and the frequency of "si-ta" is obviously high, we might extract "si-ta" in some $\{Ws(AN)\}$. Of course it is not desirable because apparently "si-ta" is not a paraphrase of any action noun. The other extreme case is a long expression like "saku-zitu tou-kyou de tai-ho sa-re-ta"(got arrested in Tokyo yesterday). It is not a paraphrase of action noun "tai-ho"(arrest) because it expresses too detailed information than "tai-ho" expresses. Thus we have to exclude too long and too short strings.

For this, firstly, we take longer strings more seriously as defined by the following formula,

$$\log(n - 1) \ . \tag{5}$$

where $n$ is obviously the length($Cs(n)$). By (5), the importance of $Cs(n)$ in which $n < 3$ is less than or equal to zero. Thus every $Cs(n)$, $n < 3$, are virtually ruled out. It is reasonable for Japanese because 1) action nouns almost always consists of two Chinese characters and 2) Paraphrases in longer Web sentences for action nouns in mobile sentences are expected to be longer than two characters. A longer $Cs$ gets a high score but the score is saturated because of log function.

Secondly, undesirably long and redundant strings are expected to have low frequencies. Thus, the combination of length and frequency would exclude too long and too short strings at the same time. In addition, it is not necessary to search very long character strings because the target is the paraphrase of one action noun. Thus we only focus on character strings that are shorter than 30 characters.

**Ranking Algorithm.** Based on these considerations, we propose a paraphrase extraction system which corresponds to Step 3 of Figure 2 as shown below.

---

**Step A:** Scan every sentence of $\{Ws(AN)\}$ backward from the end of sentence to extract character strings: $Cs(n)$ of any $n$ less than 30.
**Step B:** Calculate $BB(Cs(n))$ for every $Cs(n)$ extracted at Step A.
**Step C:** Pick up $Cs(n)$ whose $BB(Cs(n))$ increases from the resultant set of strings at Step A. We denote a set of character strings chosen by this method as $\{CBB(n)\}$.
**Step D:** Sorting all strings in $\{CBB(n)\}$ on the descending order of evaluation function $Para(CBB(n))$.

---

**Fig. 5.** Sorting algorithm to sort paraphrase candidates

The remaining problem is the definition of evaluation function $Para(CBB(n))$ that we use at Step D of Figure 5. For this we propose the following definitions,

$$Para(CBB(n)) = BB(CBB(n)) \times \log(n - 1) \times \mathrm{freq}(CBB(n)) \ . \tag{6}$$

where $\mathrm{freq}(CBB(n))$ is the frequency of the string $CBB(n)$ in $\{Ws(AN)\}$. As stated in 4.3, if two factors: 1) decreasing of frequency of $Cs(n)$ and 2) increasing of $\log(n - 1)$ as $n$ becomes longer, are well combined, we can exclude too long $Cs(n)$. Since as stated in 4.3, we can exclude too short $Cs(n)$ by $\log(n - 1)$, we can exclude too short and too long $Cs(n)$ and pick up frequently occurred character strings.

## 4.4    Word Based Extraction

The above described character based extraction method is easily translated into a word based extraction method by replacing character with word. For this, since Japanese is an agglutinative language, we have to do morphological analysis and POS tagging. We use the evaluation function defined by (6) except that the length used in (5) is counted in character because we try to exclude one or two character length string. If we count the length in word, we may exclude longer strings which may be a good paraphrase.

## 4.5    Experimental Results and Evaluation

We evaluated the paraphrases resulted in sorting algorithm described in Figure 2 and Figure 5 combined with definitions of *Para* defined in (6) and word based method described in 4.4. For evaluation, we test whether the resultant candidates of paraphrases are correct paraphrases in any context. This is done by hand because this correctness is known based on deep semantic analysis including even some consideration about contexts. Since we have 1159 action nouns among 4566 types of sentence ending expressions, we cannot evaluated every candidate of paraphrase for every ranked character string $CBB(n)$ by hand. Then we firstly evaluate precisely the ten most frequently used Japanese action nouns in our corpus as shown below.

> Ha-ppyou (announce), Tai-ho (arrest), Kai-dan (have a talk), Hyou-mei(express, demonstrate), Si-bou (die), Ke-ttei (decide), Kyou-chou (coordinate), Han-mei (proved to be, turn out to be, discover), Gou-i (agree), Ken-tou (examine).

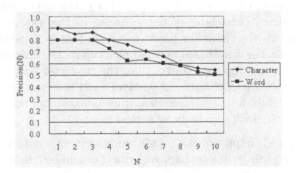

**Fig. 6.** Precision($N$) for 10 highest ranked candidates of character based method (Character) and word based method with morphological analyzer (Word)

If we closely look at these action nouns, all of them have several meanings. However they are roughly similar. We expect to extract paraphrases that are similar but have a little bit distinct meaning of the original action nouns. Actually, we test by hand the ten highest ranked candidates for each of these ten *ANs* resulted in by the above described sorting algorithms and calculate precisions against the first to $N$-th candidates defined by the following formula, where $C(i)$ means the number of correct paraphrases within $i$ highest candidates.

$$\text{Precision}(N) = \frac{1}{N} \sum_{i=1}^{N} C(i) \ . \tag{7}$$

The results of the methods described above are shown in Figure 6. When the character based method is used, nine out of the ten highest ranked candidates are correct paraphrases. If we take three highest ranked candidates of the character based method, almost 87% are still correct. Moreover about 55% of ten highest ranked candidates are correct. It is not shown in Figure 6, but the Precision($N$) gradually degraded and Precision(20) still maintains 50%. The results of word based method is lower. This result indicates that the expressions resulted in by our aligned sentences and extraction algorithm based on character string are high quality candidates of paraphrase.

**Table 2.** Accuracy of top 3 candidates

|                 | First | Second | Third |
| --------------- | ----- | ------ | ----- |
| Character based | 0.82  | 0.68   | 0.59  |
| Word based      | 0.83  | 0.81   | 0.57  |

In order for more comprehensive result, we test the highest to third highest ranked candidates of paraphrases for most frequent 100 action nouns. The results of their accuracy for each of first, second and third candidates are shown in the Table 2, where "accuracy" means the ratio of the extracted correct paraphrases over all the extracted candidates of paraphrases. In this case, the word based method outperforms the character based method. This is because in middle to low frequency cases, the word based method picks up more grammatically reasonable paraphrases at the top three ranks than the character based method which is aimed at extracting paraphrases exhaustively and may produce ungrammatical candidates. Finally we show some examples of extracted paraphrases where we can replace "sth" with any noun or action noun.

*happyou (announce).* happyou-sita (made anouncement), suru-to happyou-sita (announced to do sth.), sita-to happyou-sita (announced that sth. has been done), seisiki-ni happyou-sita (formally announced), akiraka-ni sita (disclosed), kouhyou-sita (publicly announced), to kata-tta (talked "..."),

*kettei (decide).* kime-ta (decided), suru-koto-wo kime-ta (decided to do sth.), kettei-sita (made a decision), suru-koto-wo kettei-sita (decided to do sth.), wo kettei-sita (made a decision to do sth.),

*hanmei (discover).* waka-tta (discovered), de waka-tta (discovered by sth (=evidence)), akira-ka ni na-tta (proven to be), hamei-sita (turn out to be)

# 5   Conclusions

We collected and aligned Web news articles and news articles for mobile phones over two years. Using this aligned corpus, we extract character strings of paraphrases of action nouns appearing at the end of mobile sentences based on the combination of branching factor, frequency and length. The samples of the result show high precision and indicate semi-automatic paraphrase extraction to be realistic in practical use. Our future work is to extract other types of paraphrases like sentence end with post positional particles from our aligned sentences.

# References

1. Brazilay, R. and McKeown, K.: Extracting paraphrases from a parallel corpus. In Proceedings of ACL-EACL2001 (2001) 50-57.
2. Feng, H., Chen, K., Kit, C. and Deng, X.: Unsupervised Segmentation of Chinese Corpus Using Accessor Variety. In Proceedings of the First International Joint Conference on Natunal Language Processing (2004) 255-261.
3. Ibrahim, A., Katz, B. and Lin, J.: Extracting Structural Paraphrases from Aligned Monolingual Corpora. In IWP2003 ACL2003 (2003) 57-64.
4. Inui, K. and Hermjakob, U. (eds.): Proceedings of the Second International Workshop on Paraphrasing: Paraphrase Acquisition and Applications (IWP2003) ACL2003 (2003).
5. Inui, K., Fujita, A., Iida, R. and Iwakura, T.: Text Simplification for Reading Assistance: A Project Note. In IWP2003 ACL2003 (2003) 9-16.
6. Lin, D. and Pantel, P.: DIRT-discovery of inference rules from text. In Proceedings of the ACM SIGKDD Conference on Knowledge Discovery and Data Mining (2001) 323-328.
7. Matsumoto, Y., et al. : Manual of morphological analyzer "Chasen". Ver.2.2.9, Nara Advanced Institute of Technology (2002).
8. Sato, S. and Nakagawa, H. (eds.): Proceedings of Workshop: Automatic Paraphrasing: Theories and Applications, NLPRS2001 (2001).
9. Tanaka-Ishii, K., Yamamoto, M. and Nakagawa, H.: Kiwi: A Multilingual Usage Consultation Tool based on Internet Searching. In Proceedings of the Interactive Posters / Demonstrations, ACL-03 (2003) 105-108.

# Improving Transliteration with Precise Alignment of Phoneme Chunks and Using Contextual Features

Wei Gao, Kam-Fai Wong, and Wai Lam

Department of Systems Engineering and Engineering Management,
The Chinese University of Hong Kong,
Shatin, N.T., Hong Kong, China
{wgao, kfwong, wlam}@se.cuhk.edu.hk

**Abstract.** Automatic transliteration of foreign names is basically regarded as a diminutive clone of the machine translation (MT) problem. It thus follows IBM's conventional MT models under the source-channel framework. Nonetheless, some parameters of this model dealing with *zero-fertility* words in the target sequences, can negatively impact transliteration effectiveness because of the inevitable *inverted* conditional probability estimation. Instead of source-channel, this paper presents a *direct* probabilistic transliteration model using contextual features of phonemes with a tailored alignment scheme for phoneme chunks. Experiments demonstrate superior performance over the source-channel for the task of English-Chinese transliteration.

## 1 Introduction

Automatic transliteration of foreign names, e.g. names of people, places, organizations, etc., is recognized as an important issue in many cross language applications. Cross-lingual information retrieval involves query keyword translation from the source to target language and document translation in the opposite direction. For similar reasons, machine translation and spoken language processing, such as cross-lingual spoken document retrieval and spoken language translation, also encounters the same problem of translating proper names. Contemporary lexicon-based translation is ineffective as proper name dictionaries can never be comprehensive. New names appear almost daily and become unregistered vocabulary in the lexicon. This is known as the *Out-Of-Vocabulary* (OOV) problem. The lack of translation for OOV names can impact the performance of applications adversely and sometimes seriously.

Based on pronunciations, foreign names can usually be translated, or more appropriately transliterated, into target languages, which were originally hand-coded with rules of thumb by human translators. De facto standard has been established, but is often inconsistently used. The rules are subjected to the interpretation of individual producers. In Mandarin Chinese, for instance, the name

S. H. Myaeng et al. (Eds.): AIRS 2004, LNCS 3411, pp. 106–117, 2005.

of "Bin Laden" can be translated as /ben la deng/[1], /bin la deng/, /ben la dan/ and /bin la dan/. Sometimes dialectical features can further ambiguate the standard. For these reasons, rule-based transliteration approach has been undermined in English-to-Chinese machine translation applications. Thus, an effective data-driven transliteration model is required.

In this paper, we present a statistical phoneme-based method for forward transliteration of foreign names from English to Chinese. Grapheme-to-phoneme transformation and Pinyin-to-Hanzi conversion applied in the phoneme-based methods are extensively studied areas. We focus on the intermediate tasks for transliterating phoneme pairs. The rest of the paper is organized as follows: Section 2 summarizes related work; Section 3 explains the drawbacks of source-channel model in our task; Section 4 illustrates our methods in detail; Section 5 presents and analyses experimental results; Section 6 concludes this paper.

## 2   Related Work

Several approaches have been proposed for automatic name transliteration between various language pairs. Based on the source-channel framework, [7] described a generative model, in which they adopted finite state transducers and a channel decoder to transform transliterated names in Japanese back to their origins in English. The source-channel model was later on applied or extended by a number of other tasks: backward transliteration from Arabic to English in [12], forward transliteration from English to Korean in [8], and forward transliteration from English to Chinese in [13].

The aim of work by [13] is closest to ours. Their fundamental equation derived from the IBM statistical machine translation (SMT) model proposed by [2]:

$$\hat{C} = \underset{C}{\operatorname{argmax}}\, p(C|E) = \underset{C}{\operatorname{argmax}} \left\{ p(E|C) \times p(C) \right\} \tag{1}$$

where $E = e_1^{|E|}$ denotes a $|E|$-phoneme English word as the observation on channel output, and $C = c_1^{|C|}$ represents $E$'s $|C|$-phoneme Chinese translation by pinyin as the source of channel input. As shown in Fig. 1, the channel decoder reverses the direction to find the most probable input pinyin sequence $\hat{C}$ given an observation $E$. The posterior probability $p(C|E)$ is indirectly maximized by optimizing the combination of the transliteration model $p(E|C)$ and the language model $p(C)$. $p(E|C)$ was trained from name pairs represented by International Phonetic Alphabet (IPA) symbols for English names (obtained from a speech synthesis engine) and pinyin notations for their Chinese counterparts (obtained from a Hanzi-Pinyin dictionary). It proceeded by Expectation-Maximization ($EM$) iterations of standard IBM SMT model training method by using GIZA++ toolkit, bootstrapping from Model-1 through Model-4 [13]. Language model $p(C)$ was trained by using pinyin symbol trigrams and applying

---

[1] Mandarin pinyin is used as phonetic representation of Chinese characters throughout this paper. For simplicity, we ignore the four tones in the pinyin system.

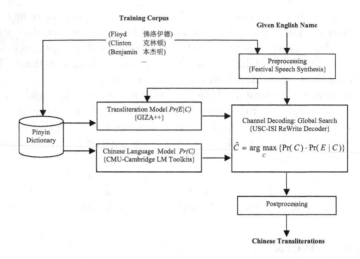

**Fig. 1.** English-to-Chinese transliteration system [13] based on IBM SMT model

*Good-Turing smoothing* with *Katz back-off* in CMU-Cambridge language modeling toolkits [3]. Search was done by use of USC-ISI ReWrite Decoder [5]. The method demonstrates pinyin error rates in edit distance of around 50% [13].

## 3    Drawbacks of Source-Channel

When the IBM SMT model is applied in our task, i.e. English-to-Chinese transliteration, it has several limitations:

1. $p(E|C)$ is approximated by the Markov chains [11] under Markov assumption (zero order or first-order) on state transition as well as conditional independence assumption on observation. Markov assumption hypothesizes that the transition probability to a state, i.e. phoneme of Chinese pinyin, depends only on its previous one state at most. Longer history may suffer from data sparseness and renders the model computationally expensive with the increase of state space; conditional independence assumption assumes that an observation unit, i.e. English phoneme, depends only on the state that generates it, not on its neighboring observation units. With these hypotheses, it is hard to extend the model by using additional dependencies, such as flexible features of neighboring phonemes. Albeit the trigram language model $p(C)$ is combined, (1) cannot be be optimal unless both $p(E|C)$ and $p(C)$ use the true probability distributions. Yet, the used models and training methods in machine translation empirically testified that they only provided poor approximations of the true distributions [9, 13].

2. Because of the *inverted* conditional probability $p(E|C)$, only a target language phoneme can be associated with a contiguous group of source language phonemes, but not vice visa, i.e. never could one English phoneme be con-

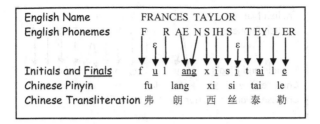

| English Name | FRANCES TAYLOR | | | | | | |
| English Phonemes | F  R AE N S IH S  T EY L ER | | | | | | |

| Initials and Finals | f | u  l | ang | x i | s i | t | ai | l  e |
| Chinese Pinyin | fu | | lang | xi | si | tai | | le |
| Chinese Transliteration | 弗 | | 朗 | 西 | 丝 | 泰 | | 勒 |

**Fig. 2.** English-to-Chinese transliteration example in [13], considering unaligned symbols as zero-fertility

verted to a group of pinyin symbols. The example in [13] exposes this obvious limitation (see Fig. 2[2]): /u/ and the second /i/ in the third line have to be considered as spuriously produced from nothing or from a mute $\varepsilon$. Under the IBM models, such inserted symbols are known as *zero-fertility* "words". They are deleted by source-channel during training and reproduced by decoder by considering inserting one of them before each target symbol of each remaining unaligned source phoneme in terms of the number of possible zero-fertility symbols [5]. Although adding zero-fertility symbols may increase the probability of hypothesis, incorrect transliterations are still abundant as such insertions are frequent.

3. Due to smoothing, the language model may not assign zero probability to an illegal pinyin sequence, e.g. one containing two consecutive initials [13]. Such sequences need to be manually corrected by inserting certain finals between them until a legitimate pinyin sequence is obtained. Moreover, the training of language model is independent of transliteration model and their combination sometimes can yield unpredictable results.

## 4   Direct Transliteration Model

Instead of source-channel, we aim to estimate the posterior probability directly. We rectify the angle of observation to avoid the use of the reversed conditional probability. Figure 3 shows the application of the alignment scheme of our approach to the previous example in Fig. 2, where we look possible combinations of pinyin symbols as initial-final clusters converted from single English phonemes. The condition of the probability to be estimated thus turns out to be $E$ instead of $C$. The distribution can be approximated directly by Maximum Entropy (*Max-Ent*) approach [1]. Under *MaxEnt* model, the language model can be considered as an optional feature [9]. Its absence could be compensated if other cutting edge features could be chosen.

---

[2] Lowercase letters denote pinyin symbols. Capital letters are English phonemes represented by computer-readable IPA notations — ARPABET.

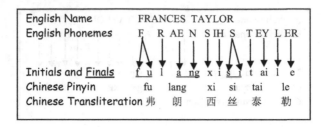

**Fig. 3.** Our phoneme alignment scheme in direct transliteration modeling

### 4.1   Baseline of Direct Model

We introduce the pinyin mapping units of each $e_i$ denoted by $cmu_i$, which can be individual pinyin symbols or clusters of initial and final. In an alignment, each English phoneme aligns to only one $cmu$. Thus, the transliteration model $p(C|E)$ can be approximated by:

$$p(C|E) \approx \prod_{i=1}^{|E|} p(cmu_i|e_i) \; . \tag{2}$$

The unknown $cmus$ (clusters) can be discovered on the fly during $EM$ training for computing the *Viterbi* alignments and symbol-mapping probabilities by using GIZA++, where we can make the source and target consistent with the actual transliteration direction, i.e. from English to Chinese.

Equation (2) gives poor approximation as no contextual feature is considered. From our perspective, the transliteration is to classify each phoneme of a given English name into its most probable $cmu$ according to the frames of various features. We then yield a better approximation:

$$p(C|E) \approx \prod_{i=1}^{|E|} p(cmu_i|h_i) \tag{3}$$

where $h_i$ denotes the history or context of $e_i$, which is described as follows:

$$h_i = \{e_i, e_{i+1}, e_{i+2}, e_{i-1}, e_{i-2}, cmu_{i-1}, cmu_{i-2}\} \; . \tag{4}$$

History of an English phoneme is defined as its left-two and right-two neighboring phonemes plus the two $cmus$ at pinyin side, to which its left-two phonemes align. For each $e$ in a given pair of $\{e_1, e_2, \ldots, e_n\}$ and $\{cmu_1, cmu_2, \ldots, cmu_n\}$, its conditional transliteration probability to produce the $cmu$ with respect to its contextual history $h$ can be computed by:

$$p(cmu|h) = \frac{p(h, cmu)}{\sum_{cmu' \in \Omega} p(h, cmu')} \tag{5}$$

where $\Omega$ is the set of all $cmus$ mapped from $e$ and observed in the training data, and $p(h, cmu)$ is the joint probability distribution of observing $h$ and $cmu$

**Table 1.** Baseline model feature templates

| Category | Contextual Feature Templates | # of Possible Features |
|---|---|---|
| 1 | $e_i = \mathcal{X}$ and $cmu_i = \mathcal{Z}$ | $|\mathcal{V}_\mathcal{E}| \cdot |\mathcal{V}_\mathcal{C}|$ |
| 2 | $cmu_{i-1} = \mathcal{X}$ and $cmu_i = \mathcal{Z}$ | $|\mathcal{V}_\mathcal{C}|^2$ |
| 3 | $cmu_{i-2}cmu_{i-1} = \mathcal{X}\mathcal{Y}$ and $cmu_i = \mathcal{Z}$ | $|\mathcal{V}_\mathcal{C}|^3$ |
| 4 | $e_{i-1} = \mathcal{X}$ and $cmu_i = \mathcal{Z}$ | $|\mathcal{V}_\mathcal{E}| \cdot |\mathcal{V}_\mathcal{C}|$ |
| 5 | $e_{i-2} = \mathcal{X}$ and $cmu_i = \mathcal{Z}$ | $|\mathcal{V}_\mathcal{E}| \cdot |\mathcal{V}_\mathcal{C}|$ |
| 6 | $e_{i+1} = \mathcal{X}$ and $cmu_i = \mathcal{Z}$ | $|\mathcal{V}_\mathcal{E}| \cdot |\mathcal{V}_\mathcal{C}|$ |
| 7 | $e_{i+2} = \mathcal{X}$ and $cmu_i = \mathcal{Z}$ | $|\mathcal{V}_\mathcal{E}| \cdot |\mathcal{V}_\mathcal{C}|$ |

simultaneously, which can be trained using maximum likelihood estimation. We use the *MaxEnt* model to solve the joint probability distribution [10]:

$$p(cmu, h) = \pi\mu \prod_{j=1}^{k} \alpha_j^{f_i(h,cmu)} \tag{6}$$

where $\pi$ is a normalization constant, $\{\mu, \alpha_1, \alpha_2, \ldots, \alpha_k\}$ are the model parameters and $\{f_1, f_2, \ldots, f_k\}$ are features which are all binary. Each parameter $\alpha_j$ corresponds to a feature $f_j$. A feature takes the following form:

$$f_1(h_i, cmu_i) = \begin{cases} 1 & \text{if } e_i=/\text{F}/ \ \& \ e_{i-1}=\text{START} \ \& \ e_{i+1}=/\text{R}/ \ \& \ cmu_i=/\text{fu}/ \\ 0 & \text{otherwise} \end{cases}$$

or

$$f_2(h_i, cmu_i) = \begin{cases} 1 & \text{if } e_i=/\text{S}/ \ \& \ e_{i+1}=/\text{T}/ \ \& \ cmu_{i-1}=/\text{IH}/ \ \& \ cmu_i=/\text{si}/ \\ 0 & \text{otherwise .} \end{cases}$$

The general feature templates we used in experiments are listed in Table 1, where $\mathcal{X}$, $\mathcal{Y}$ and $\mathcal{Z}$ can be any individual English phoneme or Chinese pinyin $cmu$, and $|\mathcal{V}_\mathcal{E}|$ and $|\mathcal{V}_\mathcal{C}|$ are the number of elements in the respective sound vocabulary of English and Chinese.

### 4.2   Improving the Baseline Model

**Deficiencies of the Baseline Model.** There are two critical problems in the baseline model that can be improved:

1. The search space for finding the *Viterbi* alignment from all possible alignments is extremely large. Take the name pair in Fig. 3 for example, there is no means to prevent ill-formed *cmu*s, e.g. the first phoneme /F/ maps to /f/ and the second one /R/ maps to /ul/, which is an unfavorable alignment but cannot be avoided if such mappings dominate the training data. This could produce many illegal pinyin sequences and give rise to a large number of probable *cmu*s. They turned out adding uncertainties to phonetic transcriptions.

2. Because of compound pinyin finals, two consecutive English phonemes may map to a single pinyin symbol, such as mapping from /AE N/ to /ang/ in the example (see Fig. 2), which is not allowed in the baseline. This linguistic knowledge need not be imparted ad-hoc in the model. We can decompose compound finals into multiple basic finals, e.g. from /ang/ into /a/ and /ng/, to reduce the size of target phonetic vocabulary. The original mapping is broken into /AE/-to-/a/ and /N/-to-/ng/.

**Precise Alignment of Phoneme Chunks.** We introduce alignment indicators between a pair of sound sequences of $E$ and $C$. Within 39 English phonemes (24 consonants, 15 vowels) and 58 pinyin symbols (23 initials and 35 finals), there are always some indicative elements for alignment, i.e. indicators. For $E$, they are all the consonants, the vowel at the first position and the second vowel of two contiguous vowels; for $C$ correspondingly, they are all the initials, the final at the first position and the second final of two contiguous finals. Also, we define the following variables: $\tau(S)$ is defined as # of indicators in sequence $S$ ($S \in \{E, C\}$); $t(E, C) = \max\{\tau(E), \tau(C)\}$ represents the maximum # of indicators in $E$ and $C$; $d(E, C) = |\tau(E) - \tau(C)|$ is the difference of the # of indicators in $E$ and $C$.

We chunk $E$ and $C$ by tagging the identified indicators and compensate the one with fewer indicators by inserting $d$ number of mute $\varepsilon$ at its $\min\{\tau(E), \tau(C)\}$ possible positions ahead of its indicators. $\varepsilon$ is practically an indicator defined for alignment. This ensures that both sequences end up with the same number of indicators. The $t$ chunks separated by indicators in $E$ should align to the corresponding $t$ chunks in $C$ in the same order. They are called alignment chunks. There are $\|A\| = \binom{d}{t} = \dfrac{t!}{(t-d)!d!}$ number of possible alignments at chunk level with respect to different positions of $\varepsilon$.

This method can guarantee each chunk contains two sound units at most. Thus, in a pair of aligned chunks, only three mapping layouts between phoneme elements are possible:

1. $e$-to-$c_1c_2$: The alignment would be $e$-to-$c_1c_2$ where $c_1c_2$ is considered as an initial-final cluster ($cmu$);
2. $e_1e_2$-to-$c1c_2$: The alignment at phoneme level would be extended to $e_1$-to-$c_1$ and $e_2$-to-$c_2$. Note that no new alignment is generated under this condition. Thus, the total number of alignments remains unchanged;
3. $e_1e_2$-to-$c$: By adding an additional $\varepsilon$ at $C$ side, the alignment at phoneme level would be extended to $e_1$-to-$c$ and $e_2$-to-$\varepsilon$ or $e_1$-to-$\varepsilon$ and $e_2$-to-$c$. In this case, one more new alignment will be produced and we update $\|A\| = \|A\| + 1$.

EM **Training for Symbol-Mappings.** We then applied *EM* algorithm [4, 7] to find the *Viterbi* alignment for each training pair as follows:

1. *Initialization:* For each English-Chinese pair, assign equal weights to all alignments generated based on phoneme chunks as $\|A\|^{-1}$.

2. *Expectation Step:* For each of the 39 English phonemes, count the instances of its different mappings from the observations on all alignments produced. Each alignment contributes counts in proportion to its own weight. Normalize the scores of the mapping units it maps to so that the mapping probability sums to 1.
3. *Maximization Step:* Re-compute the alignment scores. Each alignment is scored with the product of the scores of the symbol mappings it contains. Normalize the alignment scores so that each pair's scores sum to 1.
4. *Repeat* step 2-3 until the symbol-mapping probabilities converge, meaning that the variation of each probability between two iterations becomes less than a specified threshold.

With the improved alignment, the mappings crossing chunks are avoided. Thus, the *EM* training becomes more precise and produces significantly fewer possible alignments compared to the baseline.

## 5    Experiments and Evaluation

### 5.1    Data Set

We obtained the beta release v.1.0 of LDC's Chinese-English bi-directional named entity list compiled from Xinhua's database, from which we chose the English-to-Chinese proper name list of people as raw data. The list contains 572,213 foreign people's names and their Chinese transliterations. Note that although the list is in English, it contains names originated from different languages (e.g. Russian, German, Spanish, Arabic, Japanese, Korean, etc.). One assumption is that the Chinese translations were produced based on their English pronunciations directly. The exceptions are Japanese and Korean names, which are generally translated in terms of meaning as opposed to pronunciation, and we consider them as noise. We resorted to CMU's pronunciation dictionary and LDC's Chinese character table with pinyin to convert the names into a parallel corpus of sequences of English phonemes and pinyin symbols. We ended up with 46,305 pairs, which were then used as our experimental data pool.

### 5.2    Performance Measurement

There is no standard for measuring machine transliteration. Some tests require human judgment. The performance was evaluated with two levels of accuracy, i.e. character-level accuracy (*C.A.*) and word level accuracy (*W.A.*) in [6]:

$$C.A. = \frac{L - (i + d + s)}{L} \tag{7}$$

$$W.A. = \frac{\# \ of \ correct \ names \ generated}{\# \ of \ tested \ names}. \tag{8}$$

In (7), $L$ is the length of the standard transliteration of a given foreign name, and $i$, $d$, and $s$ are the number of insertion, deletion and substitution respectively, i.e. edit distance between machine-generated transliteration and the standard. If $L < (i+d+s)$, we set $C.A. = 0$. Equation (8) is the percentage of the number of transliterations identical to the standards in all the tested names. A name often has acceptable transliteration alternatives. Hence, we will also measure how the percentage of the number of transliterations distributes over different character-level accuracy ranges, which is referred to as $C.A.$ Distribution ($C.A.D.$):

$$C.A.D. = \frac{\# \ of \ names \ with \ C.A. \in [r_1, r_2)}{\# \ of \ tested \ names} \tag{9}$$

where $[r_1, r_2)$ (denotes $r_1 \leq C.A. < r_2$) is the bound of a $C.A.$ range. We set up six $C.A.$ ranges: $[0\%, 20\%)$, $[20\%, 40\%)$, $[40\%, 60\%)$, $[60\%, 80\%)$, $[80\%, 100\%)$ and $[100\%]$. We are especially interested in the names within the ranges $[0\%, 20\%)$ and $[80\%, 100\%)$ since the former could be considered as "completely incorrect" and the latter "acceptable".

## 5.3    Experiments and Results

In each trial of our experiments, individual translation name pairs, hereafter referred to as instances, were randomly selected from the data pool to build 10 subsets. Each respectively accounts for 10% to 100% (step=10%) of the total instances in the entire pool. In each subset, we used 90% of the instances for training and the remaining 10% for open test. Also the same number of instances (10%) were randomly selected from the training data for close test.

**Experiments.** The baseline model was trained and tested as follows:

1. Using *EM* iterations in GIZA++ to obtain *Viterbi* alignment of each pair of names in the training set. The bootstrapping settings were the same as [13] (see Sect. 2). Note that the direction of estimation is from $E$ to $C$ directly;
2. Aligned training instances were then passed to *GIS* (Generalized Iterative Scaling) algorithm for training the *MaxEnt* models [1, 10]. This fulfilled training the models that can transliterate phoneme sequences of given English names into pinyin sequences;
3. Tests were conducted on the trained *MaxEnt* models. "Beam search" [10] was used where a beam size of 5 was adopted. For each given name, only top-1 transliteration was accepted.

The experiments on the improved model were conducted under similar settings except that the tailored alignment scheme and the *EM* training (see Sect. 4.2) were applied in the step 1.

To investigate the influence of data sizes on performance, the above procedure was applied to the 10 subsets with different data sizes as described previously. And the performance of the model was measured by the average accuracy ($C.A.$ and $W.A.$) of 50 trials of the experiments. $C.A.D.$ was measured with average distributions of 50 trials on 100% data size only.

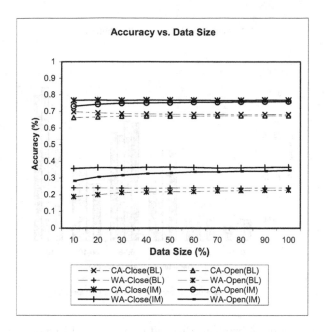

**Fig. 4.** Comparison of baseline (BL) and improved (IM) models on $C.A.\&W.A.$ vs. Data Size

**Results and Discussions.** Figure 4 shows the average $C.A.$ and $W.A.$ of the baseline model (BL) and the improved model (IM) over different data sizes. IM significantly outperforms BL on all tests. In open tests with 100% data, for example, IM demonstrates improvements on $C.A.$ by 8.56% and on $W.A.$ by 11.84%. Recall that in the IM model, we chopped longer pinyin symbols, e.g. compound finals, into smaller sound units, i.e. basic finals, and aligned chunks of English phonemes with corresponding chunks of pinyin symbols, prohibiting alignments across chunk borders. This could produce: 1. more precise mappings between English phonemes and $cmus$; 2. less possible $cmus$ for each English phoneme, reducing uncertainties; 3. less $cums$ forming illegal pinyin syllables, leading to more legitimate pinyin sequences. The figure also shows that with enough instances, the models could achieve almost equal performance in open tests to closed ones.

Figure 5 shows the average percentage of the number of transliterations distributed over their $C.A.$ values (on all data). For $C.A.$ ranging from 0% to 80%, BL produced more transliterations throughout the four ranges than IM. In the remaining $C.A.$ ranges, IM produced more high-quality transliterations (see $C.A. \geq 80\%$) and considerably more correct transliterations (see $C.A. = 100\%$) than BL.

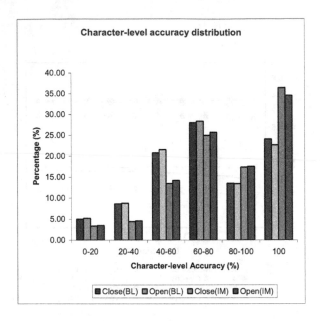

**Fig. 5.** Comparison of baseline (BL) and improved (ML) models on *C.A.D.*

**Table 2.** Results of comparisons on different systems

| Systems/Accuracy | | SC | BL | IM |
|---|---|---|---|---|
| *C.A.* | Close | 66.35% | 68.18% | 76.97% |
| | Open | 65.15% | 67.18% | 75.08% |
| *W.A.* | Close | 20.73% | 23.47% | 36.19% |
| | Open | 18.27% | 21.49% | 32.50% |

## 5.4   Comparisons with Source-Channel System

We compared our work with the source-channel (SC) system described in [13]. Their method (the first translation system) was replicated with the only exception that we obtained phoneme sequences of foreign names via a lookup of CMU's pronunciation dictionary, whereas they adopted the Festival text-to-speech system for English pronunciations. Then we tested the SC system, our BL system and IM system using the entire data pool with 41,674 instances for training and the remaining 4,631 for testing. The language model of SC system was trained on the 41,674 pinyin sequences in the training portion, similarly using trigram by CMU-Cambridge toolkits as [13]. The results are shown in Table 2. Our BL and IM system outperformed the source-channel approach by about 3% and 10% respectively in all tests using the same data set.

# 6     Conclusion

We modeled English-to-Chinese transliteration as direct phonetic mapping from English phonemes to a set of basic pinyin symbols plus dynamically discovered mapping units from training. Contextual features of each phoneme are taken into consideration in the model. An effective algorithm for precise alignment of phoneme chunks was presented, which demonstrated improvements on performance. Comparisons show that our approaches significantly outperforms traditional source-channel model. Future work will include incorporating different features, such as additional contexts, target language model, or even the composition of direct and inverted transliteration model under *MaxEnt* framework.

## Acknowledgement

The work described in this paper is partially supported by CUHK under the Strategic Grant initiative (Project Account No.: 4410001). We are especially grateful to Prof. Helen Meng for her support and help on obtaining LDC corpora.

## References

1. Berger, A.L., Della Pietra, S.A., and Della Pietra, V.J.: A Maximum entropy approach to natural language processing. Computational Linguistics **22** (1996) 39–27
2. Brown, P.F., Della Pietra, S.A., Della Pietra, V.J., and Mercer, R.L.: The mathematics of statistical machine translation: Parameter Estimation. Computational Linguistics **19** (1993) 261–311
3. Clarkson, P., and Ronsenfeld, R.: Statistical language modeling using the CMU-Cambridge toolkit. In Proc. of the 5th EuroSpeech (1997) 2707–2710
4. Gao, W., Wong, K.F., and Lam, W.: Phoneme-based transliteration of foreign names for OOV problem. In Proc. of IJCNLP (2004) 374–381
5. Germann, U., Jahr, M., Knight, K., Marcu, D., and Yamada, K.: Fast decoding and optimal decoding for machine translation. In Proc. of ACL (2001) 228–235
6. Kang, I.H., and Kim, C.C.: English-to-Korean transliteration using multiple unbounded overlapping phoneme chunks. In Proc. of COLING (2000) 418–424
7. Knight, K., and Graehl, J.: Machine transliteration. In Proc. of ACL (1997) 128–135
8. Lee, J.S., and Choi, K.S.: English to Korean statistical transliteration for information retrieval. Computer Processing of Oriental Languages **12** (1998) 17–27
9. Och, F.J., and Ney, H.: Discriminative training and maximum entropy models for statistical machine translation. In Proc. of ACL (2002) 295–302
10. Ratnaparkhi, A.: A maximum entropy model for Part-Of-Speech tagging. In Proc. of EMNLP (1996) 133–141
11. Rabiner, L.R.: A tutorial on Hidden Markov Models and selected applications in speech recognition. In Proc. of IEEE **77** (1989) 257–286
12. Stalls, B.G., and Knight, K.: Translating names and technical terms in Arabic text. In Proc. of COLING/ACL Workshop on Computational Approaches to Semitic Languages (1998)
13. Virga, P., and Khudanpur, S.: Transliteration of proper names in cross-lingual information retrieval. In Proc. of ACL Workshop on Multi-lingual Named Entity Recognition (2003)

# Combining Sentence Length with Location Information to Align Monolingual Parallel Texts

Weigang Li, Ting Liu, and Sheng Li

Information Retrieval Laboratory,
School of Computer Science and Technology,
Box 321, Harbin Institute of Technology,
Harbin, P.R. China, 150001,
{lee, tliu, lis}@ir.hit.edu.cn
http://ir.hit.edu.cn/

**Abstract.** Abundant Chinese paraphrasing resource on Internet can be attained from different Chinese translations of one foreign masterpiece. Paraphrases corpus is the corpus that includes sentence pairs to convey the same information. The irregular characteristics of the real monolingual parallel texts, especially without the strictly aligned paragraph boundaries between two translations, bring a challenge to alignment technology. The traditional alignment methods on bilingual texts have some difficulties in competency for doing this. A new method for aligning real monolingual parallel texts using sentence pair's length and location information is described in this paper. The model was motivated by the observation that the location of a sentence pair with certain length is distributed in the whole text similarly. And presently, a paraphrases corpus with about fifty thousand sentence pairs is constructed.

## 1   Introduction

Paraphrases are alternative ways to convey the same information [1]. And the paraphrase phenomenon is a common language phenomenon. The paraphrasing technology has been applied for various applications of natural language processing, such as question answering [2, 3, 4, 5, 6], information extraction [7], machine translation [8], information retrieval [9], multidocument [10], and it can improve the whole performance of these applications. There are many kinds of naturally occurred paraphrases resource. Barzilay extracted them from different translations of the same masterpiece [1]. Shinyama extracted paraphrases from news articles [11]. But the language they processed most is English or Japanese, the Chinese paraphrases is seldom researched. And in this paper, we mainly referred to the Barzilay's method to build a Chinese paraphrases corpus. The practical alignment technology is necessary for doing it.

Alignment technology is mainly applied for the bilingual texts. There have been a number of papers on aligning bilingual texts at the sentence level in the last century, e.g., [12, 13, 14, 15]. On clean inputs, such as the Canadian Hansards and the Hong Kang Hansards, these methods have been very successful. Church

S. H. Myaeng et al. (Eds.): AIRS 2004, LNCS 3411, pp. 118–128, 2005.

[16] and chen [17] proposed some methods to resolve the problem in noisy bilingual texts. Cognate information between Indo-European languages pairs are used to align noisy texts. But these methods are limited when aligning the languages pairs which are not in the same genre or have no cognate information. Fung [18] proposed a new algorithm to resolve this problem to some extent. The algorithm uses frequency, position and recency information as features for pattern matching. Wang [19] adapted the similar idea with Fung [18] to align special domain bilingual texts. Their algorithms need some high frequency word pairs as features. When processing the texts that include less high-frequency words, these methods will perform weakly and with less precision because of the scarcity of the data problem.

The real monolingual parallel texts always include some noisy information. They have the following characteristics as follows:

1) The monolingual parallel texts are aligned in chapter;

2) There are no strict aligned paragraph boundaries in real monolingual parallel text;

3) Some paragraphs may be merged into a larger paragraph since the translator's individual idea;

4) There are many complex translation patterns in real text;

5) There exist different styles and themes.

The tradition approaches to alignment fall into two main classes: lexical and length. All these methods have limitations when facing the real monolingual parallel texts according to the characteristics mentioned above. We proposed a new alignment method based on the sentences length and location information. The basic idea is that the location of a sentence pair with certain length is distributed in the whole text similarly. The local and global location information of a sentence pair is fully combined together to determine the probability with which the sentence pair is a sentence bead.

In the first of the following sections, we describe several concepts. The subsequent section reports the mathematical model of our alignment approach. Section 4 presents the process of anchors selection, algorithm implementation is shown in section 5. The experiment results and discussion are shown in section 6. In the final section, we conclude with a discussion of future work.

## 2    Several Conceptions

It is necessary to clarify several concepts for understanding the alignment process. As shown below:

1) Alignment anchors: Brown [12] firstly introduced the concept of alignment anchors when he aligned Hansard corpus. He considered that anchors are some aligned sentence pairs which divided the whole texts into small fragments.

2) Sentence bead: And at the same time, Brown [12] called each correct aligned sentence pair a sentence bead. Sentence bead has some different styles, such as (0:1), (1:0), (1:1), (1:2), (1: more), (2:1), (2:2), (2: more), (more: 1), (more: 2), (more: more).

3) Sentence pair: Any two sentences in the monolingual parallel text can construct a sentence pair.

4) Candidate anchors: Candidate anchors are those that can be possible alignment anchors. In this paper, all (1:1) sentence beads are categorized as candidate anchors.

# 3   Alignment Mathematical Model

The alignment process has two steps: the first step is to integrate all the origin paragraphs into one large paragraph. This can eliminate the problem induced by the vague paragraph boundaries. The second step is the alignment process. After alignment, the monolingual parallel texts become sequences of aligned fragments. And the unit of a fragment can be one sentence, two sentences or several sentences.

In this paper the formal description of the alignment task was given by extending the concepts of bipartite graph and matching in graph theory.

## 3.1   Bipartite Graph

Bipartite graph: Here, we assumed G to be an undirected graph, then it could be defined as $G =< V, E >$. The vertex set of V has two finite subsets: $V_1$ and $V_2$, also $V_1 \cup V_2 = V$, $V_1 \cap V_2 = \emptyset$. Let E be a collection of pairs, when $e \in E$, then e=$\{v_i, v_j\}$, where $vi \in V_1, vj \in V_2$. The triple G was described as, $G =< V_1, E, V_2 >$, called bipartite graph. In a bipartite graph G, if each vertex of $V_1$ is joined with each vertex of $V_2$, or vice versa, here an edge represents a sentence pair. The collection E is the set of all the edges. The triple $G =< V_1, E, V_2 >$ is called complete bipartite graph. We considered that: $|V_1| = m, |V_2| = n$, where the parameters m and n are respectively the elements numbers of $V_1$ and $V_2$. The complete bipartite graph was usually abbreviated as Km, n as shown in Figure 1.

## 3.2   Matching

Matching: Assuming $G =< V_1, E, V_2 >$ was a bipartite graph. A matching of G was defined as M, a subset of E with the property that no two edges of M have a common vertex.

## 3.3   Best Alignment Matching

The procedure of alignment using sentence length and location information can be seen as a special matching. We defined this problem as "Best Alignment Matching" (BAM).

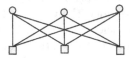

**Fig. 1.** K$_{3,3}$ complete bipartite graph

BAM: If $M = \langle S, EM, T \rangle$ is a best alignment matching of $G = \langle S, E, T \rangle$, then $E_M$ must meet the following conditions:

1) All the vertexes in the complete bipartite graph are ordered;
2) The weight of any edges in $E_M$ d(si, tj) has: d(si, tj)< D (where D is alignment threshold); at the same time, there are no edges sk, tr which made k<i and r>j, or k>i and r<j;
3) If we consider: $|S|$=m and $|T|$=n, then the edge sm, tn belonged to $E_M$;

Best alignment matching can be attained by searching for the smallest weight of edge in collection E, until the weight of every edge d(si, tj) is equal or more than the alignment threshold D. Generally, the alignment threshold D is determined according to experience.

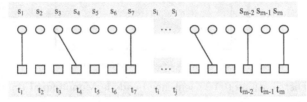

**Fig. 2.** Sketch map of $K_{m,n}$ BAM under alignment threshold D

If each sentence in the text S (or T) corresponds with a vertex in $V_1$(or $V_2$), the text S or T can be denoted by S(s1, s2, s3, ..., si, sj, ..., sm) or T(t1, t2, t3, ..., ti, tj, ..., tn). Considering the form merely, each element in S combined with any element in T can create a complete bipartite graph. Thus the alignment task can be seen as the process of searching for the BAM in the complete bipartite graph. As shown in Figure 2, the edge e = {si, tj} belongs to $E_M$; this means that the i-th sentence in text S and the j-th sentence in text T can make an alignment anchor. Each edge is corresponding to an alignment value. In order to ensure the monolingual parallel texts are divided with the same fragment number, we default that the last sentence in the monolingual parallel texts is aligned. That is to say, {sm, tn}$\in E_M$ was correct, if $|S|$=m and $|T|$=n in the BAM mathematical model.

We stipulated the smaller the alignment value is, the more similar the sentence pair is to be a candidate anchor. The smallest value of the sentence pair is found from the complete bipartite graph. That means the selected sentence pair is the most probable aligned (1:1) sentence bead. Alignment process is completed until the alignment anchors become saturated under alignment threshold value.

Sentence pairs extracted from all sentence pairs are seen as alignment anchors. These anchors divide the whole texts into short aligned fragments. At the same time, these anchors themselves are extracted as correct sentence pairs independently. The definition of BAM ensures that the selected sentence pairs cannot produce cross-alignment errors, and some cases of (1:more) or (more:1) alignment fragments can be attained by the fragments pairs between two selected alignment anchors.

## 4    Alignment Anchors Selection

All (1:1) sentence beads are extracted from different styles of monolingual parallel texts. Their distribution states are similar as presented in Figure 3. The horizontal axis denotes the sentence number in one Chinese translation text, and the vertical axis denotes the sentence number in another Chinese translation text.

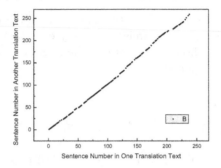

**Fig. 3.** Distribution of (1:1) sentence beads in Monolingual Parallel texts

Statistical results show that more than 85% sentence beads are (1:1) sentence beads in monolingual parallel texts and their distributions obey an obvious law well. DeKai, Wu offered that (1:1) sentence beads occupied 89% in English-Chinese as well [15]. If we select these sentence beads as candidate anchors, the alignment method will be general on any languages pairs. Length and location information of sentence pair is used fully to calculate the alignment weight of each sentence pair. Finally, the sentence pair with high value will be filtered by the similarity of the two sentences in a sentence pair.

In order to calculate the alignment value of sentence pair of si, tj, four parameters are defined:

Whole text length ratio: P0 = Ls / Lt;
Upper context length ratio: Pu[i, j] = Usi / Utj;
Nether context length ratio: Pd[i, j] = Dsi / Dtj
Sentence length ratio: Pl[i, j] = Lsi / Ltj;
Where
si the i-th sentence of S;
tj the j-th sentence of T;
Ls the length of one translation text S;
Lt the length of anther translation text T;
Lsi the length of si;
Ltj the length of tj;
Usi the upper context length above sentence si;
Utj the upper context length above sentence tj;
Dsi the nether context length below sentence si;
Dtj the nether context length below sentence tj;
Figure 4 illustrates clearly the relationship of all variables.

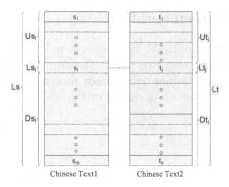

**Fig. 4.** Sketch map of variables relationship

If si and tj can construct a (1:1) alignment anchor, P[i, j] must be less than the alignment threshold, where P[i,j] denotes the integrated alignment value between si and tj. We construct a formal alignment function on every sentence pair:

$$P[i,j] = \alpha(Pu[i,j] - P0)^2 + (Pl[i,j] - P0)^2 + \alpha(Pd[i,j] - P0)^2 + (1/Sim)$$

Where, the parameter $\alpha$ is balanced coefficient, and it can adjust the weight of sentence pairs' length and the weight of context lengths well. The longer the text is, the more insensitive the effect of the context length is. So $\alpha$'s value should change in order to balance the whole proportion. The short text is vice versa. In this paper we define:

$$\alpha = (Ls/Lsi + Lt/Ltj)/2$$

And the similarity of the two sentences in a sentence pair is calculated through:

$$Sim = \frac{Intersection(L_1, L_2) \times 2}{L_1, L_2} \times \frac{min(L_1, L_2)}{max(L_1, L_2)}$$

Where $L_1$ and $L_2$ are the lengths of the two sentences; $min(L_1,L_2)$ is the length of the shorter sentence, $max(L_1,L_2)$ is the length of longer sentence; Intersection($L_1,L_2$) is the common length of the two sentences.

According to the definition of BAM, the smaller the alignment function value of P[i, j] is, the more the probability of sentence pair si, tj being a (1:1) sentence bead is. In this paper, we adopt a greedy algorithm to select alignment anchors according to all the alignment function values of P[i, j] which are less than the alignment threshold. This procedure can be implemented with a time complexity of O(m*n). To obtain further improvement in alignment accuracy the similarity is used to filter the wrong sentence pairs independently. And calculation approach of the similarity is same with the method mentioned above.

After the above similarity filtering, although the alignment recall is reduced, the alignment precision is improved greatly. Here, those candidate alignment anchors whose similarities exceed the similarity threshold will become the final alignment anchors. These final anchors divide the whole monolingual parallel texts into aligned fragments.

# 5    Algorithm Implementation

According to the definition of BAM, the first selected anchor will divide the whole monolingual parallel texts into two parts. We stipulated that the sentences in the upper part of one translation text cannot match any sentence in the nether part of anther translation text. As shown in Figure 5.

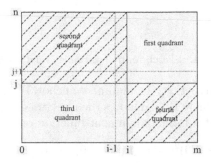

**Fig. 5.** Anchors selection in Monolingual Parallel texts

After the first alignment anchor was selected, the second candidate anchor must be selected in the first quadrant or the third quadrant and exclusive from the boundary. It is obvious that the cross alignment error will happen if the candidate anchor exists in the second quadrant or fourth quadrant. For example, if the $(i, j)$ is the first selected alignment anchor, and the $(i-1, j+1)$ is the second selected alignment anchor, the cross alignment appears. We can limit the anchors selection field to prevent the cross-alignment errors.

In addition, in order to resolve the problem that the first and the last sentence pair is not a (1:1) sentence bead, we use a virtual sentence length as the origin alignment sentence bead when we initialize the alignment process. The implementation of alignment algorithm is described as followed:

1) Load the monolingual parallel texts;

2) Identify the sentences boundaries and number each sentence;

3) Calculate every sentence pair's alignment value;

4) Search the sentence pair that is corresponding to the smallest alignment function value;

5) If the smallest alignment function value is less than the alignment threshold and the go to step 6), and if the smallest value is equal to or more than the threshold, then go to step 7);

6) If the similarity of the sentence pair is more than a certain threshold, the sentence pair will become an alignment anchor and divide the monolingual parallel text into two parts respectively, then limit the search field of the next candidate anchors and go to the step 4);

7) Output the aligned texts, and go to the end.

# 6    Results and Discussion

Because the translations of most of masterpiece are aligned in chapter and the sentence number of every chapter are less than 500, our algorithm works well on the monolingual parallel texts with the sentence number under 500. Part of translations in "The Sorrows of Young Werther" and "Cien Años de Soledad" are selected as test set. The concrete information is shown in Table 1 and 2.

The alignment experiments are performed under the condition of with similarity filtering and without similarity filtering. The precision and recall are defined:

Precision = The correct aligned sentence pairs number / The total number of alignment sentence pairs in monolingual parallel texts

Recall = The correct aligned sentence pairs number / The total alignment sentence pairs in standard test texts

The comparison results are presented in Table 3.

With similarity filtering, the alignment precision is improved greatly. We take a statistic on all the errors and find that most errors are partial alignment errors. Partial alignment means that the alignment location is correct, but a half pair of the alignment pair is not integrated. The result shows that similarity filtering can resolve the problem in some extent. The recall is so low because there are some correct aligned fragment pairs with more than one sentence which cannot match the correspondent sentence pair in the standard set.

**Table 1.** Concrete information of the test data

|                    |                 | Translation 1 | Translation 2 |
|--------------------|-----------------|---------------|---------------|
| The Sorrows        | Size            | 18.40K        | 18.40K        |
| of Young Werther   | Sentence number | 238           | 260           |
| Cien Años de       | Size            | 49.5K         | 49.2K         |
| Soledad            | Sentence number | 639           | 559           |

**Table 2.** Distribution of different alignment style

|                              | Total sentence number | 1:1 | 1:2 | 2:1 | other |
|------------------------------|-----------------------|-----|-----|-----|-------|
| The Sorrows of Young Werther | 222                   | 185 | 24  | 8   | 5     |
| Cien Años de Soledad         | 520                   | 436 | 44  | 19  | 21    |

**Table 3.** Comparison results with similarity

|               |                             | Precision(%) | Recall(%) |
|---------------|-----------------------------|--------------|-----------|
| The Sorrows   | Without similarity filtering | 70.9         | 75.1      |
| of Young Werther | With similarity filtering | 85.5         | 72.3      |
| Cien Años de  | Without similarity filtering | 68.5         | 74.2      |
| Soledad       | With similarity filtering    | 83.2         | 70.6      |

**Table 4.** Comparison results between two methods

|  | Precision(%) | Recall(%) |
|---|---|---|
| Method based on length | 35.0 | 33.9 |
| Our method | 85.5 | 72.3 |
| Combination method | 91.2 | 85.6 |

In order to verify the validity of our algorithm, we implement the classic length-based sentence alignment method using dynamic programming. And combining the traditional alignment method with our method, the results are shown in Table 4.

Because the origin monolingual parallel texts have no obvious aligned paragraph boundaries, the error extension phenomena happen easily in the length-based alignment method. Its alignment results are so weaker that it cannot be used. If we omit all of the origin paragraphs information and merge all the paragraphs in the monolingual parallel text into one larger paragraph respectively. The length-based alignment method rated the precision of 35.0%. This is mainly because different translators have different translation styles and different comprehension on the same foreign texts. But our method rated 160 (1:1) sentence pairs as alignment anchors which divide the monolingual parallel text into aligned fragments. Then the length-based classic method was applied to these aligned fragments and got a high precision.

Figure 6 shows 160 selected anchors distribution which is in the same trend with all the (1:1) sentence beads. Their only difference is the sparse extent of the aligned pairs. We can make a conclusion that Our method performs very well to align the real monolingual parallel texts.

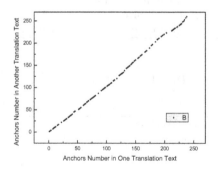

**Fig. 6.** Distribution of alignment anchors

## 7    Conclusion

This paper proposed a new method for fully aligning real monolingual parallel texts using sentence length and location information, described concretely in section 3 and 4. The model was motivated by the observation that the location of a sentence pair with certain length is distributed in the whole text similarly. It uses the (1:1) sentence beads instead of the high frequency words as the candidate anchors. Local and global location characteristics of sentence pairs are involved to determine the probability which the sentence pair is an alignment anchors.

Every sentence pair corresponds to an alignment value which is calculated according to the formal alignment function. Then the process of BAM is performed to get the alignment anchors. This alignment method can restrain the errors extension effectively in comparison to the traditional length-based alignment method. Furthermore, it has shown strong robustness, even if when it meets ill-quality texts that include incorrect sentences. To obtain further improvement in alignment accuracy sentence similarity filtering was performed. The algorithm need not segment the Chinese sentence require little cost to implement.

Additionally, we can adjust the alignment and similarity thresholds dynamically to get high precision alignment anchors, for example, applying the first test set, even if we get only 107 (1:1) sentence beads but the precision is 98.13%. We found that this method can perform the function of paragraph alignment very well and ensure the alignment precision simultaneously.

Of these pairs about half of total number of (1:1) sentence beads can be even extracted from the monolingual parallel text directly to build a large scale paraphrase corpus if the original monolingual parallel text is abundant. And the rest text can be used as spare resource. Now, we have obtained about 50,000 Chinese paraphrase pairs with high quality.

In the future, we hope to do further alignment on the basis of current work and extend the method to align other languages pairs.

## Acknowledgments

This research was supported by National Natural Science Foundation (60203020) and Science Foundation of Harbin Institute of Technology (hit.2002.73).

## References

1. Barzilay, R., McKeown, K.: Extracting paraphrases from a parallel corpus. In: Meeting of the Association for Computational Linguistics. (2001) 50–57
2. Lin, D., Pantel, P.: Discovery of inference rules for question answering. Natural Language Engineering 1 (2001)
3. Rinaldi, F., Dowdall, J., Kaljurand, K., Hess, M., Mollá, D.: Exploiting paraphrases in a question answering system. In Inui, K., Hermjakob, U., eds.: Proceedings of the Second International Workshop on Paraphrasing. (2003) 25–32
4. France, F.D.: Learning paraphrases to improve a question-answering system. In: EACL-Natural Language Processing for Question Answering. (2003)
5. Tomuro, N.: Interrogative reformulation patterns and acquisition of question paraphrases. In Inui, K., Hermjakob, U., eds.: Proceedings of the Second International Workshop on Paraphrasing. (2003) 33–40
6. Takahashi, T., Nawata, K., Kouda, S., Inui, K., Matsumoto, Y.: Effects of structural matching and paraphrasing in question answering. IEICE Transactions on Information and Syste (2003)

 7. Shinyama, Y., Sekine, S.: Paraphrase acquisition for information extraction. In Inui, K., Hermjakob, U., eds.: Proceedings of the Second International Workshop on Paraphrasing. (2003) 65–71
 8. Kanayama, H.: Paraphrasing rules for automatic evaluation of translation into Japanese. In Inui, K., Hermjakob, U., eds.: Proceedings of the Second International Workshop on Paraphrasing. (2003) 88–93
 9. Jacquemin, C.: Syntagmatic and paradigmatic representations of term variation. In: 37th Annual Meeting of the Association for Computational Linguistics (ACL'99), Proceedings, Maryland, pages (1999) 341–348
10. Barzilay, R., Elhadad, N., McKeown, K.R.: Inferring strategies for sentence ordering in multidocument news summarization. Journal of Artificial Intelligence Research **17** (2002) 35–55
11. Shinyama, Y., Sekine, S., Sudo, K., Grishman, R.: Automatic paraphrase acquisition from news articles (2002)
12. Brown, P.F., Lai, J.C., Mercer, R.L.: Aligning sentences in parallel corpora. In: Meeting of the Association for Computational Linguistics. (1991) 169–176
13. Gale, W.A., Church, K.W.: A program for aligning sentences in bilingual corpora. Computational Linguistics **19** (1993) 75–102
14. Simard, M., Foster, G.F., Isabelle, P.: Using cognates to align sentences in bilingual corpora. In: Proc. of the Fourth International Conference on Theoretical and Methodological Issues in Machine Translation: Empiricist vs. Rationalist Methods in MT, Montreal, Canada (1992) 67–81
15. Wu, D.: Aligning a parallel english-chinese corpus statistically with lexical criteria. In: Meeting of the Association for Computational Linguistics. (1994) 80–87
16. Church, K.W.: Char_align: A program for aligning parallel texts at the character level. In: ACL93. (1993) 1–8
17. Chen, S.F.: Aligning sentences in bilingual corpora using lexical information. In: Meeting of the Association for Computational Linguistics. (1993) 9–16
18. PASCALE, F., MCKEOWN, K.: Aligning noisy parallel corpora across language groups: Word pair feature matching by dynamic time warping (1994)
19. Bin, W., Qin, L., Xiang, Z.: Automatic chinese-english paragraph segmentation and alignment. Journal of Software **11** (2000) 1547–1553

# Effective Topic Distillation with Key Resource Pre-selection

Yiqun Liu[1], Min Zhang[2], and Shaoping Ma[2]

[1] State Key Lab of Intelligent Tech. and Sys., Tsinghua University,
Beijing, 100084, China
liuyiqun03@mails.tsinghua.edu.cn
[2] State Key Lab of Intelligent Tech. and Sys., Tsinghua University,
Beijing, 100084, China
{z-m, msp}@tsinghua.edu.cn

**Abstract.** Topic distillation aims at finding key resources which are high-quality pages for certain topics. With analysis in non-content features of key resources, a pre-selection method is introduced in topic distillation research. A decision tree is constructed to locate key resource pages using query-independent non-content features including in-degree, document length, URL-type and two new features we found out involving site's self-link structure analysis. Although the result page set contains only about 20% pages of the whole collection, it covers more than 70% of key resources. Furthermore, information retrieval on this page set makes more than 60% improvement with respect to that on all pages. These results were achieved using TREC 2002 web track topic distillation task for training and TREC 2003 corresponding task for testing. It shows an effective way of getting better performance in topic distillation with a dataset significantly smaller in size.[1]

## 1 Introduction

Currently, Web Information Retrieval (IR) presents a technical challenge due to the size exploding of web document collection, which contains over 20 billion pages as of February, 2003[1]. The number of pages indexed by web search engines is increasing at a high speed, for example, Google indexed over 3.3 billion pages in September, 2003, which is about 7 times as many as what it indexed in the year of 2000[2]. How to achieve better performance with fewer pages indexed is becoming more and more interesting in web IR research. Many web search engines have adopted some techniques to identify the quality of web pages independent of a given user request, in order that they can index more high quality pages with limited resources. But these approaches such as PageRank[3] only use link structures of the web and a better estimate should require additional non-content sources of information both within a page and across different pages.

---

[1] Supported by the Chinese Natural Science Foundation (NO. 60223004, 60321002, 60303005)

S. H. Myaeng et al. (Eds.): AIRS 2004, LNCS 3411, pp. 129–140, 2005.

Topic distillation is a web search task to find high-quality web pages (called key resources) for a particular topic. These pages can offer users with credible information or a good entry point to several useful pages. About 78% search engine queries are related to this task according to query log analysis of Broder [4]. If key resources can be pre-selected from the whole collection, a large number of web search requests can be satisfied by indexing these pages only (because only these pages may be returned as results in topic distillation). It is possible to find several non-content features to determine whether one web page is a key resource page or not. In fact, previous TREC experiments[5][6][7] show that some non-content sources of information such as URL and link structure enhance retrieval effectiveness. Which non-content features are useful in selecting key resources and how to use them are the key components of our work.

The remaining part of the paper is constructed as follows: Section 2 gives a brief review of related works in non-content features of web pages. Section 3 compares differences between key resource pages and ordinary pages in these features. A decision tree-based key resource pre-selection algorithm is demonstrated in section 4. Section 5 describes experiment results of both key resource pre-selection and related content retrieval process. Finally come discussion and conclusion.

## 2    Related Work

Non-content features are sources of non-content information both within a page and across different pages, such as URLs and links. Existing studies on web page non-content features are mostly related to site finding. A site finding task is one where the user wants to find a particular site and his query names the site. It belongs to navigational search whose percentage in web search queries is over 20% according to Broder[4].

Many efforts have been made to find several non-content information sources in web site finding and several of them have proved effective. Westerveld et al. defined a web page's URL-type according to its URL length and number of dashes ('/'). They computed the probability of a page being an entry page given the type of its URL: root, subroot, path, or file. Combination of this feature and some content features helps them to obtain the best result in TREC 2001's home page finding task[8]. His colleague Kraaij et al. further proved the effectiveness of document length and in-degree in selection of home pages in 2002[9]. Craswell et al. also found that, in optimal conditions, all of the query-independent methods they studied (in-degree, URL-type, and two variants of PageRank) offered a better than random improvement on a content only baseline in site finding[10].

Key resource page has the function of providing credible information on a certain topic. It means that key resource page is different from ordinary entry page, although it is defined as entry of one key resource site according to TREC 12 web track's guideline[6]. It means that effective non-content features in site finding should be re-studied in topic distillation and new features should be introduced to separate key resource pages from ordinary pages.

# 3    Non-content Features of Key Resources

This section compares differences between ordinary and key resource pages in some non-content features. The features to be discussed are in-link count (in-degree), document length, URL-type, in-site out-link number and in-site out-link anchor rate; features involving self-link analysis of the site are designed specially for key resource separation. Key resource training set used here is composed of relevant qrels of TREC 2002's topic distillation task (See Section 5).

Some common-used features are discussed at first, followed by new features we proposed involving in-site out-link analysis.

## 3.1    Study of Common-Used Non-content Features

**In-Link Count (In-Degree).** Analysis of both .GOV and key resource training set shows that entry pages are different from ordinary pages in the distribution of in-degree.

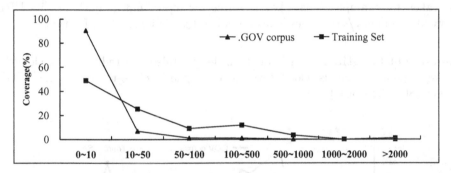

**Fig. 1.** In-degree distributions of key resource training set and .GOV. The category axis represents in-degree

It can be concluded from Fig. 1 that key resource pages have many more in-links than ordinary pages. The plots show that 51.03% key resource pages have more than 10 in-links; while only 9.45% ordinary pages have in-degree over 10. Key resources are high quality pages which interests many web users so a large number of web sites create hyper-links to them for convenience of these users.

**URL-Type.** URL-type proves to be stable and effective in site finding task according to previous TREC experiments [10][14][15]. We followed Kraaij et al. and classified URLs (after stripping off a trailing index.html, if present) into four categories: ROOT, SUBROOT, PATH and FILE. Our experiment gives the following statistics from .GOV and the training set.

There are 57.27% key resource pages in the "non-FILE" URL-type page set, which contains 11.10% pages of .GOV corpus according to Fig. 2. It means that key resource pages are likely to be "non-FILE" type. Key resource pages have

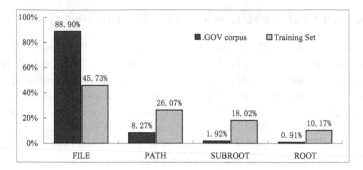

**Fig. 2.** URL-type distributions of key resource training set and .GOV

URLs ending with "/" or "index.html" because their authors usually use them as index pages for web sites.

The statistics also show that not all key resource pages are "non-FILE" URL-type. In fact about half of them are FILEs. NIH marijuana site entry page can be taken for example: It is a key resource for topic "marijuana", but its URL www.nida.nih.gov/drugpages/marijuana.html is obviously "FILE" type.

**Document Length.** Average document length (also referred to as page length) of key resource pages is 9008.02 words; it is quietly close to that of the .GOV pages (8644.62 words).

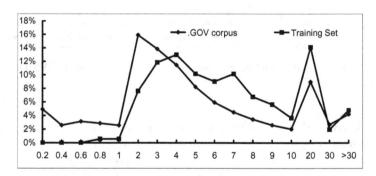

**Fig. 3.** Page length distributions of key resource training set and .GOV, the category axis represents "document length" in K words

The distributions in document length of .GOV and key resource pages are compared in Fig. 3. The figure shows that two page sets have similar distribution of page length, except that key resource page length would not be too short. Only 1.12% of key resource pages are shorter than 1000 words. While in ordinary pages, the percent is 16.00%. It means that pages with too few words cannot be key resource pages. This feature can be applied to reduce redundancy in the key resource page set.

## 3.2  New Features Involving In-Site Out-Link Analysis

In-site out-links of a certain Page A are links from A to other pages in the site where A is located. For example, the hyperlink from AIRS 2004 homepage to AIRS 2004 call for paper page is an in-site out-link of the former page. The link is located in the former page and is at the same time an in-link for the latter one.

**In-Site Out-Link Number.** Key resource pages have the function of linking to other informative pages of the same site. The function is like entry page's "navigation function" described in Craswell et al [11]. Key resources need more in-site out-links to finish the navigation function according to the statistics in Fig 4. Average in-site out-link number of key resource set is more than 37 compared with that of less than 18 in ordinary page set.

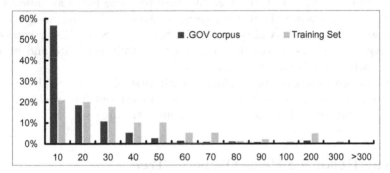

**Fig. 4.** In-site out-link number of key resource training set and .GOV. The category axis represents in-site out-link anchor number

A large number of in-site out-links is given to one key resource page, so it can connect directly to these informative pages in its site/sub-site. Further analysis in Section 4 shows that this attribute doesn't work as effective as in-degree and URL-type in pre-selection process, however, it helps separate key resource pages where traditional features fail.

**In-Site Out-Link Anchor Rate.** Key resource pages' in-site out-link anchors can be regarded as a brief review of the other pages' content in the same site. For a certain page A, in-site out-link anchors are anchors describing links from A to other pages in the same site/sub-site as A. In-site out-link anchors are located on A; that is quite different from in-link anchors which are frequently used in content analysis.

In-site out-link anchor rate is defined as:

$$rate = \frac{WordCount(in-site\ out-link\ anchor\ text)}{WordCount(fulltext)} . \tag{1}$$

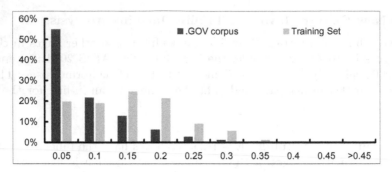

**Fig. 5.** In-site out-link anchor rate of key resource page training set and .GOV. The category axis represents in-site out-link anchor rate

According to Fig. 5, key resource training set has more pages with a high rate. It can be explained by the fact that key resource pages are always index pages for a site/sub-site. They are representatives of sites and in-site out-link anchors work as a summary of these sites' other pages. So one key resource page should have lots of in-site out-link anchors to introduce its site and its in-site out-link anchor rate is reasonably high.

This feature's distribution is similar with that of in-site out-link number. There are less than 24% pages in .GOV with in-site out-link anchor rate over 0.1; the percentage is about 61% in key resource training set.

## 4     Key Resource Page Decision Tree

Decision tree learning is adopted to combine non-content features discussed in Section 3. It is a method for approximating discrete-valued functions that is robust of noisy data and capable of learning disjunctive expressions. We choose decision tree because it is usually the most effective and efficient classifier when we have a small number of features; it also provides us with a metric to estimate feature quality in the form of information gain (ID3) or information ratio (C4.5).

ID3 algorithm proposed by Quinlan (referring to Mitchell [12]) is performed to construct the decision tree. This algorithm uses information gain to decide which attribute should be selected as the decision one for the root node of the decision tree. Information gain is a statistical property that measures how well a given attribute separates the training examples according to their target classification.

For this particular problem of selecting key resource pages, non-content features with continuous values should be discretized in advance. Attributes range should be partitioned into two or several intervals using a single or a set of cut points (thresholds). One fixed threshold is chosen in our method and after discretization process each non-content feature has boolean values.

According to ID3 algorithm described in [12], in-degree with the most information gain should be chosen at the root node (cf. Table 1). Then the process of selecting a new attribute and partitioning the training examples is repeated

**Table 1.** Information gain with different attributes

| Attribute | Information gain |
|---|---|
| In-degree | 0.2101 |
| URL-type | 0.1981 |
| Document length | 0.0431 |
| In-site out-link number | 0.0796 |
| In-site out-link anchor rate | 0.0988 |

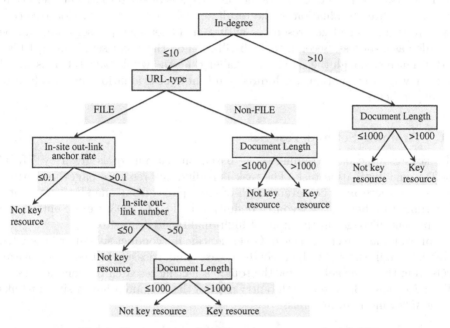

**Fig. 6.** Key resource decision tree constructed with ID3

for each non-terminal descendant node, with only the examples associated with the node.

In the construction process, non-content attributes are ranked by their information gains while choosing the root node, as shown in Table 1.

In-degree and URL-type are good classifiers to separate key resources from ordinary pages and their information gains are almost the same. Information gain of document length is quite low, but according to Section 3.3 it can be used to keep out redundancy.

The two features obtained from in-site out link analysis are important in decision tree learning. Sub tree with these two features is used to select key resources from the pages whose URL type are FILE and in-degree are less than 10. The percentage of such kind of pages in .GOV is 68.53%, and there are 29.38% of key resource pages among them. Without in-site out-link analysis, it is impossible to get these key resource pages separated.

When one example set is mainly composed of pages with the same target attribute value or all attributes have been tested, the construction process ends. Finally the decision tree shown in Fig. 6 is constructed. With this decision tree, any page in .GOV page set can be judged whether it belongs to the key resource set or not.

# 5    Experiments and Discussions

There are two methods to evaluate the effectiveness of key resource pre-selection. First is the direct evaluation: if the result set selected by the decision tree covers a large number of key resources with a small set size, pre-selection can be regarded as a success. Second, topic distillation on the result set should get high performance although the set size is smaller than the whole page set: It is called indirect way of evaluation. Both direct and indirect evaluations are involved in our experiments.

## 5.1    Training Set and Test Set

The key resource page training set is based on relevant qrels given by TREC 2002's topic distillation task. The task is to find key resource pages for certain topics but there may be several high-quality pages from one single web site. According to a better key resource definition in TREC 2003, we use entry page to represent all pages in this kind of high-quality site/sub-site.

For example, the entry page of ORI scientific misconduct sub-site whose URL is http://ori.dhhs.gov/html/misconduct/casesummaries.asp is used to replace all qrels from the same sub-site for the topic "scientific research misconduct cases" (Topic No. 559). Because it is the entry page of the site and it has in-site out-links to the following relevant qrels:

> http://ori.dhhs.gov/html/misconduct/elster.asp
> http://ori.dhhs.gov/html/misconduct/french.asp
> http://ori.dhhs.gov/html/misconduct/hartzer.asp
> (another 21 qrels from the same site are omitted here)

TREC 2003's topic distillation task is to find as many key resource site entry pages as possible. The task's 50 queries are from log analysis of web search engines and NIST gets credible qrels with pooling technology. The topics and relevant qrels are directly used as the test set.

## 5.2    Key Resource Coverage of the Result Set

When fixed threshold values vary in the discretization process described in Section 4, result sets with different size and key resource coverage are built correspondingly. Statistics of several result sets are shown in Fig. 7, from which two important conclusions can be drawn:

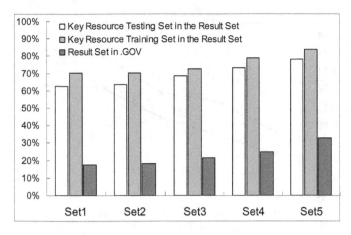

**Fig. 7.** Key Resource coverage and the result set size

1. With the decision tree constructed in Section 4, it is possible to cover about 70% test set pages (key resource pages) with about 20% of .GOV size. The result set can cover a high percentage of key resources with a small amount of pages because non-content features which we adopt are fit for the job of selecting key resources. There are still about 30% key resource testing set pages which the result set doesn't include. It means the up-limit of this key resource pre-selection method is about 70%. However, topic distillation performance is less than 13% with the measure of precision at 10 currently according to TREC 11 and TREC 12 reports[5][6]. The up-limit doesn't depress the performance too much.
2. Key resource coverage increases with the key resource set amount. It is obvious that 100% coverage is got with the result set equaling .GOV, so there must be a trade off between the result set size and key resource testing set coverage.

## 5.3    Topic Distillation on the Key Resource Set

Experiments in this Section are based on a key resource set with 24.89% pages in .GOV and 73.12% key resources (Set 4 in Fig. 7). BM2500 weighting and default parameter tuning described in[14] are performed in all experiments. Topics and relevant qrels in TREC 2003 topic distillation task are used for testing. In-link anchor retrieval is evaluated together with full text retrieval both on .GOV corpus and on the key resource set, because in-link anchor proves to be an effective source for topic distillation in [6]. These results are also compared with TREC 2003 best run to validate effectiveness of our method in Fig. 8.

In-link anchor retrieval performs much better than full text retrieval on .GOV (46% improvement in R-precision and 42% in P@10). It accords with the previous conclusion by Craswell et al[11] that BM25 ranking applied to link anchor documents significantly outperforms the same ranking method applied to document content. However, even anchor text retrieval on .GOV gets much worse

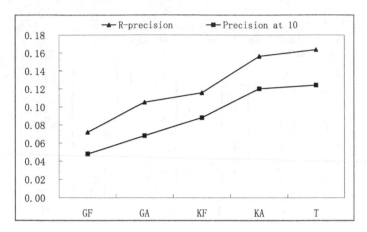

**Fig. 8.** Text retrieval on .GOV and Key resource set compared with TREC 2003 best run, *GF*: Full text of .GOV corpus, *GA*: In-link anchor text of .GOV corpus, *KF*: Full text of Key resource page set, *GA*: In-link anchor text of Key resource page set, *T*: Best run in TREC 2003 web track

performance than TREC 2003 best run. It can be explained by the fact that these results don't make use of any kind of non-content information.

Fig. 8 also proves that retrieval on key resource site get much better performance than that on .GOV. Similar with retrieval results on .GOV, anchor text retrieval on key resource set outperforms full text retrieval by about 35% in both evaluation metrics. However, from Fig. 8, we can see both full text and anchor text retrieval gain much progress on key resource set. Anchor text retrieval on key resource set achieves almost the same ranking as TREC 2003 best run (cf. Table 2).

These experiments are not biased towards the key resource pre-selection method. TREC 2003 topic distillation topics and relevant qrels are not used for training in any way. BM2500 ranking based on link anchor of the key resource set gets the 2nd highest R-precision and 4th highest Precision at 10 according to TREC 2003 topic distillation task results. Further improvement is expected if better stemming and parameter tuning technology are introduced.

**Table 2.** Anchor text retrieval on different data set compared with TREC 2003 topic distillation best results

| Evaluation metric | Precision at 10 | R-Precision |
|---|---|---|
| Whole page set | 0.0680 | 0.1050 |
| Key resource page set | 0.1200 | 0.1562 |
| Best run in TREC 2003 (highest P@10)[6] | 0.1280 | 0.1485 |
| Best run in TREC 2003 (highest R-precision)[6] | 0.1240 | 0.1636 |

The key resource set works well with only 20% amount of the whole page set. It proves that key resource set has included a large number of high quality pages. This set can be used for topic distillation instead of the whole page set.

# 6   Conclusions and Further Work

In this paper, an effective topic distillation method with key resource pre-selection is proposed. Experiments prove that information retrieval on this pre-selected key resource page set gets much better performance than that on the whole collection. We can conclude that:

1. A small sub-set of web pages which contains most key resource pages can be built with key resource pre-selection technology. This sub-set contains only about 20% of the whole collection, but covers more than 70% key resources.
2. Decision tree learning is fit for the job of selecting key resources using non-content features. ID3 also provides us with a credible metric of non-content feature effectiveness.
3. Pre-selection of key resource pages works well for topic distillation according to experiment results. Full text retrieval on key resource page set gets more than 60% improvement comparing with that on the whole collection. Anchor text retrieval works as well as TREC 2003's best run on this sub set.

This research may help web search engines to index fewer pages without losing performance in topic distillation. It can also be used to evaluate web page quality query-independently based on whether it is a key resource or not.

However, a great many aspects of the key resource set are to be investigated in future work: Will we get better results if existing link analysis methods such as HITS and PageRank are performed on this set? In-site out-link analysis has proved effective in key resource selection, how well does retrieval on in-site out-link anchor instead of in-link anchor in the key resource set? Is it possible to find the best trade off point between key resource set size and key resource coverage?

# References

1. Lyman, Peter and Hal R. Varian, "How Much Information", 2003. Retrieved from http://www.sims.berkeley.edu/how-much-info-2003 on April 2th, 2004
2. Danny Sullivan, Search Engine Sizes. In search engine watch website; September 2, 2003. Online at: http://searchenginewatch.com/reports/article.php/2156481
3. S. Brin and L. Page. The anatomy of a large-scale hypertextual web search engine. In Computer Networks and ISDN Systems (1998) 30(1–7): 107–117
4. Andrei Broder: A taxonomy of web search. SIGIR Forum(2002), Volume 36(2):3–10
5. N. Craswell, D. Hawking: Overview of the TREC-2002 web track. In NIST Special Publication 500-251: The Eleventh Text REtrieval Conference (TREC 2002)
6. N. Craswell, D. Hawking: Overview of the TREC-2003 web track. In NIST Special Publication 500-255: The twelfth Text REtrieval Conference (TREC 2003):78–92

7. E. M. Voorhees and D. K. Harman, editors. The Tenth Text Retrieval Conference (TREC-2001), volume 10. National Institute of Standards and Technology, NIST, 2001
8. T. Westerveld, D. Hiemstra, W. Kraaij. Retrieving Web Pages Using Content, Links, URLs and Anchors. In NIST Special Publication 500-250: The Tenth Text REtrieval Conference (TREC 2001) (2001) 663–672
9. W. Kraaij, T. Westerveld, and D. Hiemstra. The importance of prior probabilities for entry page search. In 25th ACM-SIGIR conference on research and development in information retrieval (2002) 27–34
10. Nick Craswell and David Hawking. Query-independent evidence in home page finding. In ACM Transactions on Information Systems (2003) 21(3): 286–313
11. Nick Craswell, David Hawking and Stephen Robertson. Effective Site Finding using Link Anchor Information. In 24th ACM-SIGIR Conference on Research and Development in Information Retrieval. (1998) 250–257
12. Tom M. Mitchell. Chapter 3: Decision Tree Learning, in Machine Learning (ISBN 0-07-115467-1). Pages 55-64, McGraw-HILL INTERNATIONAL EDITIONS, 1997
13. Van Rijbergen.: Information Retireval. Butterworths, London, 1979
14. S. E. Robertson, S. Walker, M.M. Hancock-Beaulieu, and M. Gatford. Okapi at TREC-3. In NIST Special Publication 500-225: Overview of the Third Text Retrieval Conference (1994) 109–127

# Efficient PageRank with Same Out-Link Groups[*]

Yizhou Lu[1,**], Xuezheng Liu[2], Hua Li[1], Benyu Zhang[3], Wensi Xi[4],
Zheng Chen[3], Shuicheng Yan[3], and Wei-Ying Ma[3]

[1] School of Mathematical Sciences, Peking University,
100871 Beijing, P.R. China
[2] Department of Computer Science and Technology,
Tsinghua University, 100084 Beijing, P.R. China
[3] Microsoft Research Asia, 49 Zhichun Road,
100080 Beijing, P.R. China
[4] Virginia Polytechnic Institute and State University,
Blacksburg, 24060 VA. U.S.A.

**Abstract.** Traditional PageRank algorithm suffers from heavy computation cost due to the huge number of web pages. In this paper, we
propose a more efficient algorithm to compute the pagerank value for
each web page directly on the same out-link groups. This new algorithm
groups the pages with the same out-link behavior (SOLB) as a unit. It is
proved that the derived PageRank is the same as that from the original
PageRank algorithm which calculates over single webpage; while our proposed algorithm improve the efficiency greatly. For simplicity, we restrict
the group within a directory and define metrics to measure the similarity of the pages in same out-link behavior. We design the experiments
to group from 0.5 liked to exact SOLB pages; the results show that such
group offers similar rank scores as traditional PageRank algorithm does
and achieves a remarkable 50% on efficiency.

## 1 Introduction

Link analysis techniques [4, 7] have received many attentions these years in the
field of Web Search. The modern searching engines need to sort the search result
and offer personalized search service with the help of link analysis algorithms.
One of the most used link analysis algorithm is the PageRank algorithm, which is
proposed by Brin and Page [7] in 1998. The PageRank algorithm used a random-
surfer model to simulate the users' browsing behavior. It can also be explained
as a Markov Chain. The principal eigenvector of the Markov probability matrix
is considered to be the rank score of web pages.

In practice, the power method [3] is the basic idea for page rank score computation of the search engines [7]. This is also the convenient method for computing the principal eigenvector of large-scale matrices. However, it is still a time

---

[*] This work is done at Microsoft Research Asia.
[**] contact email: luyizhou@pku.edu.cn

S. H. Myaeng et al. (Eds.): AIRS 2004, LNCS 3411, pp. 141–152, 2005.

consuming task, due to the huge amount of pages in the growing world wide web. The search engines need a even quicker PageRank computation method for fresher rank score, as well as the personalized PageRank computation [7, 8].

Some PageRank acceleration algorithms have been proposed in recent years. From the viewpoint of numerical analysis, Arasu[1] suggested that using iterative methods, like Gauss-Seidel method[3] and SOR (successive over-relaxation) method[3], on the PageRank computation. Those methods fail to exploit the structure information of web graph. Meanwhile, these algorithms require fairly amount of operations on the matrix elements, and are not practical for web search engine.

Kamvar et.al proposed the BlockRank algorithm[8],which computes the local rank score of each block and the rank score of each block individually. Each page's rank score is weighted by the block score of their nested block and the weighted rank score vector is defined as the start point in the iterative power method. The algorithm take advantage of block structure information, however, they need the power iteration on the whole graph. Furthermore, the web has more attributes to be exploited. Recently, Yizhou and Benyu et.al proposed the PowerRank algorithm[9] in which several graphs are constructed in different granularity of the original web graph (pages, hosts, domains ... ). The rank score vector in the thicker graph is filtered and expanded to the thinner graph according to the rank score and a global rank score vector obtained with the combination of them. Such algorithm is a framework to make use of the hierarchy structure and the power-law distribution [5] of the web graph, which offers a little different result to that of the PageRank algorithm's.

In this article, we mined the web graph's implicit attribute on page's SOLB (same out-link behavior) character. We call two pages $P_i$, $P_j$ are SOLB pages, if any other page $P$ is always both page $P_i$'s and page $P_j$'s out-linked page. The relationship between PageRank algorithm and the SOLB character follows the theorem: "If clustered the SOLB pages of a web graph and performed the PageRank algorithm on such granularity graph, the rank score of each cluster is equal to the sum of its nested pages' rank score. Mathematically, $pr(C) = \sum_{P \in C} pr(P)$. Thus using same out-link groups as computation units will not reduce the accuracy of PageRank algorithms but can greatly improve the efficiency.

To reduce the computing complexity of clustering SOLB pages in the whole web, we make a restriction that only the SOLB pages within the same directory will be clustered. The reason is also because the fairly amount of the links within the same directory can be regarded as noise or spam link. With the restriction, we make a trade-off between the computing complexity and the scale reduction. Meanwhile, we introduce two measurements (the cosine similarity and the set similarity) for the similarities between the pages' out-link behavior. In our experiment, we cluster from 0.5 liked SOLB pages to exact SOLB pages in each directory. The experimental results show that the iteration computing on the same out-link group offers a remarkable more than 50% acceleration. The rank score vector is very similar to the PageRank score vector even the threshold set to 0.5 (measured by the L1-norm[1] ).

The rest of this article is organized as follows. Section 2 will firefly introduce the PageRank algorithm. Section 3 will give out the detailed descriptionof our theorem. The design and results of our experiments are presented in the section 4. We make a conclusion in section 5.

## 2    Preliminary and Related Works

The PageRank algorithm simulates user's behavior of browsing web pages. With probability $c$, user randomly selects a link from current page and jump to the page it links to; with probability $1 - c$ the user jumps to a web page uniformly and at random from the collection. The rank score of a web page is defined as the probability a user is browsing the page at long enough time. This model defines a Markov chain on the web graph, with the transition matrix,

$$M = cA + (1 - c)E .$$

Here, the value of $A_{ij}$ is zero if the $i$th page has link to $j$th page; is $\frac{1}{\text{outDegree}(i)}$, if the $i$th page has no link to $j$th page. $E$ is a uniform probability matrix with all element is $\frac{1}{N(P)}$ ($N(P)$ is the number of pages in the collection). The rank score vector $\boldsymbol{pr}$ of all collected pages is happened to be the static transition probability vector. Its computing follows the iterative format:

$$\boldsymbol{pr}^{n+1} = M^T \boldsymbol{pr}^n . \tag{1}$$

This iteration converges to the principal eigenvector of matrix $M^T$.

### 2.1    BlockRank

Suppose the web can be split to $K$ blocks that index by the upper case characters I, J,... Ignored the links between blocks, each block can be treated as a graph. The local rank score vector $\boldsymbol{pr}_\text{I}$ is the static transition probability vector of markov chain defined on Block I. The obtained local rank scores are used to weight the edges between blocks, while calculating the block's importance. The relation between block I and J follows: $B_{\text{IJ}} = \sum_{i \in \text{I}, j \in \text{J}} A_{ij} \cdot \boldsymbol{pr}_i$ .

If define $L$ to be $N(P) \times K$ matrix, whose column are local rank vector $\boldsymbol{l}_J$. And Define $S$ to be $N(P) \times K$ matrix that has the same structure as $L$, but all non-zero element replaced by 1:

$$L = \begin{pmatrix} \boldsymbol{pr}_1 & \boldsymbol{0} & \cdots & \boldsymbol{0} \\ \boldsymbol{0} & \boldsymbol{pr}_2 & \cdots & \boldsymbol{0} \\ \vdots & \vdots & \ddots & \boldsymbol{0} \\ \boldsymbol{0} & \boldsymbol{0} & \cdots & \boldsymbol{pr}_K \end{pmatrix} .$$

Then we have $B = L^T A S$.

---

[1] L1-norm $\|\boldsymbol{x}\|_1$ of vector $\boldsymbol{x} = (x_1, x_2, \ldots, x_n)$ is defined as: $\|\boldsymbol{x}\|_1 = \sum_{i=1}^{n} |x_i|$.

And the transition matrix of the block graph follows: $M_{\text{blcok}} = cB + (1 - c)E_{\text{block}}$ . The block importance follows: $pr_{\text{block}}^{n+1} = M_{\text{block}}^{T} pr_{\text{block}}^{n}$ . And the approximated global PagaRank vector, can be obtained from the block importance and each block's local rank vector: $pr_{\text{approximated}}(j) = pr_{J}(j) \cdot pr_{\text{block}}(J)$ , which is the start point of PageRank algorithm in the step 4 of BlockRank[8].

We notice the original BlockRank adopt $E_{\text{block}}$ as a uniform transition matrix with same dimension as B, any elements in $E_{\text{block}}$ is equal to $\frac{1}{K}$ . It means user will jump to other block with uniform probability 1-$c$. But, intuitively, the block level transition probability of a random surfer should be related with the block's size. Thus, we suggest matrix $E_{\text{block}}$ should be obtained by "zipping" matrix $A$ corresponding to the size of blocks. It is, $E_{\text{block}} = S^{T}AS$ instead of uniform matrix with all element $\frac{1}{K}$.

## 2.2     PowerRank

The theorem in [9] reveals, in the original PageRank algorithm, the low-ranked pages are likely to be pages with less in-links. Because of the power-law attribute of the web, which is experimentally proved by lots of paper [2, 5, 6], there are great mount of such pages in the whole web. That indicates the PageRank scores can be taken as guideline to prune less in-linked pages from the web.

Furthermore, experimently, they found the power-law attribute exists in each granularity of the web. The rank score on each granularity can also guide the pruning process. Based on those, they designed the PowerRank, a multi-granularity algorithm, corresponding to the web's hierarchy structure. The pseudocode of the algorithm is described as:

```
Input: [Graph(1), ...  , Graph(k)];
M(1) = AdjacencyMatrix(Graph(1));
For i = 1 : k-1
    [HighRankId(i), LowRankId(i)] = DetectLowRank (M(i), a%);
    M(i+1) = expand (HighRankId(i), Graph(i+1));
    M'(i)  = expand (LowRankId(i), Graph(i+1));
    LocalRank(i) = PageRank (M'(i));
End for
R(k)= PageRank (M(k));
For i = k-1 : 1
    R(i) = combine (LocalRank(i), R(i+1));
End for
PR = R(1)
```

*Algorithm 1: Pesudocode of the PowerRank algorithm*

## 3     Theorem Result

First, we give the explanation of the symbols used in the following proof which would be helpful for clear expressions and better comprehension.

## 3.1    Notations

We use the upper case of character 'C' to denote the SOLB pages' cluster, and the upper case 'P' to denote the page. The "$pr$" denotes the PageRank score vector, and the superscript of it denotes the iteration number for computing this vector. The notation 'A' is used for the adjacency matrix. We append "zip" on the subscript of those symbols, if they are corresponding to the clustered web graph. "Zip" is also the name of one famous compression software. We adopt this term to illustrate the dimensions of adjacency matrix and rank score vector are reduced after SOLB pages clustered.

The more detailed symbols definitions are listed in the table 1 and table 2.

**Table 1.** Symbols in the Pages and Directories denotation

| Notations | Explanation |
|---|---|
| $N(P)$ | Number of pages in the web graph |
| $N(C)$ | Number of Clusters of SOLB pages in the web graph |
| $C_i, i = 1, \ldots, N(C)$ | $i$th Cluster in the graph |
| $n_i$ | Number of pages in $i$th Cluster |
| $P_y^x$ | $y$th page in the $x$th cluster |
| $P_1^j, \ldots, P_{n_j}^j$ | $n_j$ pages in the $j$th Cluster $C_j$ |
| $S_k^j$ | Out-Linked pages set of page $P_k^j$, $\{P_y^x | \text{out-linked pages by } P_k^j\}$ |
| $\|S_k^j\|$ | Number of pages in set $S_k^j$ |

**Table 2.** Symbols in the PageRank Computation

| Notations | Explanation |
|---|---|
| $pr^n$ | $n$th PageRank score vector |
| $pr_{\text{zip}}^n$ | $n$th zipped PageRank score vector(Cluster Rank score vector) |
| $pr^n(P_y^x)$ | PageRank Score of Page $P_y^x$ in the $n$th iteration |
| $pr_{\text{zip}}^n(C_j)$ | Rank score of Cluster $C_j$ in the $n$th iteration |
| $A$ | Adjacency matrix of the web graph |
| $A_{\text{czip}}$ | Column dimension of $A$ is reduced corresponding to the Clusters |
| $A_{\text{zip}}$ | Column and the row dimensions of $A$ are reduced |
| $A(:, \{_t^j\})$ | The column in the matrix $A$ correspond to the page $P_t^j$ |
| $A_{\text{czip}}(:, \{_t^j\})$ | The column in the matrix $A_{\text{czip}}$ correspond to the page $P_t^j$ |
| $M(:, \{_t^j\})$ | The column in the matrix $M$ correspond to the page $P_t^j$ |
| $M_{\text{czip}}(:, \{_t^j\})$ | The column in the matrix $M_{\text{czip}}$ correspond to the page $P_t^j$ |
| $M_{\text{zip}}(:, C_j)$ | The column in the matrix $M_{\text{zip}}$ correspond to the cluster $C_j$ |

## 3.2    The Rank Scores of the Aggregated Clusters

Suppose we have clustered SOLB pages in the web graph, then we have $N(C)$ clusters $,C_i, i = 1, \ldots, N(C)$. As the definition of the SOLB attribute, if a page $P_s^j$ in cluster $C_j, s \in \{1, \ldots, n_j\}$, links to one page $P_y^x$, then for all $s \in \{1, \ldots, n_j\}$, $P_s^j$ links to the page $P_y^x$. Hence, $\|S_1^j\| = \|S_2^j\| = \ldots = \|S_{n_j}^j\|$, and we define this uniform value as $\|S^j\|$. And $P_1^j, \ldots, P_{n_j}^j$ have same value of matrix entry $A(:, l)$ for all $l = 1, \ldots, N(P)$. That is, $A(P_1^j, l) = A(P_2^j, l) = \ldots = A(P_{n_j}^j, l)$, for each column $l$. Specially, if $P_1^j$ links to $P_y^x$, $A(P_1^j, P_y^x) = A(P_2^j, P_y^x) = \ldots = A(P_{n_j}^j, P_y^x) = \frac{1}{\|S^j\|}$.

We can reduce the column dimension of the original matrix, corresponding to the cluster $C_i, i = 1, \ldots, N(C)$. Similarly, we can reduce row dimension of $A_{\text{czip}}$. We sum link weight from $C_j$ to $P_t^i$, (all page in the cluster $C_i$), to be the link weight from $C_j$ to $C_i$. For example, if each page $P_t^i$ in the cluster $C_i$ has links from $C_j$, then the element $(C_i, C_j)$ of $A_{\text{zip}}$ is $\frac{n_i}{\|S^j\|}$:

$A_{\text{czip}}$ :

$$
\begin{array}{c}
\quad\quad P_1^i \cdots P_{n_i}^i \\
\begin{matrix} \vdots \\ \vdots \\ C_j \end{matrix}
\begin{matrix} \rightarrow \\ \rightarrow \end{matrix}
\begin{pmatrix}
\cdots & \cdots & \cdots & \cdots \\
\cdots & \cdots & \cdots\cdots & \cdots \\
\cdots & \vdots & \vdots & \vdots & \cdots \\
\cdots & \frac{1}{\|S^j\|} & \cdots\cdots & \cdots \\
\cdots & \cdots & \cdots\cdots & \cdots
\end{pmatrix}
\end{array}
$$

$A_{\text{zip}}$ :

$$
\begin{array}{c}
\quad\quad C_i \cdots\cdots \\
\begin{matrix} \vdots \\ \vdots \\ C_j \end{matrix}
\begin{matrix} \rightarrow \\ \rightarrow \end{matrix}
\begin{pmatrix}
\cdots & \cdots & \cdots & \cdots \\
\cdots & \cdots & \cdots\cdots & \cdots \\
\cdots & \vdots & \vdots & \vdots & \cdots \\
\cdots & \frac{n_i}{\|S^j\|} & \cdots\cdots & \cdots \\
\cdots & \cdots & \cdots\cdots & \cdots
\end{pmatrix}
\end{array}
$$

The dimension of uniform probability matrix $E$ is also reduced with the same pattern. We denote $cA + (1-c)E$ as $M$ , $cA_{\text{czip}} + (1-c)E_{\text{czip}}$ as $M_{\text{czip}}$, $cA_{\text{zip}} + (1-c)E_{\text{zip}}$ as $M_{\text{zip}}$ . Then we have:

**Theorem.** *If clustering the SOLB pages in the web graph, and combining the pages' link to be the cluster link, the PageRank on this granularity follows:*

$$pr_{\text{zip}}^{n+1} = M_{\text{zip}}^T pr_{\text{zip}}^n . \tag{2}$$

*The iterative process converges, and the rank score of cluster $C_i$, $PR(C_i)$, is equal to the sum of pages' rank score in it, that is $pr(C_i) = \sum_{t=1}^{n_i} pr(P_t^i)$ .*

To prove it, we give out a lemma firstly.

**Lemma.** *After each iteration of original PageRank, $pr^{n+1} = M^T pr^n$, we sum the rank score of pages in each cluster $C_i$ as $C_i$'s rank score. That defines an iteration representation for cluster rank score vector and it is same to the iteration format (2).*

*Proof.* Suppose $P_s^j, s = 1, \ldots, n_j$, is a page in the cluster $C_j$. Here, $j = 1, \ldots, N(C)$. From the iterative format (1), we have the following $n_j$ equations.

$$pr^{n+1}(P_1^j) = M(:, \{_1^j\}) \cdot pr^n \tag{3}$$

$$\cdots$$

$$pr^{n+1}(P_{n_j}^j) = M(:, \{_{n_j}^j\}) \cdot pr^n . \tag{4}$$

Suppose, in the $n$th iteration, the cluster's score has been the sum of its nested pages' score. And because the pages in same cluster are SOLB pages, the $n_j$ equations can be written as:

$$pr^{n+1}(P_1^j) = M_{\text{czip}}(:, \{_1^j\}) \cdot pr_{\text{zip}}^n \tag{5}$$

$$\cdots$$

$$pr^{n+1}(P_{n_j}^j) = M_{\text{czip}}(:, \{_{n_j}^j\}) \cdot pr_{\text{zip}}^n . \tag{6}$$

In the $n + 1$ iteration, we have the following expression of cluster $C_j$'s rank score:

$$pr_{\text{zip}}^{n+1}(C_j) = \sum_{t=1}^{n_j} pr^{n+1}(P_t^j)$$

$$= \sum_{t=1}^{n_j} M_{\text{czip}}(:, \{_t^j\}) \cdot pr_{\text{zip}}^n$$

$$= M_{\text{zip}}(:, C_j) \cdot pr_{\text{zip}}^n \tag{7}$$

The expression (7) reveals that, for the rank score vector of all clusters: $pr_{\text{zip}}^{n+1} = M_{\text{zip}}^T pr_{\text{zip}}^n$ . $\qquad \square$

*Proof of Theorem.* It has been proved that the iterative format (1) is converged. That is, for any $\varepsilon > 0$, if $n$ is large enough, $\|pr^{n+1} - pr^n\|_1 < \varepsilon$. From the above lemma, which defined an iterative format same to (2), we know , for any $\varepsilon > 0$, if n is large enough, $\|pr_{\text{zip}}^{n+1} - pr_{\text{zip}}^n\|_1 < \varepsilon$. The format (2) converges. Furthermore, from the construction of the new iterative format in the above lemma, we know that the rank score of cluster $C_i$, $pr(C_i)$, is equal to the sum of its nested pages' score. That is $pr(C_i) = \sum_{t=1}^{n_i} pr(P_t^i)$. $\qquad \square$

### 3.3    Distribute Clusters' Scores to SOLB Pages

In this section, we defined a matrix-vector multiplication format to obtain the pages' rank score, after we have computed the directory's rank score. Firstly, we review the formula for single page $P_s^j$'s PageRank computation.

From the equation (5) - (6), the PageRank score of page $P_s^j$, in the $n$th iteration, follows:

$$pr^n(P_s^j) = M_{\text{czip}}(:, \{_s^j\}) \cdot pr_{\text{zip}}^n \tag{8}$$

$$= \sum_{k, C_k \to P_s^j} \frac{1}{\|S^k\|} \cdot pr_{\text{zip}}^n(C_k) \tag{9}$$

$s = 1, \ldots, n_j$ .

Here, on the expression (9), the term $\frac{1}{\|S^k\|}$ represents the weight of links from other cluster $C_k$ to page $P_s^j$ which nests in the cluster $C_j$, while the term $pr_{\text{zip}}^n(C_k)$ represents the Rank Score of $C_k$ in the $n$th iteration.

From the expression (8), we can derive an equivalent vector-matrix expression:

$$pr^{n-\text{threshold}} = M_{\text{czip}}^T \cdot pr_{\text{zip}}^{n-\text{threshold}} . \tag{10}$$

Here, $n$-threshold is the iteration number in computation. That means, after an additional vector-matrix multiple operation, the Page's Rank score can be obtained.

## 3.4    Restriction and Relaxation

The graph, with all SOLB pages clustered, is called complete SOLB clustered graph. If only parts of SOLB pages in the web are clustered, we call it the incomplete SOLB clustered graph. The result of PageRank computation on such graph has similar character as on the complete SOLB clustered graph. A corollary of the theorem in the section 3.2 supports that.

**Corollary.** *On the incomplete SOLB clustered graph, denoted clusters as $C_i', i = 1, 2, \ldots$, we have $pr(C_i') = \sum_{P_t' \in C_i'} P_t'$. That is, the rank score of incomplete cluster is also the sum of its nested pages' original rank score.*

This result is interesting and can be easily derived from the aforementioned theorem. It reveals that we can constrict an incomplete SOLB clustered graph and also obtain the exact rank result. To cluster pages just in the same directory helps to reduce the computing complexity from $O(N^2)$ to $O(mN)$. Here, $N$ is the scale of whole graph, $m$ is the average page number in a directory.

Moreover, it is necessary to measure the similarity of same out-link behavior between two pages. We introduce two measurements on pages' SOLB: cosine similarity and the set similarity. The cosine similarity defines on pages' out-link vector. If we have allocated each page a uniform id, the out-link vector $V_i$ of the page $P_i$ can be defined as:

$$V_i(k) = \begin{cases} 1 & P_i \to P_k \\ 0 & P_i \not\to P_k \end{cases} , k = 1, \ldots, N(P) .$$

Hence, the cosine based similarity can be defined as:

$$S^a(P_i, P_j) = \frac{V_i^T V_j}{\|V_i\| \cdot \|V_j\|} . \tag{11}$$

The set similarity measurement is based on the set's operation. In short, if $S_i$ is the set of out-linked pages by page $P_i$ and $\|S_i\|$ denotes number of pages in $S_i$, the similarity between page $P_i$ and $P_j$ can be defined as:

$$S^b(P_i, P_j) = \frac{\|S_i \cap S_j\|}{\sqrt{\|S_i\| \cdot \|S_j\|}} . \tag{12}$$

With these measurements, we are able to relax "SOLB" condition to "SOLB liked" condition. We can cluster more than 0.9 SOLB liked pages in the same directory. Under those cluster thresholds, we follow the same formula (12) to distribute rank score to the nested pages after the rank score of cluster obtained.

# 4   Experiments

We perform experiments on a web data collection which contains about 80 million web pages. To obtain a more connected graph, we selected several core pages, and expand through the forward links and the backward links several times. Such treated graph contains 5.3 Million pages, and about 0.7 Million SOLB pages.

With the aforementioned cosine similarity, we prepare two experiments to show how the similarity threshold selection makes influence on the performance. In the first experiment, the similarity for clustering SOLB pages is from 0.9 to 1, with interval 0.01. In the second, the similarity is set from 0.5 to 1, with interval 0.1. For each experiment, we report the L1-norm of minus vector of the result vector and the PageRank vector. (By [8], L1-norm can be the measurement for the similarity of rank order. Smaller the L1-norm of minus vector, more similar the two rank orders.) We also give charts to compare the computing time, in which the time of original PageRank algorithm is set to 1.

## 4.1   Experiment Result

In the first experiment, we obtained 11 rank score vectors with different threshold. From the Fig. 1 (a), the L1-norm of minus vector decrease while the threshold increasing from 0.9 to 1. With the threshold 0.9, which graph is different to the original graph, the L1-norm of minus vector is 3e-6. That means the result with the threshold 0.9 is very near to the PageRank result, because the stop error in our experiment is 1e-6.

The Fig. 1 (b) is the comparison of computing time with same threshold. If the threshold sets to 0.9, our algorithm saves 10% computing time than the PageRank. The computing time increases while the threshold increased. If the threshold set to 1, our algorithm saves 7% computing time than PageRank.

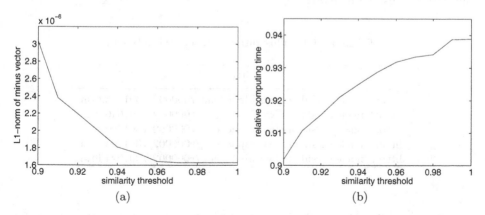

$$(a) \qquad\qquad (b)$$

**Fig. 1.** L1-norm and time comparison of experiment I

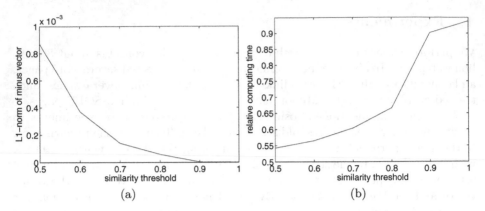

**Fig. 2.** L1-norm and time comparison of experiments II

**Table 3.** SOLB illustration - a site of lyrics collection

| URL |
| --- |
| www.lyricalline.com/writeandwrong/songs/mr2.html |
| www.lyricalline.com/writeandwrong/songs/nc5.html |
| www.lyricalline.com/writeandwrong/songs/bsc1.html |
| www.lyricalline.com/writeandwrong/songs/cws1.html |
| www.lyricalline.com/writeandwrong/songs/jam1.html |

We obtain 6 rank score vectors in the second experiments. With smaller similarity threshold, the Fig. 2 give their L1-norm distance to the PageRank score vector and the computing time for obtaining them.

In the Fig. 2 (a), if the threshold is 0.5, the L1-norm of minus vector is 0.8e-3; while the threshold is 1, the L1-norm of minus vector is 1.7e-6. Here, the stop error of experiments II is set to 1e-6 and those L1-norms also decrease while the similarity threshold increases.

**Table 4.** SOLB illustration - structured web data

| URL |
| --- |
| http://nucleus.cshl.org/worldpac/eng/r000002/r001323.htm |
| http://nucleus.cshl.org/worldpac/eng/r000002/r001646.htm |
| http://nucleus.cshl.org/worldpac/eng/r000002/r001700.htm |
| http://nucleus.cshl.org/worldpac/eng/r000002/r001703.htm |
| http://nucleus.cshl.org/worldpac/eng/r000002/r001924.htm |

The computing time with threshold 0.5 is less than 55% of PageRank. Especially, if the threshold is 0.8, the L1-norm of minus vector is 5e-5 which is

**Table 5.** SOLB illustration - unskilled web site

| URL |
| --- |
| www.putuo.sh.cn/socialservice/sqfw/0301a07.htm |
| www.putuo.sh.cn/socialservice/sqfw/0401a04.htm |
| www.putuo.sh.cn/socialservice/sqfw/0601a01.htm |
| www.putuo.sh.cn/socialservice/sqfw/0701b22.htm |
| www.putuo.sh.cn/socialservice/sqfw/0701b27.htm |

very near to the stop error 1e-6 and the computing time is about 65% of PageRank. Also, from the Fig. 2 (b), the computing time increases with the threshold increasing.

### 4.2    Case Study

We will give three examples sites in our data set to illustrate how SOLB pages exist in the web graph. The first illustration comes from the site: www.lyricalline.com. These pages provide lyrics and enable people to critique them. In each page, the lyric occupy the main space of the page, and two links link to the critique page.

The second illustration is catalog of a digital library: "http: //nucleus.cshl.org/ worldpac/eng/wphome.htm". In the page of each entry of the catalog, it is just have three links from help page, home page, and the anchor in the top of this page. Furthermore, the SOLB pages widely exist in early digital library sites, or the manually maintained sites in scientific field, such as the netlib repository.[2]

We also find the SOLB pages in the site constructed by less skilled designer. The web site, of the social service of Putuo district, Shanghai, "http: // www.putuo. sh.cn /socialservice/ index0.htm", is an example. Pages in the site have just one link to the home page, and the left space filled with the text information.

## 5    Conclusion

We find a theorem result on the web graph's implicit attribute: clusters of SOLB pages. Taken the clusters as computing units, we can efficiently improve the PageRank computation. We also define metrics to measure the similarity of pages in same out-link behavior. The experimental results show that, even if we cluster the 0.5 liked SOLB pages, the rank result is very near to the original PageRank and the required computing time is reduced to 50% of the original.

The work was done in Microsoft Research Asia. Thank to the abundant web data they offered.

---

[2] http://www.netlib.org

# References

1. A. Arasu. PageRank Computation and the Structure of the Web: Experiments and Algorithms. 11th International WWW Conference, May 2002.
2. A. Medina, I. Matta, J. Byers. On the Origin of Power Laws in Internet Topologies. ACM Computer Communication Review, vol. 30, no. 2, pp. 18–28, Apr. 2000.
3. G. H. Golub, and C. F. Van Loan. Matrix Computations. The Johns Hopkins University Press, Baltimore, 1996.
4. J.Kleinberg. Authoritative sources in a hyperlinked environment. Journal of the ACM, 46(1999), pp.604-632
5. M. Faloutsos, P. Faloutsos, C. Faloutsos. On Power-Law Relationships of the Internet Topology. Proceedings of ACM SIGCOMM, Aug. 1999.
6. Q. Chen, H. Chang, R. Govindan, et al. The Origin of Power Laws in Internet Topologies Revisited. Proceedings of IEEE INFOCOM '02, 2002.
7. S. Brin, L. Page, R. Motwami, and T. Winograd. The PageRank citation ranking: bringing order to the web. Stanford University Technical Report, 1998.
8. S. D. Kamvar, T. H. Haveliwala, C. D. Manning and G. H. Golub. Exploiting the Block Structure of the Web for Computing. Stanford University Technical Report, 2003
9. Yizhou Lu, Benyu Zhang, Wensi Xi, Chen Zhen, et al. The PowerRank Web Link Analysis Algorithm. Proceedings of 13th International WWW Conference, May 2004.

# Literal-Matching-Biased Link Analysis

Yinghui Xu and Kyoji Umemura

Toyohashi University of Technology,
xyh@ss.ics.tut.ac.jpumemura@tutics.tut.ac.jp
Information and Computer Science Department, Software System Lab.
1-1, Hibarligaoka, Tempaku, Toyohashi 441-8580, Aichi Japan

**Abstract.** The PageRank algorithm, used in the Google Search Engine, plays an important role in improving the quality of results by employing an explicit hyperlink structure among the Web pages. The prestige of Web pages defined by PageRank is derived solely from surfers' random walk on the Web Graph without any textual content consideration. However, in the practical sense, user surfing behavior is far from random jumping. In this paper, we propose a link analysis that takes the textual information of Web pages into account. The result shows that our proposed ranking algorithms perform better than the original PageRank.

## 1 Introduction

Most of search engines have tried some form of hyperlink analysis because it significantly improves the relevance of search results. The two best-known algorithms that perform link analysis are HITS [10] and PageRank [2]. The latter, used in Google, employs the hyperlink structure of the Web to build a stochastic irreducible Markov chain with a transition matrix P. The transition matrix was built on the assumption that a "random surfer" who is given a Web page at random and keeps clicking on successive links, never hits "back" but eventually gets bored and starts on another random page. The irreducibility of the chain guarantees that the long-run stationary vector , known as the PageRank vector, exists. The calculated PageRank vector is then used to measure the importance of Web resources. From the standpoint of the PageRank algorithm, the regular PageRank of a page can be regarded as the rate at which a surfer would visit that page. The PageRank model is based only on the hyperlink structure without considering the textual information that is carried along the edge between connected pages, and the transition probability from a given page to its outgoing links is weighted with equal chance. The transition probability might become more accurate if we consider some practical aspects of human search behavior. We might describe it as follows: a user goes somewhere based on his interests. He arrives there and begins thinking about the displayed content. He may find some content of interest that is somewhat similar to what he is looking for, but that still does not satisfy his requirements based on some textual information in the page. Therefore, the user will continue searching. Where will he go? The information obtained through reading a page and the literal information

S. H. Myaeng et al. (Eds.): AIRS 2004, LNCS 3411, pp. 153–164, 2005.

of outgoing links will help the user weight further options because additional searching might provide a more literal match between the words in the user's mind and the words on the page. Therefore, we propose to modify the underlying Markov process by giving different weights to different outgoing links from a page based not only on the hyperlink structure but also the textual clues on the Web graph. The intuition here is that inbound links from pages with a similar theme to a more significant influence on its PageRank than links from unrelated pages. Although such considerations have been discussed by some researchers, an actual system and its experimental results have not been reported yet. In this paper, we propose a link analysis model that combines the literal information resources extracted from Web pages themselves with the hyperlink structure for calculating the query-independent page importance. This paper is organized as follows: Section 2 will introduce the PageRank algorithm and its several variations. We propose our algorithm in Section 3. Experiment results and analysis are presented in Section 4. Lastly, we present our conclusions in Section 5.

## 2     PageRank and Its Variations

First, let us make a brief observation about Google's PageRank algorithm. Imagine a Web surfer jumps from Web page to page, choosing with uniform probability which link to follow at each step. In order to reduce the effect of dead-end or endless cycles, the surfer will occasionally jump to a random page with some small probability $d$, or when the page has no out-links. To reformulate this in graphic terms, consider the web as a directed graph, where nodes represent web pages, and the edges between nodes represent links between web pages. Let N be the number of verticals on the Web graph, $F_i$ be the set of pages $i$ links to, and $B_i$ be the set pages which link to page $i$. If averaged over a sufficient number of steps, the probability the surfer is on page $j$ at some point in time is given by the formula: $PR(j) = (1 - d)/N + d \times \sum_{i \in B_j} (PR(i)/|F_i|)$. The PageRank score for page $j$ is denoted as $PR(j)$. Because the equation shown above is recursive, it must be iteratively evaluated until $PR(j)$ converges. Typically, the initial distribution for $PR(j)$ is uniform. PageRank is equivalent to the primary eigenvector of the transition matrix $A$:

$$A = (1 - d) \left[ \frac{1}{N} \right]_{N \times N} + d \times M, \quad M = \begin{cases} 1/|F_i| \ there \ is \ an \ edge \ from \ i \ to \ j \\ 0 \ otherwise \end{cases}.$$

Note the original PageRank is based on purely explicit link structure among Web pages, and the assumption that outgoing links from a page are equally important when calculating the transition probability is naive. Two kinds of representative approaches for the integration of themes in the PageRank technique have been proposed recently: On the one hand, the "intelligent surfer" introduced by Matthew Richardson [12] and on the other hand the topic-sensitive PageRank proposed by Tather Haveliwala [7]. M. Richardson and P. Domingo employ a random surfer model in order to explain their approach to the implementation of themes in the PageRank technique. Instead of a surfer who follows links completely at random, they suggest a more intelligent surfer who, on the one hand, follows only links which are related to an original search query and, on the other hand, after "getting bored", only jumps to a page which relates to the original query. Thus, an "intelligent surfer" only surfs relevant pages that contain the term of an initial query. To model how "intelligent" behavior of a random surfer influences PageRank, every term occurring on the Web has to be processed through a separate PageRank calculation, and each calculation is solely based on links between pages that contain that term. Apparently, computing time requirements are indeed a serious issue with this approach after viewing "intelligent surfer PageRank". The approach of Taher Havilewala seems to be more practical for actual usage. Similar to Richardson and Domingos, Havilewala also suggests the computation of different PageRanks for different topics. Compared with "intelligent surfer PageRank", the Topic-Sensitive PageRank does not consist of hundreds of thousands of PageRanks for different terms, but rather of a few PageRanks for different topics. Topic-Sensitive PageRank is based on the link structure of the whole Web, according to which the topic sensitivity implies that there is a different weighting for each topic. However, there are two serious limitations in Haveliwala's work on Topic-Sensitive PageRank. One is that the computation of PageRank for topics in the Web is very costly and the other is that it is difficult to determine what the "topic" of a search should be. Having a Topic-sensitive PageRank is useless as long as one does not know in which topics a user is interested in.

## 3    Our Approach

In this paper, we concentrate on investigating a way to take the textual information of Web pages into account for link analysis with a more accurate Markov transition model. Hyperlinks in a page might serve different roles. We divide the hyperlinks in a page into two types, informative links and referential links. We are interested in the links with a literal matching between pages (informative links), because the purpose of such types of links is to point to similar, more detailed, or additional information. As for the referential links, they are the links in a page that have no literal matching with its target. Users surfing on those referential links are conducting the same behavior as drawing an unseen ball from an urn while the link transfer on those informative links are guided by a literal matching with their corresponding target.

### 3.1    Virtual Document

One major challenge to our proposed model is how to introduce the appropriate textual information units of Web pages for the literal matching function. It is well known that the similarity measure of two connected pages using content might fail to reflect the correct relationship between them due to the heterogeneous

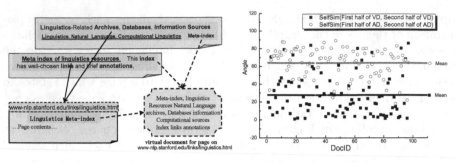

**Fig. 1.** a sample of VD

**Fig. 2.** Comparison of homogeneity bewteen AD and VD

characteristics of unstructured Web pages. An IR system is purely dependent on page content and this has proved to be weakness [13]. To avoid such problem issues, an information unit, called a virtual document (VD), is generated for Web pages. What is the VD? There may be several possible ways to define VD. It is dependent on information processing tasks. The concept of the VD was introduced by Glover [6], and they used VD for Web pages classifying and describing task. In our approach, the VD of a given page is comprised of the expanded anchor text (EAT) from pages that point to him and the title words on the page itself. The definition is shown as follows:

$AnchorText\,(i,j)$ : *set of terms in and around the anchor of the link from page i to j.*
$TitleText\,(j)$ : *set of terms in the title tag of page j.*
$VD\,(j)$ : *set of terms in the virtual document j.*
$VD\,(j) = \bigcup_{i} (AnchorText\,(i,j)) \cup TitleText\,\{(j)\}$ .

The solid building method refers to three steps: At first, we create a link text table that includes triple elements. . It represents that the page with to the page with has description text (DT). The DT is extracted based on DOM tree structure. The left and right sibling node with text properties of the anchor tag "a" node and the text information under it are all extracted as the description data. Considering the case that sibling node with text properties around an anchor tag node may be over several lines long and deviate from the main author's motivation due to poor page structure, in our experiment, only the text information around the anchor tag within one phrase is kept as the description data of anchor link. Next, we also extract text information of the title tag. Lastly, the VD of are organized by the integrated textual information from all associated description texts of its ingoing hyperlinks and title text of its own. Fig. 1 is a sample illustration of a VD.

**Features of Virtual Document.** From the definition of the VD of a page, the information resources that it contains is from the expanded anchor text of pages that point to it and its title. Such combined resources make the VD share characteristics of both anchor and title. Recent studies [5][11] examined several aspects of anchor text. They showed evidence that anchor text provides a summarization of the target document within the context of the source document, and the process of creating anchor text for document might be a good approximation of the type of summarization presented by users to search system in most queries. As Eiron argued in his paper[5], a searching procedure may be looked at as one way of building an optimized query formulation through an anchor text collection. What is more, a page may have several ingoing links that all carry descriptions about their own impression of that page. Such collective wisdom of what the page is about is a useful implicit resource for characterizing the page. As for the title information on a page, Jin, Haupmann, and Zhai[8] found that document titles also bear a close resemblance to queries, and that they are produced by a similar mental process. Thus, it is natural to expect that both titles and anchor texts capture some notion of what a document is about. Moreover, the size of VD collection is much smaller than that of the total collected data. Thus, processing it is much faster than processing the total document data.

**Stopword Removal from VD Space.** A collection of VD sets can be looked at as a Web-based concept database. Just as an ordinary text database has certain words that appear frequently such as "the" and "if", so does a VD. In particular, there are many commonly used words in VD such as "click here" or "home", or "next", and these types of terms will decrease the quality of a VD and have a negative influence on the literal matching function. In our experiment, words with an entropy [3][9] value larger than 0.65 (the maximum word entropy in our experiment is 0.78, and the thresholds are set based on human judgment) are regarded as Stopwords and it will be removed from the VD space. The entropy calculating formula is as follows:

$$VDTF(w,j) = \#\{w|w \in VD(j)\}; \quad P(w,j) = VDTF(w,j) \Big/ \sum_{k=1}^{N} VDTF(w,k)$$

$$VDET(w) = -\sum_{j=1}^{N} P(w,j)\log_N P(w,j); \quad VDTW(w) = 1 - VDET(w).$$

*where:* $N$ is the *number of virtual documents in virtual document collection.*

The rational for an entropy-based Stopword filter is that terms used in many pages usually carry less information to users, and terms appearing in fewer pages likely carry more information of interest.

**Definition 1.** *Given an actual web page $i$, we have*

$$Page(i): \ set \ of \ terms \ in \ the \ content \ of \ page \ i.$$
$$tf(w,i) = \#\{w|w \in Page(i)\}.$$
$$df(w) = \#\{i|w \in Page(i)\}.$$

**Definition 2.** *Given a page* p *and its outgoing sets , the transition odds from* p *to* $q_k$ *are determined by:*

$$TranOdds\,(p \rightarrow S_k)$$
$$= SIM\_IR\,(VD\,(S_k)\,,p) + SIM\_JACCARD\,(VD\,(p)\,,VD\,(S_k))\,.$$
$$where:$$
$$SIM\_IR\,(VD\,(S_k)\,,p) = \begin{cases} TF \bullet IDF \\ OKAPI\_BM25 \end{cases}.$$
$$SIM\_JACCARD\,(VD\,(p)\,,VD\,(S_k)) = \frac{|VD(p) \cap VD(S_k)|}{|VD(p) \cup VD(S_k)|}.$$

**Homogeneity of Virtual Document.** As we mentioned in Section 3.1, due to the heterogeneous characteristic of unstructured Web pages, a general similarity metric cannot capture the correct relationship between two connected pages. Our intuition about the VD space tells us that the information resources in VD are somewhat homogeneous. To test the homogeneity of VD, 100 Web pages were selected randomly from the Web corpuses that satisfy the condition that the length of their VDs was more than 18 words. The method used to measure the homogeneity of a document was: we divided the document into two sections and measured the similarity of the two sections to each other. Here we use the cosine measure for the similarity metric. Fig. 2 plots the comparative results of homogeneity between an actual document (AD) and its VD. The X-axis is the document ID and the Y-axis is the angle between the vectors of the two sections of its correspondent document. It shows clearly that the variation in the two sections in its VD is much smaller than that of the actual page content. The average angle between the two parts of the AD is around 64 degrees and most points are above the average, whereas the average angle for VD is around 27 degrees, and most points are below the average. Such analyzed results are consistent with our intuition. This homogeneous characteristic makes VD a more reasonable information resource than actual page content for literal matching calculations.

## 3.2    Algorithms

In this section we present formally how literal mining is combined with link analysis based on the Markov transition model in our approach. The main purpose of us is to introduce a more accurate transition probability calculation method for modeling actual user search behavior. Thus, we aim to assign a reasonable link weights through literal information on the connected pages. In our intuition, surfer behavior will be influenced by not only the suffer purpose but also the page content that author wrote. We assume that a surfer prefers a relevant target. The mechanisms that define how a user chooses his target after viewing the current page, on one hand, should in some way reflect how likely the surfer could satisfy his need through the page being viewed, or indicate how much influence on the surfer that page's contents impose. When a user continues his search (clicking), he does have a reasonable idea of what the activated page will looks like. We assume that there is something present in the user's mind (VD) about

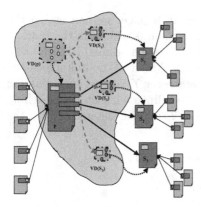

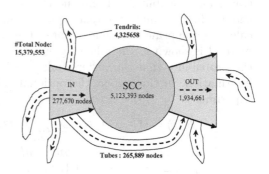

**Fig. 3.** Computing elements relationship illustration of our algorithm

**Fig. 4.** Bow tie structure of experiment corpus

the destination page that they would like to visit. In our approach, the VD is used to represent the user's idea of the actual destination page. In addition, we assume that information searching is a process that gradually approaches the user's desired outcome. Accordingly, user's minds are somewhat consistent in their searching path. Based on the above considerations, we propose two factors to evaluate the likelihood that a user will surf from $p$ to $S_k$, denoted by: $TranOdds\,(p \rightarrow S_k)$. The first factor is to measure how likely the VD of the activated target page $S_k$ can be generated from the page p being viewed, denoted by: $SIM\_IR\,(p, VD\,(S_k))$. The higher the similarity between a page and the VD of its target page, the more likely the surfer will accept the recommendation from the page's author. Most state-of-the-art similarity equations can be used to calculate the similarity between $p$ and $S_k$. In our experiment, OKAPI BM25 and TFIDF schemes are performed. The second factor is to measure the similarity between two connected VDs denoted by: $SIM\_JACCARD\,(VD\,(p), VD\,(S_k))$. This value is used to indicate the consistency of surfer's view of two connected nodes in the Web graph. Because the unique word list of a VD is usually short, a set operation is feasible and an easy way to measure the similarity between them. The relationship of computing elements in our algorithm is plotted in Fig. 3. Before introducing equations for calculating these two factors, we provide some definitions (definition 1 and 2) that we will use in our approach. Hyperlinks in a page might serve different roles. In our proposed link analysis model, we divide the hyperlinks in page into two types, informative links, termed InforLink, and referential link, termed as referLink. We are interested in the links with literal matching between pages, because the purpose of such kind of links is to point to similar, more detailed, or additional information. As for the referential links, they are the links in a page that have no literal matching with its target. User surfing on referential links are conducted the same behaviors as drawing a ball from an urn while surfing on informative links are guided by the literal matching with their corresponding targets. Based on the calculated values that indicate transition likelihood for all possible connections on a page, we can assign the transition probability to them and regard them as the link weight in the Markov chain. Then we can use the same processing method as the original PageRank to calculate the principle eigen-vector of the transition Matrix. The link allocation method is shown in the following equations. Parameter $\Upsilon$ is used for adjusting the probability that the surfer tends to follow those links with literal matching information. In this paper, the optional value of $\Upsilon$ will be determined by the experiment.

$$PR(j) = (1 - \lambda) \, 1/N + \lambda \sum_{i \in B_j} PR(i) \, prob(i \to j),$$

$B(i):$ $set\ of\ pages\ which\ link\ to\ page\ i;$

$$LinkType(i \to j) = \begin{cases} 1, & if: A \wedge B \wedge C \\ 0, & otherwise \end{cases}$$

$$\begin{cases} A : VD(j) \neq \emptyset; \ B : \ Page(i) \neq \emptyset; \\ C : \{w | w \in (VD(j) \cap Page(i))\} \neq \emptyset; \end{cases}$$

$InforLink(i) = \{j | LinkType(i \to j) = 1\},$

$referLink(i) = \{j | LinkType(i \to j) = 0\},$

$F(i) = InforLink(i) + referLink(i).$

$$prob(i \to j) = \begin{cases} \gamma \times \dfrac{TranOdds(i \to j)}{\sum\limits_{k \in InforLink(i)} TranOdds(i \to k)} \\ (1 - \gamma) \times \dfrac{1}{\# referLink(i)} \end{cases}, \gamma = 0.7$$

$where:$

$r:$ $the\ probability\ that\ transition\ follows\ informative\ link.$

## 4    Experiment Results

We ran experiments on NTCIR 100G Web corpus [4]. There are 10,894,819 pages in the crawled Web corpus. As for the link structure of the Web corpus, "Bow Tie" structure [1] is used to describe its graph features. Over 15 million nodes expose the large-scale structure of the Web graph as having a central, strongly connected core (SCC), a sub-graph (IN) with directed paths leading into the SCC, a component (OUT) leading away from the SCC, and relatively isolated tendrils attached to one of the three large sub-graphs. The statistical information on the NTCIR Web corpus is shown in Fig. 4. The experimental steps are: Firstly, SCC shown in Figure 4 was constructed based on the link structure of the Web corpus. In our experiment, there are 5 million nodes extracted iteratively from the Web graph, which is around 30 percent of the overall Web corpus. Next, link analysis based on our approach and PageRank are both performed on the SCC of the Web corpus. Nodes in the OUT sets of the Web graph are processed iteratively based on the calculated page rank value of their source nodes in the second step. Finally, around 7 million nodes obtained their page rank value. We note that not all page entities in the Web corpus have their own page rank value. There are around 3 million pages without page rank value through link analysis. The reason is that not all Web pages of the nodes in the

Web graph were downloaded into the page repository. The Web corpus that we used in our experiment is just a simulated Web environment. To measure the efficiency of our proposed link analysis, several evaluation experiments were conducted. At first, query-dependent, full-text-based information retrieval was conducted on the whole corpus. Only the top 1,000 return results of a given query were extracted for evaluation. In the Web corpus, 778 relevant documents from 42 queries were provided for conducting general precision-recall evaluation of information retrieval system. As we pointed out above, due to the incomplete link structure of Web corpus, one-fourth of the pages in the Web repository has no its correspondent page rank value. This leaves only 350 relevant documents from 38 queries left for the page sets that have page rank value. To make the comparison experiment reasonable, we remove the pages that have no page rank value from the result sets of the IR baseline. Therefore, all our evaluation experiments will be performed only on the document sets that are all have a page rank value.

## 4.1    Original PageRank Versus Our Approach

Our first experiment was to perform a comparison between the IR precision for those documents that are returned by content information retrieval based on their original PageRank score and the IR precisions based on the score of our proposed model. We choose $\Upsilon$ to be 0.7 from observing the behavior of our system. To find the best literal matching model for link analysis, the comparison experiments were performed among all possible matching schemes. The results are figured out in Fig. 5. As we expected, Literal-Matching-Biased link analysis all performed better than the original PageRank for both Ave. Precision-Recall and R-precision among the top 10 return documents. In addition, we found that the literal matching model using OKAPI-BM25 for with the factor together do the best job. Therefore, the following experiments were performed based on this literal matching scheme.

## 4.2    Direct Rank Comparison

Direct comparison experiments of the rank value for the right answer (relevant file) in the results sets between PageRank and our approach were performed. Among 350 relevant documents from 38 queries, we received 214 "win", 120 "fail", and 16 "equal". The rough metric for win, fail and equal is defined as:

$R$ : $return\ document\ sets\ for\ a\ given\ query.$
$\tau_2$ : $document\ in\ R\ sort\ by\ original\ pagerank\ value.$
$\tau_3$ : $document\ in\ R\ sort\ by\ pagerank\ value\ of\ our\ approach.$
$\tau_k\ (i)$ : $rank\ of\ i\ in\ \tau_k$
$$\begin{cases} win & : & \tau_2\ (i) > \tau_3\ (i) \\ fail & : & \tau_2\ (i) < \tau_3\ (i) \quad , i \in R. \\ equal & : & \tau_2\ (i) = \tau_3\ (i) \end{cases}$$

To observe the different degrees clearly, the plot based on the summation of rank difference, $(\tau_2\ (i) - \tau_3\ (i))$, of relevant files for each query is shown in Fig. 6. It

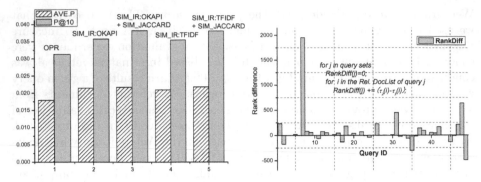

**Fig. 5.** Comparison between IR prec. based on ori. PR score and the IR prec. based on our proposed link analysis score under all possible matching schemes

**Fig. 6.** Rank difference of relevant file for each query

shows clearly that the rank values based on our approach are less than those of the original PageRank in most cases. From the results of this direct rank comparison, we can deduce that our approach does make sense for achieving more accurate page importance.

### 4.3 Ideal Re-ranking Experiment

Evaluating the usefulness of query-independent evidence in boosting search effectiveness is complicated by the need to combine the query-independent score with a query-dependent score. There is a risk that a spurious negative conclusion could result from a poor choice of the combining function. Accordingly, we introduced an ideal re-ranking experiment, which is used to gauge the improvement likelihood. The way that we use this here is: first, locate the right answer (relevant file) in the IR baseline, and then compare the IR rank value with its page rank value. If the IR rank value of the relevant file is smaller than the page rank value, we will keep it as it was, but if the IR rank of relevant file is larger than its page rank, we will make some adjustment based on the scheme showed in figure 7. The lower rank boundary is the IR rank, and the upper rank boundary is the PR rank. After this rank adjustment through our ideal re-ranking method, the traditional IR precision-recall evaluation method was used to compare their efficiency. Comparison results shows in Figure 8 and Table 1. It is evident that the query-independent analysis (original PageRank) and our approach, both boost search effectiveness significantly. Compared with original PageRank, however, our approach does a good job on the lower recall and our approach obtained 8.5 percent improvement at the lowest recall. Therefore, the ideal re-ranking experiment results indicate that our proposed model promises to determine more accurately the page importance on the Web.

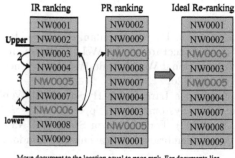

IR ranking    PR ranking    Ideal Re-ranking

Move document to the location equal to page rank. For documents lies between rank bounds (lower → upper), Simply increase their rank by 1. If meet relevant file during transfer, keep it as it was.

**Fig. 7.** Re-ranking scheme on ideal case

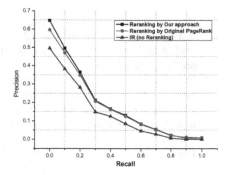

**Fig. 8.** Recision recall comparison based on ideal re-ranking experiment

**Table 1.** Precision recall comparison results among our approach, Original PageRank and IR baseline

|     | Our approach | Origina PR | IR baseline |
| --- | --- | --- | --- |
| 0   | 0.6467 ( +30.6% +8.5% ) | 0.596 ( +20% +0% ) | 0.4949 (+0% +0%) |
| 0.1 | 0.4955 ( +29.5% +5.3% ) | 0.4703 ( +23% +0% ) | 0.3825 (+0% +0%) |
| 0.2 | 0.3647 ( +30% +4.6% ) | 0.3486 ( +24% +0% ) | 0.2813 (+0% +0%) |
| 0.3 | 0.2113 ( +42% +2.3% ) | 0.2065 (+39% +0%) | 0.1482 (+0% +0%) |
| 0.4 | 0.1645 ( +0.31% +1.0% ) | 0.1628 (+30% +0%) | 0.1252 (+0% +0%) |
| 0.5 | 0.1298 ( +53% +3.2% ) | 0.1257 (+49% +0%) | 0.0846 (+0% +0%) |
| 0.6 | 0.0826 ( +84% +2.4% ) | 0.0806 (+80% +0%) | 0.0448 (+0% +0%) |
| 0.7 | 0.0529 ( +88% +2.3% ) | 0.0517 (+84% +0%) | 0.028 (+0% +0%) |
| 0.8 | 0.0213 ( +222% +3.9% ) | 0.0205 (+210% +0%) | 0.0066 (+0% +0%) |
| 0.9 | 0.0084 ( − -23% ) | 0.0109 (+7.3% +0%) | 0 (+0% +0%) |
| 1.0 | 0.0075 ( − -9% ) | 0.0083 ( − +0%) | 0 (+0% +0%) |

## 5   Conclusion and Future Work

In this paper, we proposed a unified model, which combines literal matching with link structure analysis. We provide a sound mechanism for unifying literal mining and link analysis for Web information retrieval. The experiment results showed that our approach will potentially do a better job for the IR task than the original PageRank. The future research that we plan to undertake is as follows:

- We are planning to do more investigation on the literal matching kernel used in our approach.
- The $LinkType(i \rightarrow j)$ factor in our approach could be used later to distinguish more than just informative and referential links, and in particular, to distinguish several subtypes of informative links.
- Finding an appropriate practical re-ranking scheme to evaluate the effectiveness of our approach.

# References

1. Broder, A., Kumar, R., Maghoul, F., Raghavan, P., Rajagopalan, S., Stata, R., Tomkins, A., and Wiener, J. (2000): Graph Structure in the Web: Experiments and Models. In Proceedings of the 9th International World Wide Web Conference on Computer Networks. Amesterdam, pp. 309-320.
2. Brin, S. and Page, L. (1998): The Anatomy of a Large-Scale Hypertextual Web Search Engine. Computer Networks and ISDN Systems 30, 1-7, pp. 107-117.
3. Christopher D. Manning and Schutze, H. (1999): Foundation of Statistical Natural Language Processing. The MIT Press, Cambridge Massachusetts, London, England.
4. Eguchi, K., Oyama, K., Ishida, E., Kando, N., and Kuriyama, K. (2003): System Evaluation Methods for Web Retrieval Tasks Considering Hyperlink Structure, The 12th International World Wide Web Conference, No.poster-344, Budapest, Hungary.
5. Eiron, N., and McCurley, K. S. (2003):, Analysis of Anchor Text for Web Search. In Proc. of the 26th Annual International ACM SIGIR'03 Conference on Research and Development in Information Retrieval, pages 459-460, Toronto, Canada, August.
6. Glover, E., Tsioutsiouliklis, K., Lawrence, S., Pennock, D., and Flake, G. (2002): Using Web Structure for Classifying and Describing Web Pages. In Proc. 11th WWW, pages 562–569.
7. Haveliwala, T. (2002): "Topic-sensitive PageRank", in Proceedings of the eleventh international conference on World Wide Web, pp. 517-526, ACM Press.
8. Jin, R., Hauptmann, A. G., and Zhai, C. (2002): Title Language Model for Information Retrieval. In Proc. of the 25th Annual International ACM SIGIR Conference on Research and Development in Information Retrieval, pages 42-48, Tampere, Finland, ACM.
9. Kao, H.-Y. and Lin, S.-H. (2004): Mining Web Information Structure and Content Based on Entropy Analysis. IEEE Transactions on Knowledge and Data Engineering, Vol 16, No. 1.
10. Kleinberg, J. M. (1998): Authoritative Sources in a Hyperlinked Environment. Proceedings of the Ninth Annual ACM-SIAM Symposium on Discrete Algorithms.
11. Kraft, R., and Zien, J. (2003): Mining Anchor Text for Query Refinement, in Proceeding of the Thirteenth International Conference on World Wide Web, New York, USA, May 17-22.
12. Richardson, M. and Domingos, P. (2002): The Intelligent Surfer: Probabilistic Combination of Link and Content Information in Pagerank. In Advances in Neural Information Processing Systems, Cambridge, MA, MIT press pp. 1441-1448.
13. Westerveld, T., Kraaij, W., and Hiemstra, D. (2001): Retrieving Web Pages Using Content, Links, URLs and Anchors, In Voorhees and Harman, pages 52-61.

# Multilingual Relevant Sentence Detection Using Reference Corpus

Ming-Hung Hsu, Ming-Feng Tsai, and Hsin-Hsi Chen

Department of Computer Science and Information Engineering
National Taiwan University, Taipei, Taiwan
{mhhsu, mftsai}@nlg.csie.ntu.edu.tw
hhchen@csie.ntu.edu.tw

**Abstract.** IR with reference corpus is one approach when dealing with relevant sentences detection, which takes the result of IR as the representation of query (sentence). Lack of information and language difference are two major issues in relevant detection among multilingual sentences. This paper refers to a parallel corpus for information expansion and translation, and introduces different representations, i.e. sentence-vector, document-vector and term-vector. Both sentence-aligned and document-aligned corpora, i.e., Sinorama corpus and HKSAR corpus, are used. The factors of aligning granularity, the corpus domain, the corpus size, the language basis, and the term selection strategy are addressed. The experiment results show that MRR 0.839 is achieved for similarity computation between multilingual sentences when larger finer grain parallel corpus of the same domain as test data is adopted. Generally speaking, the sentence-vector approach is superior to the term-vector approach when sentence-aligned corpus is employed. The document-vector approach is better than the term-vector approach if document-aligned corpus is used. Considering the language issue, Chinese basis is more suitable to English basis in our experiments. We also employ the translated TREC novelty test bed to evaluate the overall performance. The experimental results show that multilingual relevance detection has 80% of the performance of monolingual relevance detection. That indicates the feasibility of IR with reference corpus approach in relevant sentence detection.

## 1 Introduction

Relevance detection on sentence level aims to identify relevant sentences from a collection of sentence set given a specific topic specification. Since it is an elementary task in some emerging applications like multi-document summarization and question-answering, it has attracted many researchers' attentions recently. The challenging issue behind sentence relevance detection is: the surface information that can be employed to detect relevance is much fewer than that in document relevance detection. TREC (Harman, 2002) organized a relevance/novelty detection track starting from 2002 to evaluate the technological development of this challenging problem.

S. H. Myaeng et al. (Eds.): AIRS 2004, LNCS 3411, pp. 165–177, 2005.

In the past, several approaches were proposed to identify sentence relevancy. Word matching and thesaurus expansion were adopted to recognize if two sentences touched on the same subject in multi-document summarization (Chen, *et al.*, 2003). Such an approach has been employed to detect relevance between a topic description and a sentence (Tsai and Chen, 2002). Zhang *et al.* (2002) employed an Okapi system to retrieve relevant sentences with queries formed by topic descriptions. Allan *et al.* (2003) focused on the novelty detection algorithms and showed how the performance of relevant detection affects that of novelty detection. Instead of using an IR system to detect relevance of sentences directly, a reference corpus approach has been proposed (Chen, Tsai and Hsu, 2004). In this approach, a sentence is considered as a query to a reference corpus, and two sentences are regarded as similar if they are related to the similar document lists returned by IR systems.

The above approaches focus on monolingual relevance sentence detection only. As all know, large scale multilingual data have been disseminated very quickly via Internet. How to extend the applications to multilingual information access is very important. Chen, Kuo and Su (2003) touched on multilingual multidocument summarization. To measure the similarities between two bilingual sentences is their major concern.

This paper extends the reference corpus approach (Chen, Tsai and Hsu, 2004) to identify relevant sentences in different languages. The computation of the similarities between an English sentence and a Chinese sentence, which is the kernel of multilingual relevant sentence detection, will be studied by referencing sentence-aligned and document-aligned parallel corpora. Section 2 introduces the basic concepts of the reference corpus approach, and a multilingual Okapi-based IR system used in our experiments. Section 3 shows its extension to multilingual relevance detection. Section 4 presents our reference corpora and evaluation criteria. Section 5 shows and discusses the experimental results. Section 6 further compares the performance differences between monolingual and multilingual relevance detection on TREC evaluation data. Section 7 concludes the remarks.

## 2   Relevance Detection Using Reference Corpus

To use a similarity function to measure if a sentence is on topic is similar to the function of an IR system. We use a reference corpus, and regard a topic and a sentence as queries to the reference corpus. An IR system retrieves documents from the reference corpus for these two queries. Each retrieved document is assigned a relevant weight by the IR system. In this way, a topic and a sentence can be in terms of two weighting document vectors. Cosine function measures their similarity, and the sentence with similarity score larger than a threshold is selected. The issues behind the IR with reference corpus approach include the reference corpus, the performance of an IR system, the number of documents consulted, the similarity threshold, and the number of relevant sentences extracted.

The reference corpus should be large enough to cover different themes for references. Chen, Tsai, and Hsu (2004) consider TREC-6 text collection as a reference corpus. Two IR systems, i.e., Smart and Okapi, were adopted to measure the effects of the performance of an IR system. Their experimental results show that Okapi-based relevance detector outperforms Smart-based one. Thus Okapi system is adopted in the latter experiments.

We modify Okapi-Pack[1] from City University (London) to support Chinese information retrieval in the following way. A Chinese word-segmentation system is used for finding word boundaries. Unknown words may be segmented into a sequence of single Chinese characters. While indexing, Okapi will merge continuous single characters into a word and treat it as an index term. We build a single-character word list to avoid merging a single-character word into an unknown word. Chinese stop word list is not adopted.

We adopted NTCIR3 Chinese test collection (Chen, et al., 2003) to evaluate the performance of Chinese Okapi system (called *C-Okapi* hereafter). Table 1 summarizes the performance of C-Okapi comparing to the results of the participants in NTCIR3 (Chen, et al., 2003). The first column denotes different query construction methods, where T, C, D, and N denote topic, concept, description, and narrative, respectively. The $2^{nd}$-$4^{th}$ columns, i.e., AVG, MAX, and MIN, denote the average, the maximum, and the minimum performance, respectively. C-Okapi outperforms or competes with the maximum one in T and C methods, and is above the average in the other two query construction methods. In the later experiments, we will adopt Okapi and C-Okapi for bilingual relevance detection.

**Table 1.** Performance of C-Okapi

| Topic Field | AVG | MAX | MIN | C-Okapi |
|---|---|---|---|---|
| C | 0.2605 | 0.2929 | 0.2403 | 0.2822 |
| T | 0.2467 | 0.2467 | 0.2467 | 0.2777 |
| TC | 0.3109 | 0.3780 | 0.2389 | 0.3138 |
| TDNC | 0.3161 | 0.4165 | 0.0862 | 0.3160 |

# 3   Similarity Computation Between Multilingual Sentences

In Section 2, we consult a monolingual corpus to determine the similarity between two sentences in the same language. When this approach is extended to deal with multilingual relevance detection, a parallel corpus is used instead. This corpus may be document-aligned or sentence-aligned. Figure 1 shows the overall procedure. English and Chinese sentences, which are regarded as queries to a parallel corpus, are sent to Okapi and C-Okapi, respectively. Total $R$ English and Chinese documents/sentences[2] accompanying with the relevance weights are retrieved for English and Chinese que-

---

[1] http://www.soi.city.ac.uk/~andym/OKAPI-PACK/
[2] The word documents/sentences denotes either document-aligned or sentence-aligned corpus is used.

ries. Because the corpus is aligned, the returned document (or sentence) IDs are comparable. Cosine function is used to compute the similarity, and thus the degree of relevance.

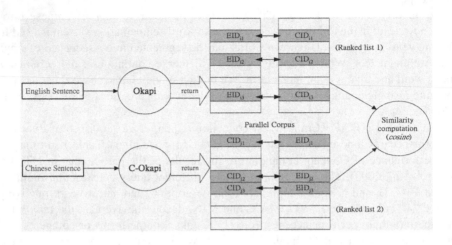

**Fig. 1.** Document-Vector/Sentence-Vector Approach

In the above, two sentences are considered as relevant if they have similar behaviors on the results returned by IR systems. The results may be ranked list of documents or sentences depending on the aligning granularity of the parallel corpus. Besides the *document-vector/sentence-vector approach* shown in Figure 1, the two vectors used in similarity computation may be in terms of relevant terms. This idea follows the corpus-based approach to query translation (Davis and Dunning, 1995) in cross language information retrieval (CLIR). In CLIR, a query in language $A$ is submitted to an $A$-$B$ parallel corpus. An IR system for language $A$ selects the relevant documents in $A$. The documents in language $B$ are also reported at the same time. The target query is composed of terms selected from the relevant documents in $B$, and finally submitted to IR system for $B$ language.

The above procedure is considered as *translation* in CLIR. Now, the idea is extended and plays the roles of both *translation* and *information expansion*. Figure 2 shows the overall flow. Similarly, an English sentence and a Chinese sentence, which will be determined relevancy, are sent to the two IR systems. $R$ most relevant documents/sentences in two languages are returned. Instead of using the retrieval results directly, we select $K$ most representative terms from the resulting documents/sentences. The two sets of $K$ terms form two vectors, so that this approach is called *term-vector approach* later. Cosine function determines the degree of relevance between the English and the Chinese sentences.

Because the $R$ most relevant documents/sentences are in two languages, we can consider either English or Chinese documents/sentences as a basis. In other words, if we select Chinese, then we map ranked list 1 (i.e., English results) into Chinese correspondent through the document-aligned/sentence-aligned chains. Similarly, ranked list 2 (i.e., Chinese results) may be mapped into English correspondent when English part is selected as a basis. Now we consider how to select the *K* most representative terms. Two alternatives shown below are adopted.

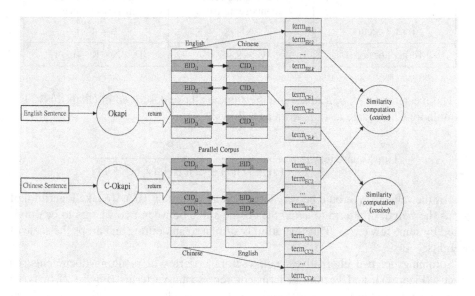

**Fig. 2.** Term-Vector Approach

### 3.1 Weighting Scheme: Okapi-FN1

An intuitive weighting scheme is the weighting function of IR system. The weighting function of a term *t* in Okapi is as follows:

$$W(t) = \log \frac{(r+0.5)(N-R-n+r+0.5)}{(R-r+0.5)(n-r+0.5)} \,. \tag{1}$$

where *N* is total number of documents/sentences in the reference parallel corpus, *R* is the number of relevant documents/sentences to a query, *n* is the number of documents/sentences in which term *t* occurs, and *r* is the number of relevant documents/sentences in which term *t* occurs.

In our experiments, top *R* (different values of *R* are tested) documents in the ranked list are parsed and the total occurrences *r* of a term *t* in the *R* documents are counted. Terms with the top *K* weights are employed for similarity computation.

### 3.2 Weighting Scheme: Log-Chi-Square

The *Chi-Square* test is used to find the terms highly relevant to the returned documents/sentences. Besides, Chi-Square test is also considered as a basis for weighting. A 2×2 contingency table shown in Table 2 is conducted for Chi-Square test.

**Table 2.** Contingency Table for Chi-Square Test

|  | Relevant documents/sentences | Non-relevant documents/sentences |
|---|---|---|
| Term $t$ occurs | $A = r$ | $B = n - r$ |
| Term $t$ not occur | $C = R - r$ | $D = N - R - (n - r)$ |

The meanings of $N$, $n$, $R$, and $r$ are the same as those described in Okapi-FN1. The formula for Chi-Square test is shown as follows:

$$\text{Chi-Square test } \chi^2 = \frac{N(AD - BC)^2}{(A + B)(A + C)(B + D)(C + D)} . \quad (2)$$

For the value of $\chi^2$ could be very large (even larger than $10^6$), we take logarithm of $\chi^2$ as the weight of a term to avoid the cosine value between two vectors to be dominated by some few terms. This operation is similar to smoothing and drops the scale of weights.

Summing up, two alternatives, i.e., vectors in terms of resulting documents/sentences (Figure 1), and vectors in terms of representative terms (Figure 2), may be considered for similarity computation in multilingual relevance detection. In the latter case, either English part or Chinese part may be considered as a basis, and each has two possible weight schemes, i.e., Okapi-FN1 and Log-Chi-Square. Thus, four possible combinations are conducted in total for the latter experiments.

## 4   Experiment Materials and Evaluation Method

Two Chinese-English aligned corpora are referenced in our experiments. One is Sinorama corpus[3], and the other one is HKSAR Corpus[4]. Sinorama consists of documents published by Sinorama magazine within 1976-2001. This magazine, which is famous for her superior Chinese-English contrast, recorded Taiwan society's various dimensions of evolvements and changes. HKSAR collects news articles released by the Information Services Department of Hong Kong Special Administrative Region (HKSAR) of the People's Republic of China. The following compares these two corpora from corpus scale, aligning granularity, average length, and so on.

Sinorama is a "sentence-aligned" parallel corpus, consisting of 50,249 pairs of Chinese and English sentences. We randomly select 500 Chinese-English pairs as test data to simulate multilingual relevance sentence detection. The remaining 49,749 pairs are considered as a parallel reference corpus. They are indexed separately as two monolingual databases, in which a Chinese sentence or an English sentence is regarded

---

[3] http://rocling.iis.sinica.edu.tw/ROCLING/corpus98/mag.htm
[4] http://wave.ldc.upenn.edu/Catalog/CatalogEntry.jsp?catalogId=LDC2000T46

as a "small document". The average length of Chinese sentences in the reference corpus is 151 bytes and that of English sentences is 254 bytes. The average length of Chinese and English test sentences is 146 and 251 bytes, respectively.

HKSAR corpus contains 18,147 pairs of aligned Chinese-English documents released by HKSAR from July 1, 1997 to April 30th, 2000. Similarly, we index all articles in the same language as a monolingual database. The average document length is 1,570 bytes in Chinese and 2,193 bytes in English. The test data used in experiments are the same sentences pairs as Sinorama.

At first, we develop an evaluation method and a set of experiments to measure the kernel operation of relevance detection only, i.e., the similarity computation between Chinese and English sentences, in Section 5. Then, we measure the overall performance of multilingual relevance detection in Section 6. As mentioned, total 500 pairs of Chinese-English sentences are randomly selected from Sinorama corpus. They are denoted as: $<C_1, E_1>, <C_2, E_2>, ..., <C_{500}, E_{500}>$, where $C_i$ and $E_i$ stand for Chinese and English sentences, respectively. Among the 500 Chinese sentences $C_1, ..., C_{500}$, $C_i$ is the most relevant to $E_i$. In other words, when we compute the similarities of all combinations consisting of one Chinese and one English sentences, $C_i$ should be the most similar to $E_i$ for $1 \leq i \leq 500$ ideally. Let $Sim(i, j)$ be the similarity function between $C_i$ and $E_j$. A match function $RM(i, j)$ is defined as follows:

$$RM(i, j) = |\{k| \, Sim(i, k) > Sim(i, j), \, 1 \leq k \leq 500\}| + 1 \,. \tag{3}$$

The match function assigns a rank to each combination. The perfect case is $RM(i, i)=1$. We call it a *perfect match* later. We also relax the case. If $RM(i, i)$ is no larger than a threshold, we consider the result of matching is "good". In our experiments, the threshold is set to 10. That is, we postulate that the first 2% of matching pairs will cover the correct matching.

Consulting the evaluation method in question answering track of TREC, we adopt **MRR** (mean reciprocal rank) score to measure the performance. Let $S(i)$ be the evaluation score for a topic $i$ (Chinese sentence). MRR is summation of $S(i)$.

$$S(i) = \begin{cases} 1 \, / \, M(i, i) & \text{if } RM(i, i) \leq 10 \\ 0 & \text{else} \end{cases} \,. \tag{4}$$

$$MRR = \frac{1}{500} \sum\nolimits_{i=1}^{500} S(i) \,. \tag{5}$$

## 5  Result Discussion

### 5.1  Using Sinorama Corpus

**Sentence Vector Approach.** Table 3 shows the experimental result of sentence-vector approach along with Sinorama corpus. Row "$RM(i, i)=1$" denotes how many topics get a "perfect match" and row "$RM(i, i) \leq 5$" denotes how manytopics get a correct match in the first 5 ranks. For example, 77.40% of test data are perfect match if 200 sentences are consulted by Okapi and C-Okapi, i.e., 200 sentences are returned for reference. In this case, the MRR is 0.839, which is the best in this experiment. When the number of returned sentences increases from 50 to 200, MRR score also increases. Then MRR score goes down until the number of returned sentences reaches about 600. After that, MRR score rises again and reaches to a stable state, i.e., 0.82-0.83. Figure 3 captures the performance change.

Analyzing the result, we find there may be two degrees of relevancy of small documents (i.e., sentences) in the corpus to a query. Documents with high relevance are easily retrieved with ranks smaller than 200. When the rank increases larger than 200, lowly relevant documents are retrieved with more non-relevant documents. That introduces noise for similarity computation. The influence reaches to the worst between ranks 500 and 600, and then goes down since the weights of vector elements are decreased. The other reason may be that some returned sentences in both vectors are complementary when smaller number of sentences is consulted, and the complementary parts show up when more sentences are consulted.

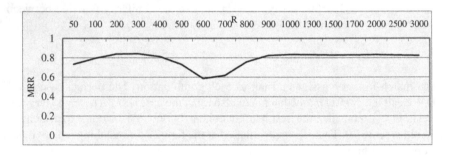

**Fig. 3.** MRR Score versus Number of Returned Sentences

**Term-Vector Approach.** Figures 4, 5, 6 and 7 show the results of term-vector approach, where terms in either English or Chinese are used, and two weighting schemes, i.e., Okapi-FN1 and Log-Chi-Square, are applied. The $x$ axis represents $k$, the number of terms used for similarity computation. The $y$ axis denotes the MRR score.

Several interesting conclusions can be made after the factors of language and weighting schemes are considered. Performances of Figures 4 and 5 are inferior to those of Figures 6 and 7. It shows that Log-Chi-Square weighting scheme is more suitable for term-vector approach than Okapi-FN1 weighting scheme. It meets our expectation that Log-Chi-Square weighting scheme properly captures concepts embedded in resulting sentences returned by Okapi and C-Okapi. Performances of the runs of smaller $k$ (=50) in the four figures show that Log-Chi-Square scheme will assign higher weights to terms which are truly relevant to the sentence (query).

Observing the differences between Figures 4 and 5, and between Figures 6 and 7, we can find that the trends of performances using terms in different languages are dissimilar. Using English terms as vector elements, the performance trend shows undulation as we saw in Figure 2, though the drops are smaller in Figures 4 and 6. On the other hand, performance trend of using Chinese terms as vector elements is monotonously increasing with $k$ when $R$ is greater than 30. It may indicate that English suffers from more noises, such as word sense ambiguity, than Chinese.

The best performance, near 0.81, appears in the case "$R=300$" of Figure 7, i.e., take Chinese as a basis and Log-Chi-Square formula. It is lower than the best performance 0.84 in sentence-vector approach. The whole performance of term-vector approach is also inferior to sentence-vector approach.

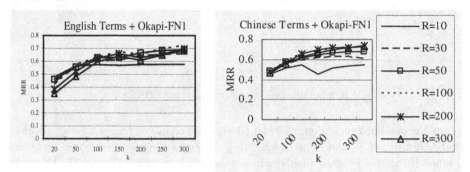

**Figs. 4 and 5.** English Terms or Chinese Terms plus Okapi-FN1 Weighting Scheme

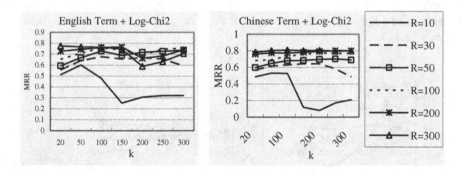

**Figs. 6 and 7.** English Terms or Chinese Terms plus Log-Chi-Square Weighting Scheme

## 5.2  Using HKSAR Corpus

**Document-Vector Approach.** Figure 8 shows the results of the application of the document-vector approach on HKSAR corpus, which is a document-aligned Chinese-English corpus. The best one has only 30% of the performance shown in Figure 3. The result shows the influence of corpus domain on reference corpus approach. Since the 500 pairs of test sentences are randomly selected from Sinorama corpus, the domain of test sentences and the reference databases are the same, i.e., content focused on major events and construction in Taiwan from 1976-2001. In contrast, the HKSAR corpus contains the news issued by HKSAR within 1997-2000. The test sentences and the reference corpus are totally different in domain of concepts so that there are rarely relevant documents in HKSAR. That introduces much more noises than useful information in ranked list. Besides the domain issue, the small size of HKSAR corpus results in poor performance in retrieval too.

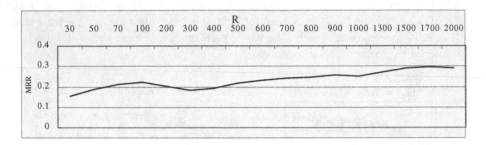

**Fig. 8** Document-Vector Approach Using HKSAR

**Term-Vector Approach.** Figures 9 and 10 show the results of term vector approach on HKSAR, using Log-Chi-Square weighting scheme. Chinese-term-based approach (Figure 10) is more robust than English-term-based approach (Figure 9). However, their performance does not compete with that of document-vector approach. As HKSAR is a "document-aligned" parallel corpus, it is more difficult to select terms suitable for information expansion. Thus, the performance goes down from Figure 8 to Figure 9 and Figure 10. The performance drop is more obvious than that between Figures 3 and 7.

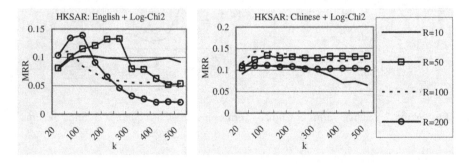

**Figs. 9 and 10.** English and Chinese Terms plus Log-Chi-Square Weighting with HKSAR

# 6   Experiments of Multilingual Relevance Detection

Besides evaluating the similarity computation, we also employ the test data in TREC 2002 Novelty track to evaluate the overall multilingual relevance detection. The test data includes 49 topics, each of which is given a set of sentences to evaluate the performance of relevance detector (Harman, 2002). All of these topics and sentences are in English. For multilingual relevant sentence detection, all topics are manually translated into Chinese. Each translated topic (in Chinese) and the corresponding given set of sentences (in English) are sent as queries to C-Okapi and Okapi respectively, so that we can compute similarity between each topic and each sentence in the given set using either document-vector or term-vector approach.

Chen, Tsai, and Hsu (2004) use logarithmic regression to simulate the relationship between total number of the given sentences and number of the relevant sentences, in TREC 2002 Novelty track. We adopt the similar approach. A dynamic percentage of sentences most similar to topic $t$ in the given set will be reported as relevant. According to the assessment of TREC 2002 Novelty track, we can compute precision, recall, and F-measure for each topic. Figure 11 shows the performance, i.e., average F-measure of 49 topics, using Sinorama and HKSAR as reference corpora, respectively. Sentence-vector approach (Section 5.1.1) and document-vector approach (Section 5.2.1) are adopted.

Apparently, using Sinorama as a reference corpus outperforms using HKSAR. This result is consistent with the evaluation in similarity computation. Chen, Tsai, and Hsu (2004) used TREC6 text collection, which consists of 556,077 documents, as reference corpus. The best performance using Sinorama for multilingual relevance detection is about 80% of monolingual relevance detection, i.e., 0.212 (Chen, Tsai, and Hsu), and using HKSAR is about 50%. Note that the human performance in monolingual relevance detection is 0.371.

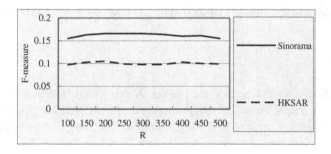

**Fig. 11.** Sentence/Document-Vector Approach Using TREC 2002 Topics

## 7   Conclusions and Future Work

This paper considers the kernel operation in multilingual relevant sentence detection. A parallel reference corpus approach is adopted. The issues of aligning granularity, the corpus domain, the corpus size, the language basis, and the term selection strategy are addressed. In the intensive experiments, the best MRR (0.839) is achieved when the test data and the reference corpus come from the same domain, the finer-grained alignment (i.e., sentence alignment), and the larger corpus are adopted. In that case, 77.40% of test data are ranked 1.

Generally speaking, the sentence-vector approach is superior to the term-vector approach when sentence-aligned corpus is employed. The document-vector approach is better than the term-vector approach if document-aligned corpus is used. In term-vector approach, Log-Chi-Square weighting scheme is better than Okapi-FN1 weighting scheme. Considering the language issue, Chinese basis is more suitable to English basis in our experiments. It shows that performance trends may depend on the characteristics of different languages.

Comparing the monolingual and the multilingual relevance detection, the latter has 80% performance of the former. It shows that IR with reference corpus approach is adapted easily to multilingual domain.

From the experiment results, we infer that if the domain of reference corpus is the same as that of query, the performance of relevance detection will be better. While the domain of a query is often unknown, large domain-coverage corpora should be more appropriate than small ones. More, we can infer that the finer-grained alignment corpus is more suitable for multilingual relevant sentence detection. In future work, we'll design more careful experiments to verify the two points and to find out other characteristics of IR with reference corpus approach.

## Acknowledgements

Research of this paper was partially supported by National Science Council, Taiwan, under the contracts NSC 93-2213-E-002-078 and NSC 93-2752-E-001-001-PAE.

## References

[1]  Allan, James; Wade, Courtney and Bolivar, Alvaro (2003) "Retrieval and Novelty Detection at the Sentence Level," *Proceedings of the 26th ACM SIGIR* , 314-321.

[2]  Chen, Hsin-Hsi; Kuo, June-Jei; Huang, Sheng-Jie; Lin, Chuan-Jie and Wung, Hung-Chia (2003) "A Summarization System for Chinese News from Multiple Sources," *Journal of American Society for Information Science and Technology*, **54**(13), 1224-1236.

[3]  Chen, Hsin-Hsi; Kuo, June-Jei; and Su, Tsei-Chun (2003). "Clustering and Visualization in a Multi-Lingual Multi-Document Summarization System," *Proceedings of 25th European Conference on Information Retrieval Research*, LNCS 2633, 266-280.

[4]  Chen, Hsin-Hsi; Tsai, Ming-Feng and Hsu, Ming-Hung (2004) "Identification of Relevant and Novel Sentences Using Reference Corpus," *Proceedings of 26th European Conference on IR Research*, LNCS 2997, 85-98.

[5]  Chen, Kuang-hua; Chen, Hsin-Hsi; Kando, Noriko; Kuriyama, Kazuko; Lee, Sukhoon; Myaeng, Sung Hyon; Kishida, Kazuaki; Eguchi, Koji and Kim, Hyeon (2003) "Overview of CLIR Task at the Third NTCIR Workshop," *Proceedings of NTCIR Workshop 3*.

[6]  Davis, Mark and Dunning, Ted (1995) "A TREC Evaluation of Query Translation Methods for Multi-lingual Text Retrieval," *Proceedings of TREC 1995*, 483-498.

[7]   Harman, Donna (2002) "Overview of the TREC 2002 Novelty Track," *Proceedings of TREC 2002*, NIST SP 500-251, Maryland.

[8]   Tsai, Ming-Feng and Chen, Hsin-Hsi (2002) "Some Similarity Computation Methods in Novelty Detection," *Proceedings of TREC 2002*, NIST SP 500-251, Maryland.

[9]   Zhang, Min; Song, Ruihua; Lin, Chuan; Ma, Shaoping; Jiang, Zhe; Yijiang Jin; Liu Yiqun; and Zhao, Le (2002) "Expansion-Based Technologies in Finding Relevant and New Information," NIST SP 500-251, Maryland.

[10]  http://trec.nist.gov/pubs/trec11/papers/NOVELTY.OVER.pdf

[11]  http://trec.nist.gov/pubs/trec11/papers/ntu.feng.final2.pdf.

[12]  http://trec.nist.gov/pubs/trec11/papers/tsinghuau.novelty2.pdf

# A Bootstrapping Approach for Geographic Named Entity Annotation

Seungwoo Lee and Gary Geunbae Lee

Department of Computer Science and Engineering,
Pohang University of Science and Technology,
San 31, Hyoja-dong, Nam-gu, Pohang, 790-784, Korea
{pinesnow, gblee}@postech.ac.kr

**Abstract.** Geographic named entities can be classified into many sub-
types that are useful for applications such as information extraction
and question answering. In this paper, we present a bootstrapping algorithm for the task of geographic named entity annotation. In the initial
stage, we annotate a raw corpus using seeds. From the initial annotation, boundary patterns are learned and applied to the corpus again
to annotate new candidates. Type verification is adopted to reduce over-
generation. One sense per discourse principle increases positive instances
and also corrects mistaken annotations. As the bootstrapping loop proceeds, the annotated instances are increased gradually and the learned
boundary patterns become gradually richer.

## 1 Introduction

The traditional named entity task can provide semantic information of key entities, and is taken as a first step by any information extraction and question
answering system. However, classifying entity names only into persons, locations and organizations (PLO) is not sufficient to answer a question *"What is
the oldest city in the United States?"* in the TREC-10 QA task [1]. Although the
answer type of the question can be determined as CITY, if location entities are
not classified into their specific sub-types, such as CITY, STATE, COUNTRY,
etc., it is impossible to filter out *'Florida'*, a state, from the text *"... Florida is
the oldest one in the United States ..."* By sub-categorizing location entities we
can effectively reduce the candidate answer space of the question and help to
point out the exact answer.

In this paper, we focus on geographic named entities (i.e., locations). Geographic named entities can be classified into many sub-types that are critical for
applications such as information extraction and question answering. As a first
step, we define their seven sub-types: COUNTRY, STATE, COUNTY, CITY,
MOUNTAIN, RIVER, and ISLAND. We attempt to identify and classify all
instances of the seven types in plain text.

Annotation of geographic named entities is a formidable task. Geographic
named entities are frequently shared with person names as well as between their

S. H. Myaeng et al. (Eds.): AIRS 2004, LNCS 3411, pp. 178–189, 2005.

sub-types. For example, *'Washington'* may indicate a person in one context but may also mean a city or state in another context. Even country names cannot be exceptions. For some Americans, *'China'* and *'Canada'* may be cities where they live. Geographic named entities such as *'Man'* and *'Center'* can also be shared with common nouns. Contextual similarity among geographic named entities is much higher than between PLO entities since they are much closer semantically. These make geographic named entity annotation task more difficult than the traditional named entity task.

The statistical machine learning approach is a current trend in linguistic area and some supervised learning approaches have achieved comparable or better performance than handcrafted rule-based approaches, especially in the named entity task [2]. However, Manual corpus annotation is tedious and time-consuming work and requires considerable human effort. Domain shift may render the existing annotated corpus useless even if it is available. On the other hand, there are many raw text corpora available for training and geographic entities for a gazetteer are also available in several web sites. These lead us to apply a bootstrapping approach using gazetteer entries as seeds. Gazetteer entries can be found in several web sites, such as the U.S. Census Bureau,[1] the Consortium for Lexical Research (CRL),[2] TimeAndDate,[3] Probert Encyclopaedia - Gazetteer,[4] The World of Islands[5] and Island Directory,[6] and we gathered entities with their types to use both as seeds and for confirming types of candidates to be annotated in the bootstrapping process.

The key idea of our bootstrapping process is as follows: in the initial stage, we annotate a raw corpus using seeds. From the initial annotation, boundary patterns are learned and applied to the corpus again to annotate new candidates of each type. Type verification is adopted to reduce over-generation, and one sense per discourse principle [3] expands the annotated instances in each document. As the bootstrapping loop continues, the annotated instances are increased gradually and also the learned boundary patterns become richer little by little.

The remainder of this paper is as follows. Section 2 presents and compares related works to our approach. Our bootstrapping approach is described in Section 3 and Section 4 gives the experimental results. Section 5 contains our remarks and future works.

## 2    Related Works

Research on analysis of geographic references recently started to appear and focused on only classifying geographic entity instances in text [4, 5]. Li et al. [4]

---

[1] http://www.census.gov/
[2] Computing    Research    Laboratory    at    New    Mexico    State    University, http://crl.nmsu.edu/Resources/resource.htm
[3] http://www.timeanddate.com/
[4] http://www.probertencyclopaedia.com/places.htm
[5] http://www.world-of-islands.com/
[6] http://islands.unep.ch/isldir.htm

suggested a hybrid approach to classify geographic entities already identified as location by an existing named entity tagger. They first matched local context patterns and then used a maximum spanning tree search for discourse analysis. They also applied a default sense heuristic as well as one sense per discourse principle. According to their experiments, the default sense heuristic showed the highest contribution.

Various unsupervised learning methods have been developed in the area of named entity recognition. These approaches relate to bootstrapping learning algorithms using natural redundancy in the data, with the help of the initial seeds. Collins and Singer [6] applied the co-training algorithm to named entity classification. They used 7 simple seed rules to extract lexical and contextual features of named entities as two competing evidences, and learned the capitalized names from the fully parsed corpus. They only focused on named entity classification by assuming consecutive proper nouns as named entity candidates. Yangarber et al. [7] presented a bootstrapping technique to learn generalized names such as disease names that do not provide capitalization any longer. This makes the identification of such names more difficult. They used manually selected seeds, and boundary patterns and entity names acted as competing evidences in the bootstrapping process. The patterns for beginning and ending boundaries are independently learned from currently annotated names and applied to the raw corpus in order to obtain new candidate names. This bootstrapping process is quite similar to our method, but annotation targets differ between each approach. We focus on the fine-grained sub-types of geographic entities whereas they deal with generalized names as well as locations. We also represent boundary patterns with character patterns and semantic information as well as lexical words, while they only use literal patterns with wildcard generalization. Their boundary patterns are used for identifying only one (left or right) boundary and the other is identified by a simple noun group matching [ADJ* N+]. But there are many instances that go beyond the boundary of the simple noun group. For example, *"Antigua and Barbuda"* and *"Bosnia and Herzegovina"* have a conjunction but indicate unique country names. Therefore, we do not restrict the entity boundary to a simple noun group.

## 3     Bootstrapping

Our bootstrapping approach is based on two competing evidences: entity instances and their boundary patterns, as in other bootstrapping approaches [6, 7, 8].

The bootstrapping flow has one initial stage and four iterative steps, as shown in Figure 1. In the initial stage, we annotate a raw corpus with the seeds automatically obtained from the gazetteer. Starting and ending boundary patterns are learned from the annotation and applied to the corpus again to obtain new candidates of each type. Then we perform type verification of each candidate entity using the gazetteer and scores of boundary patterns applied. Finally, one sense per discourse principle [3] propagates the annotated instances to the entire document, which makes the boundary patterns much richer in the next loop.

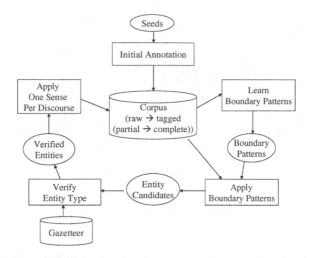

**Fig. 1.** Our bootstrapping process

The principle also helps to remove erroneous annotated instances. As the bootstrapping loop proceeds, the annotated instances are increased gradually and the learned boundary patterns also become richer little by little. These processes are explained in detail in the following subsections.

### 3.1 Initial Stage

Seed entities of each type can be obtained from the gazetteer. The gazetteer includes many ambiguous entities that should not be used as seeds. Ambiguous seeds can act as noise in the initial annotation and such noise can propagate to the next iterations. As a result, they diminish the final performance. Therefore, we automatically select only unambiguous entities by consulting the gazetteer as well as a part-of-speech dictionary.[7] Seed annotation also requires careful attention. Not all occurrences of unambiguous seeds in a corpus can be annotated as their types. For example, 'U.S.' in 'U.S. Navy' should not be annotated as COUNTRY since it is part of an organization name. For this reason, we annotate only occurrences having no proper nouns in their immediate left and right contexts. These efforts make it possible to start bootstrapping iteration with high precision (99%) and moderate recall (28%) on our test data.

### 3.2 Boundary Patterns

**Definition.** For each annotated instance, we define two boundary patterns: starting ($bp_s$) and ending ($bp_e$) independently, i.e.,

$$bp_s = [f(w_{-2})f(w_{-1})f(w_{-0})] \text{ and } bp_e = [f(w_{+0})f(w_{+1})f(w_{+2})],$$

---

[7] We also exclude entities, such as 'Man' and 'Center', shared with common nouns.

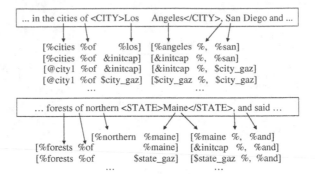

**Fig. 2.** Boundary Patterns

where $w_{\mp 0}$ are the first and the last words of an annotated instance, respectively, and $w_{\mp 1}$ and $w_{\mp 2}$ are the left and the right contexts of an annotated instance. The context is determined by a rule-based noun phrase chunking we implemented and can be grouped into two: *intra-phrase* and *inter-phrase* contexts. *Intra-phrase* context indicates a modifier or modifiee of the entity, within a base noun phrase containing the entity, and *inter-phrase* context means a modifier or modifiee of the entity, outside the base noun phrase. These can make boundary patterns richer than simple adjacent word contexts do because they reflect syntactic constraints. $f(w)$ means one of following four features: character pattern (marked with &), such as capitalization and digits, word forms (marked with %), semantic category (marked with @) based on the WordNet hierarchy,[8] and a gazetteer category (marked with $) based on the gazetteer. These features are similar to *internal sub-features* of Zhou and Su [2]. Figure 2 shows some possible starting and ending boundary patterns of entities, *'Los Angeles'* and *'Maine'*. [%northern %maine] shows *intra-phrase* context and others is *inter-phrase* contexts. The left and the right contexts have windows of maximum 2 but it is not a strict word count in the case of semantic and gazetteer category features. Multi-words like *'San Diego'* and *'vice president'* count only one and can be abstracted with their gazetteer or semantic categories, i.e., *$city_gaz* and *@person1*, respectively.

**Learning.** All possible boundary patterns are generated from the corpus partially annotated in the initial stage or in previous iterations. Then, we score the generated patterns by checking if each pattern correctly bounds the previously annotated instances. We count the followings that are similar to Yangarber et al. [7]:

- *pos(bp)*: the number of matched instances that are already annotated as the same entity type;

---

[8] We use only intermediate nodes for simplification; For example, *Mr.* and *President* are all mapped to the first sense of person (*@person1*).

– $neg(bp)$: the number of matched instances that are already annotated as a different type or that are previously filtered from the same entity type;
– $unk(bp)$: the number of matched instances that are not annotated yet.

Then, the score of a boundary pattern, $Score(bp)$, is computed as follows:

$$Score(bp) = \frac{pos(bp)}{pos(bp) + neg(bp) + unk(bp)} .$$

**Application.** To obtain new entity candidates, we apply the learned patterns to the entire corpus. Since each pattern determines only one – *i.e., starting or ending* – boundary, a candidate is identified by a pair of starting and ending patterns of the same entity type. We limit the length of each candidate to six words since a geographic entity is usually composed of a fairly small number of words. However, the entity boundary is not restricted to a simple noun chunk to allow entities such as *'Bosnia and Herzegovina'* and *'the United States of America'*.

Each candidate has starting and ending scores that are the scores of the boundary patterns used for identifying it. Generally, both the left and the right contexts can provide clues for the entity type, but it is not common for both contexts to provide reliable clues at the same time and usually only one of them supports the clue. This leads us to use a pair of thresholds, top threshold ($\theta_t$) and bottom threshold ($\theta_b$), to select only promising candidates in each iteration. At least one of the starting and ending scores of each candidate should be higher than $\theta_t$ and both should be higher than $\theta_b$.

## 3.3   Type Verification

We do not limit the entity boundary to the base noun phrase chunk in order to cover geographic entities bounded by complex noun phrases such as *'Bosnia and Herzegovina'* and *'the United States of America'*. However, this can also cause erroneous candidates such as *"Boston and Chicago"* to be generated if the left context of *'Boston'* provides a reliable starting boundary score and the right context of *'Chicago'* provides a reliable ending boundary score. Even if we have such a restriction, erroneous candidates can also be extracted by less reliable boundary patterns. They act as noise in the next iterations and less reliable boundary patterns are increased gradually as the iteration proceeds. As a result, this may cause what is called *garbage in and garbage out*. To prevent erroneous candidates from being annotated, we employ type verification of each candidate if it was identified by less reliable patterns (i.e., $Score(bp) < \theta_v$). If we know that the identified candidate is really of its classified entity type, we can be convinced that at least it can be of its entity type in some contexts although it was identified and classified by less reliable boundary patterns. To verify the type of an entity candidate, we first analyze its prefix and suffix and then consult the gazetteer. We add or remove prefixes or suffixes such as *Republic, State, Province, County, City, City of, River, Mt., Island,* etc., according to its entity type before searching the gazetteer, since the gazetteer does not have all possible forms of each entity. As a result, candidates are removed if they are less reliable and also not verified by the gazetteer.

### 3.4     Expanding and Correcting the Annotation

It is common to use the heuristic such as *one sense per discourse* in the word sense disambiguation research [3]. In our research, we set a document as a discourse because we use newspaper articles as a training and test corpus which is described in Section 4. Each article is short enough to be regarded as one discourse. This heuristic is very helpful to expand the current annotation to the entire discourse to obtain a new annotation or to correct a mistaken annotation [9, 10]. That is, it increases positive instances as well as decreases spurious instances effectively.

**Type Consistency.** Error correction is performed when the same entity candidates from a document have different entity types. By the heuristic, they are all annotated with the type of entity having the most reliable boundary score if the difference between the two boundary scores is larger than a threshold. Otherwise, they are all ignored and their annotation is delayed until the next iterations.

**Location vs. Person.** Location names, i.e., geographic named entities, are highly shared with a person's names. This is because it is common to name a location after a famous person's name, such as *'Washington'*. This makes it important to disambiguate locations from a person's names. According to the heuristic, a location candidate can be filtered out if we know that one of its instances within a document indicates a person's name. This is another error correction.

To decide whether an occurrence of a location candidate means a person's name,[9] we first check if it has a title word, such as *Mr.*, *Sen.* and *President*. Second, we check typical patterns, such as *" " PERSON says"*, *", PERSON says"*, *" " says PERSON"*, *", says PERSON"*, *"PERSON who"*, etc. Finally, we consult the gazetteer to see if it and its adjacent proper noun are all registered as a person name. For example, *'George Washington'*, an occurrence of a candidate *'Washington'*, filters out the candidate, since both *'George'* and *'Washington'* are used as a person name, based on the gazetteer.

**Location vs. part of Organization.** Another issue is how to distinguish the location from part of the organization name. This is required when propagating the annotation of an entity to other occurrences within the document by the heuristic. For example, when *'U.S.'* is annotated as COUNTRY in a document, it should not propagate to *'U.S. Navy'* in the same document since it is a part of the organization name, whereas it should propagate to *'U.S. President'*. One possible solution is to check the existence of an acronym for the suspected phrase,

---

[9] This is applied to only single-word candidates.

since it is common to use an acronym to represent a unique entity. *'U.S. Navy'* is represented as the acronym *'USN'* but *'U.S. President'* is not. To check the existence of their acronyms, we consult Web search engines by querying *"U.S. Navy (USN)"* and *"U.S. President (USP)"*, respectively, and check if there exists more than one result retrieved. The following shows some acronym examples that can be found by Web search engines and therefore whose *typewritten* sub-phrases should not be annotated: "United States-China *Policy Foundation (USCPF),*" "New York *Times (NYT)*" and *"University of* California, Los Angeles *(UCLA)*."

Another possible solution is to check if the suspected phrase beginning with a candidate can be modified by a prepositional phrase which is derived by *in* or *comma (,)* plus the candidate. For example, we can decide that *'Beijing'* in *'Beijing University'* is a part of the organization name, since the phrase *"Beijing University in Beijing"* is found by Web search engines. If *'Beijing'* in *'Beijing University'* denotes CITY, *'Beijing University'* indicates *any university in Beijing* and is not modified by the prepositional phrase duplicately. The following shows some other examples whose *typewritten* sub-phrases should not be annotated: "Shanghai *No. 3 Girls' School, Shanghai,*" "Boston *English High School in Boston*" and "Atlanta *Brewing Co. in Atlanta.*"

### 3.5   Iteration

In each iteration, we control two thresholds, $\theta_t$ and $\theta_b$, described in Subsection 3.2, of boundary scores to extract new entity candidates. We set both thresholds high to maintain high precision in the early stages and decrease them slightly to increase recall as the iteration proceeds. We decrease $\theta_b$ at first and $\theta_t$ at next, i.e., $(\theta_t, \theta_b) = (0.90, 0.90) \rightarrow (0.90, 0.85) \rightarrow \cdots \rightarrow (0.90, 0.15) \rightarrow (0.87, 0.15) \rightarrow \cdots \rightarrow (0.51, 0.15)$. This reflects our observation that the most reliable *one* – left or right – context is preferred to moderately reliable *two* – left and right – contexts. The bootstrapping stops when the two thresholds arrive at $(0.51, 0.15)$ and there are no more new boundary patterns learned.

## 4   Experiments

The algorithm was developed and tested using part of New York Times articles (June and July, 1998; 116MB; 21,000 articles) from the AQUAINT corpus. We manually annotated 107 articles among them for test and the counts of annotated instances were listed in Table 1. Others were used for training. We collected 78,000 gazetteer entries from several Web sites mentioned in Section 1. This includes non-target entities (e.g., CONTINENT, OCEAN, PERSON and ORGANIZATION) as well as various aliases of entity names. As a baseline, we could achieve high recall (94%) but poor precision (60%) by applying only the gazetteer without bootstrapping. Seeds were automatically selected from the gazetteer and their counts are also listed in Table 1.

We first performed bootstrapping on the entire training corpus and investigated the change in performance as the iteration proceeds. In each iteration,

**Table 1.** The counts of instances annotated in the test corpus and seeds obtained from the gazetteer

| Type | Test | Seeds |
|------|------|-------|
| COUNTRY | 591 | 308 |
| STATE | 425 | 225 |
| COUNTY | 60 | 2576 |
| CITY | 827 | 21819 |
| ISLAND | 25 | 3978 |
| RIVER | 24 | 2204 |
| MOUNTAIN | 15 | 3228 |
| Total | 1967 | 34338 |

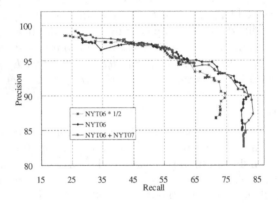

**Fig. 3.** The learning curves according to the size of a training corpus

we applied the learned boundary patterns to the test corpus and measured recall and precision using the standard MUC named entity scoring scheme. The learning curve is shown in Figure 3, marked with `NYT06+NYT07`. The figure says that recall increases (28% to 84%) while precision decreases slightly (99% to 87%). However, after approaching to the maximum recall, precision dropped steeply. The maximum value of F1 was 86.27% when recall and precision were 83% and 90%, respectively. This performance is comparable to that of Lin et al. [11] although we distinguished fine-grained types of locations whereas they did not. It is also comparable to similar approaches in the traditional named entity task, e.g. [6],[10] considering that they only attacked named entity classification and that geographic named entity annotation is much more difficult than PLO-based annotation.

Second, to investigate the effect of the size of a training corpus, we performed bootstrapping on the half of the June corpus (`NYT06*1/2`) and on the

---

[10] They achieved about 83% accuracy in noisy environment.

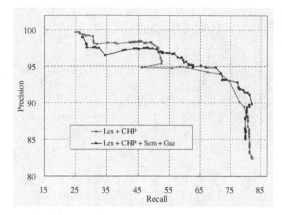

**Fig. 4.** Effects of semantic and gazetteer features

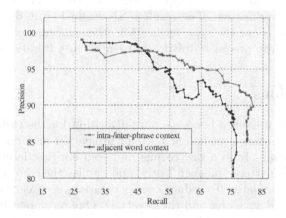

**Fig. 5.** Effects of *intra-/inter-phrase* context

June corpus (NYT06). The learning curves are also shown in Figure 3. The comparison says that the large training corpus can maintain higher precision than the small training corpus until recall approaches the maximum. This is because the large training corpus has more instances to be learned and provides much richer boundary patterns than the small training corpus.

We also explored whether semantic and gazetteer features can contribute to the performance. We performed bootstrapping on the training corpus using boundary patterns represented with only lexical and character patterns. As shown in Figure 4, semantic and gazetteer features could maintain precision a little higher although any techniques for word sense disambiguation were not employed. The effect of pattern generalization seemed to suppress noise caused by word sense ambiguity. Figure 5 explains how well the *intra-/inter-phrase* context works, compared to just adjacent word context.

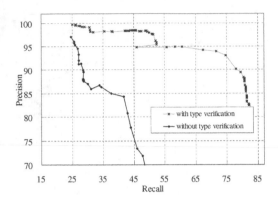

**Fig. 6.** With vs. without type verification

Finally, we performed bootstrapping without type verification to investigate the effect of the verification. The result is shown in Figure 6. The figure says that if we do not adopt the type verification, the erroneous candidates are not filtered out and drop precision drastically even in early iterations.

## 5    Conclusions

In this paper, we present a bootstrapping algorithm for the task of geographic named entity annotation including both identification and classification. The algorithm bootstraps from a raw corpus and seed entities, learns starting and ending boundary patterns independently, and applies them to the corpus again to obtain new candidates. To reduce over-generation, each candidate is verified based on its boundary scores and the gazetteer. One sense per discourse principle [3] expands the annotated instances and also corrects mistaken annotations. As the bootstrapping loop continues, the annotated instances are increased gradually and the learned boundary patterns become gradually richer. When training and testing on NYT newspaper articles, our approach achieved 86.27% F1 with 83% recall and 90% precision.

In future work, we will incorporate our heuristic knowledge used for filtering into a statistical model and include other geographic entity types as well as PERSON and ORGANIZATION. Competing entity types may help to improve overall performance, as described in Lin et al. [11]. Co-occurrence information with the default sense heuristic suggested by Li et al. [4] will also be tried for disambiguating geographic named entities effectively.

## Acknowledgements

This work was supported by 21C Frontier Project on Human-Robot Interface (by MOST).

# References

1. Voorhees, E.M.: Overview of the TREC 2001 Question Answering Track. In: Proceedings of the 10th Text Retrieval Conference (TREC 2001), Gaithersburg, MD (2001) 42–51
2. Zhou, G., Su, J.: Named Entity Recognition using an HMM-based Chunk Tagger. In: Proceedings of the 40th Annual Meeting of the Association for Computational Linguistics (ACL), Philadelphia, USA (2002) 473–480
3. Gale, W.A., Church, K.W., Yarowsky, D.: One Sense Per Discourse. In: Proceedings of the 4th DARPA Speech and Natural Language Workshop. (1992) 233–237
4. Li, H., Srihari, R.K., Niu, C., Li, W.: InfoXtract location normalization: a hybrid approach to geographic references in information extraction. In: Proceedings of the HLT-NAACL 2003 Workshop on Analysis of Geographic References, Alberta, Canada (2003) 39–44
5. Smith, D.A., Mann, G.S.: Bootstrapping toponym classifiers. In: Proceedings of the HLT-NAACL 2003 Workshop on Analysis of Geographic References, Alberta, Canada (2003) 45–49
6. Collins, M., Singer, Y.: Unsupervised Models for Named Entity Classification. In: Proceedings of the Joint SIGDAT Conference on Empirical Methods in Natural Language Processing and Very Large Corpora (EMNLP/VLC), College Park, MD, Association for Computational Linguistics (1999) 100–110
7. Yangarber, R., Lin, W., Grishman, R.: Unsupervised Learning of Generalized Names. In: Proceedings of the 19th International Conference on Computational Linguistics (COLING 2002), Taipei, Taiwan (2002) 1135–1141
8. Uryupina, O.: Semi-supervised learning of geographical gazetteers from the internet. In: Proceedings of the HLT-NAACL 2003 Workshop on Analysis of Geographic References, Alberta, Canada (2003) 18–25
9. Yarowsky, D.: Unsupervised Word Sense Disambiguation Rivaling Supervised Method. In: Proceedings of the 33rd Annual Meeting of the Association for Computational Linguistics (ACL). (1995) 189–196
10. Niu, C., Li, W., Ding, J., Srihari, R.K.: A Bootstrapping Approach to Named Entity Classification Using Successive Learners. In: Proceedings of the 41st Annual Meeting of the Association for Computational Linguistics (ACL), Sapporo, Japan (2003) 335–342
11. Lin, W., Yangarber, R., Grishman, R.: Bootstrapped Learning of Semantic Classes from Positive and Negative Examples. In: Proceedings of the ICML-2003 Workshop on The Continuum from Labeled to Unlabeled Data, Washington, DC (2003)

# Using Verb Dependency Matching in a Reading Comprehension System

Kui Xu and Helen Meng

Human-Computer Communications Laboratory,
Department of Systems Engineering & Engineering Management,
The Chinese University of Hong Kong, Hong Kong SAR, China
{kxu, hmmeng}@se.cuhk.edu.hk

**Abstract.** In this paper, we describe a reading comprehension system. This system can return a sentence in a given document as the answer to a given question. This system applies bag-of-words matching approach as the baseline and combines three technologies to improve the result. These technologies include named entity filtering, pronoun resolution and verb dependency matching. By applying these technologies, our system achieved 40% HumSent accuracy on the Remedia test set. Specifically, verb dependencies applied in our system were not used in previous reading comprehension systems. In addition, we have developed a new bilingual corpus (in English and Chinese) - the ChungHwa corpus. The best result is 68% and 69% HumSent accuracy when the system is evaluated on the ChungHwa English and Chinese corpora respectively.

## 1  Introduction

Recently, there has been increasing interest in the research of question answering (QA) systems. Researchers in the field of information retrieval (IR) have paid much attention to this topic. One branch in the study of QA system is based on the context of the reading comprehension task. This task was proposed as a method for evaluating Natural Language Understanding (NLU) technologies by a research group at MITRE Corporation [7]. In this task, MITRE Corporation developed the Remedia corpus to evaluate a reading comprehension (RC) system. In addition, another task about question answering on a large-scale is the Text REtrieval Conference Question Answering (TREC QA) track. There is a major difference between TREC QA and reading comprehension. For a given question, TREC QA systems retrieve documents in a large collection of data and then find the answer (in phrases or sentences) within the retrieved documents. Reading comprehension systems only look for answers to given questions within a given story. In this paper, we only address the QA system for reading comprehension task.

Many reading comprehension systems [7], [12], [5] assume that there are common words shared between questions and answers. A reading comprehension system measures the similarity between questions and answers by matching with

S. H. Myaeng et al. (Eds.): AIRS 2004, LNCS 3411, pp. 190–201, 2005.
© Springer-Verlag Berlin Heidelberg 2005

different features. Features can be as simple as bag-of-words or bag-of-verbs [7], [12], [5]. In related work such as the TREC QA systems, syntactic and semantic features that have been applied include syntactic parse trees, dependencies and predicate-argument structures [4], [9]. However, syntactic and semantic features are not broadly used for reading comprehension tests.

In this paper, we have developed a reading comprehension system that uses syntactic features in an attempt to improve the accuracy. Syntactic features are represented by verb dependencies in our system. As Allen described in [1], the context-independent meaning of a sentence can be represented by logical forms, which can be captured using relationships between verbs and noun phrases. In this paper, verb dependencies are defined as lexical dependencies in which the heads of dependencies are verbs. We apply verb dependencies to handle ambiguities among candidate sentences when a reading comprehension approach (e.g. BOW matching) cannot discriminate among multiple candidate sentences. We believe that matching with syntactic features like verb dependencies can perform better selection among candidate answers than the simple approach of BOW matching.

In addition, this paper reports our first attempt in developing a Chinese reading comprehension system. We begin by collecting the ChungHwa corpus, which is a bilingual corpus both in English and Chinese.

## 2    Related Work

In 1999, a group at MITRE developed a reading comprehension system, Deep Read [7]. This system used the bag-of-words (BOW) matching and automated linguistic processing to achieve 36% HumSent[1] accuracy in the Remedia test set [7]. The system applied linguistic processing such as stemming, named entity (NE) recognition, named entity filtering, semantic class identification and pronoun resolution. If multiple candidate sentences contain the maximum number of matching words in BOW matching, the first (earliest occurrence) candidate sentence will be returned as the final answer.

Riloff and Thelen [12] developed a rule-based system called Quarc and achieved 39.7% HumSent accuracy in the Remedia test set. Quarc used not only BOW matching but also a number of heuristic rules that look for lexical and semantic clues in the questions and stories. For example, the WHERE rule,

if contain(S, LOCATION), then Score(S)+=6,

can be interpreted as the following: for *where* questions, if a candidate sentence contains LOCATION, this rule will reward the candidate sentences with 6 points.

---

[1] By comparing the system answers with the human marked answers, the percentage of correct answers is used as HumSent accuracy. In other words, if the system's answer sentence is identical to the corresponding human marked answer sentence, this question scores one point. Otherwise, the system scores no point. HumSent accuracy is the average score across all questions.

Charniak et al. [5] used bag-of-verbs (BOV) matching, "Qspecific" techniques, named entities, etc. to achieve 41% HumSent accuracy. The BOV matching is a variation of BOW matching in which only verbs are examined instead of all non-stop words. The "Qspecific" techniques use different strategies for different questions. For example, the following is one of the strategies for *why* questions.

If the first word of the matching sentence is "this", "that", "these" or "those", select the previous sentence.

Ng et al. [11] used a machine-learning approach (C5 learning algorithm) to determine if a candidate sentence is the answer to a question based on 20 features such as "Sentence-contains-Person", "Sentence-is-Title" etc. This approach achieves 39.3% HumSent accuracy.

As mentioned above, some of the previous work assumed that there is a high degree of overlap between the words used in a question and those used in its correct answer. In our approach, we assume that there is a structural overlap between the syntactic structure of a question and that of the correct answer. Syntactic structures are not commonly used in previous work for reading comprehension tests. In this paper, the impact of applying syntactic structure is examined.

In addition, some rules or features [11], [12] have been applied based on the observation in the training corpus. For example, the dateline (the line that shows the date when the story happened) in a Remedia story can be the answer to a question. Riloff and Thelen [12] applied dateline rules to handle the dateline. For example:

if contain(Q, story), then Score(DATELINE)+=20

can be interpreted as the following: for *where* and *when* questions, if a question contains "story", this rule will reward the dateline with 20 points. In addition, Ng et al. [11] used "Sentence-is-Dateline" as a feature in their machine-learning approach. Such corpus-specific technologies do not have impact on the corpus that does not have datelines. In our first attempt, corpus-specific technologies are not involved. We applied a general approach that is corpus-independent.

## 3     Verb Dependencies

In the current work, verb dependencies are used to represent the syntactic structures of sentences. They can be used as auxiliary information to perform matching based on BOW matching. Verb dependencies can be obtained from parse trees of sentences. In the study of parsing technologies, lexical dependencies have been used to handle ambiguities among parse tree outputs by a PCFGs parser [6]. Research in parsing technologies [6] shows the power of lexical dependencies in improving the performance of a parser. For the same reason, if reading comprehension approaches (e.g. BOW matching) cannot discriminate among multiple candidate sentences, lexical dependencies can also be used to handle ambiguities among candidate sentences.

In the work of Collins [6], the author defines a dependency as a relation between two words in a sentence (a *modifier* and a *head*), written in:

$$< modifier \rightarrow head > .$$

If the head of a dependency is verb, we refer this type of dependency verb dependency. Dependencies can be extracted from syntactic parse trees according to the work of Collins [6]. Fig. 1, Fig. 2 and Fig. 3 show the parse trees of the following question and candidate sentences respectively:

What is the new machine called?
A new machine has been made.
The machine is called a typewriter.

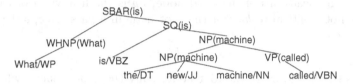

**Fig. 1.** The parse tree of "*What is the new machine called?*"

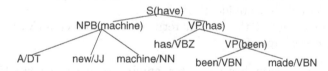

**Fig. 2.** The parse tree of "*A new machine has been made*"

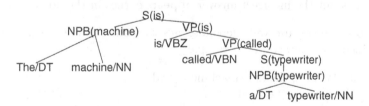

**Fig. 3.** The parse tree of "*The machine is called a typewriter*"

These parse trees are lexicalized parse trees since each node in the trees have a label and a headword. The label indicates a syntactic category (e.g. SBAR, NP and VP in Fig. 1). The headword is defined as the prime constituent of a phrase or sentence [2]. For example, the headword of a noun phrase is the noun; that of a verb phrase is the verb. Hence the headword of the noun phrase "a new machine" in Fig. 2 is "machine". Remaining words in the phrase or sentence are regarded as the left or right modifiers of the headword. More specifically, a context-free grammar rule may be represented as:

$$p \rightarrow l_n \ldots l_1 h r_1 \ldots r_m,$$

where $p$ is the parent non-terminal, $h$ is the head child of $p$, $l_i$ and $r_i$ are the left and right modifiers of $h$ respectively. If $h$ is a non-terminal, the headword of $p$ is the headword of $h$. If $h$ is a terminal, the headword of $p$ is $h$. We use the guidelines described by Collins [6] in determining the head constituent in a context-free rule. The procedure of assigning every non-terminal (every non-leave node in parse trees) a headword is called lexicalization [6]. An edge in parse trees involves two nodes: a child node and a parent node. A dependency can be written as:

$$< hc \rightarrow hp >,$$

where $hc$ is the headword of the child node, $hp$ is the headword of the parent node, $hc$ is not identical to $hp$. For the root node in a parse tree, a dependency can be written as:

$$< hr \rightarrow TOP >,$$

where $hr$ is the headword of the root node. Only the dependencies whose $hp$ or $hr$ are verbs are selected as verb dependencies. We call this selection process verb dependency extraction. The corresponding verb dependencies of parse trees in Fig. 1, Fig. 2, and Fig. 3 can be found in Table 1. In this paper, two dependencies match if and only if they are identical.

In BOW matching, if the system returns an incorrect answer, one of the following two cases happens.

- The incorrect answer has a greater number of matching words than the correct answer.
- The incorrect answer and the correct answer have an equal number of matching words but the incorrect answer appears earlier in the document.

Therefore, we can use verb dependencies to distinguish the correct answers from the incorrect answers. For example:

Question: What is the new machine called?
BOW: {be machine new call}

Suppose there are two candidate sentences that have the greatest number of matching words. The matching words in the candidate sentences are displayed in italic in the following:

Candidate sentence 1: A new machine has been made.
BOW: {*be machine new* has make}
Candidate sentence 2: The machine is called a typewriter.
BOW: {*be call machine* typewriter}

**Table 1.** Verb dependencies for the syntactic parse trees in Fig. 1, Fig. 2, and Fig. 3

| Verb dependencies for the parse tree in Fig. 1 | {<what→be>, <*machine→be*>, <*be→TOP*>} |
|---|---|
| Verb dependencies for the parse tree in Fig. 2 | {<machine→have>, <be→have>, <make→be>, <have→TOP>} |
| Verb dependencies for the parse tree in Fig. 3 | {<*machine→be*>, <call→be>, <typewriter→call>, <*be→TOP*>} |

Both candidate sentences have the maximum number (three) of matching words among all candidate sentences. The parse trees of the question and two candidate sentences are shown in Fig. 1, Fig. 2, and Fig. 3 respectively. Their corresponding verb dependencies are shown in Table 1.

The candidate sentence 2 (see Fig. 3) has the maximum number of matching dependencies against the question. Verb dependency matching selects candidate sentence 2 as the final answer, which is the true answer in this example.

## 4    The Reading Comprehension System

The flow chart of our reading comprehension system is shown in Fig. 4. In our current system, six processes are applied to identify different types of information in the story sentences and questions. These six processes are part-of-speech (POS) tagging, stemming, named entity recognition (NER), pronoun resolution, syntactic parsing and verb dependency extraction. Three out of the six processes are implemented by the use of natural language processing tools.

- The Brill POS tagger [3] is applied to perform POS tagging.
- A C programming language function (morphstr) provided by WordNet [10] is used to find the base form of nouns and verbs. This process is our stemming process.
- The Collins' parser [6] is applied to obtain the syntactic parse trees of sentences. This is the syntactic parsing process.

In addition, named entity recognition and pronoun resolution are not implemented in current system. We applied the named entity information and the pronoun resolution information that have been annotated in the corpus. These two processes will be studied in our future work. Further more, we followed the method described in Sect. 3 to implement the process of verb dependency extraction.

With these six processes, syntactic and semantic information is obtained to enrich the document sentences and questions. Such enrichment is used by the answering engine to retrieve the answer to a given question.

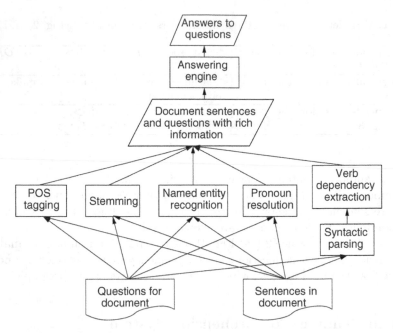

**Fig. 4.** The flow chart of our reading comprehension system

## 4.1   Answering Engine

The answering engine applied four technologies to find answers. They are BOW matching, named entity filtering, pronoun resolution and verb dependency matching. We use BOW matching as the baseline. Other three technologies are combined into the baseline incrementally to examine their impact.

**BOW Matching.** Each sentence in the story is regarded as a word set. BOW matching is conducted between the question word set and the candidate answer word set. Referring [7], a BOW matching system "measures the match by size of the intersection of the two word sets". If multiple candidate sentences contain the maximum number of matching words, the candidate sentence that appeared earlier is returned as the answer. Stop words in the word sets are removed before matching. The stop word list contains 16 words. They are: *the, of, a, an, it, and, or, do, what, where, why, who, how, when, which, all.* In addition, the nouns and verbs in the BOW are replaced by their base forms, which are the outputs of the stemming process.

**Named Entity Filtering.** Five named entity types (PERSON, ORGANIZATION, TIME, DATE and LOCATION) are used to perform answers filtering for three types of questions (*who, where, when*). The relationships are listed as the following [7]:

- For *who* questions, a candidate sentence that contains PERSON or ORGANIZATION is assigned higher priority.

- For *where* questions, a candidate sentence that contains LOCATION is assigned higher priority.
- For *when* questions, a candidate sentence that contains TIME or DATE is assigned higher priority.

**Pronoun Resolution.** For pronoun resolution, we replace five pronouns (*he, him, his, she and her*) with their referents in the word sets of candidate sentences and questions for BOW matching. In addition, other pronouns (e.g. *their, them, they, you, your, it*, etc.) and noun phrase referents are also annotated in the corpus. The system also examines the impact when all pronouns and noun phrase are replaced with their referents beside the above mentioned five pronouns.

**Verb Dependency Matching.** Verb dependencies are extracted according to the process in Sect. 3. Just like words in the word sets of questions and candidate sentences, verb dependencies are inserted into the corresponding word sets for matching.

## 5  Corpora

### 5.1  The Remedia Corpus

The Remedia corpus has been used by many researchers in previous work [5], [7], [11], [12]. It is published by Remedia Corporation. MITRE Corporation has annotated named entities, co-reference of pronouns and noun phrase anaphor and the true answer sentences on this corpus [7]. The corpus contains 55 training stories and 60 testing stories. Each story contains 20 sentences on average. There are about 20K words in this data set. For each story, five types of questions are asked: *who, what, when, where* and *why* question. Within the 60 test stories, there are 59 *who* questions, 61 *what* questions, 60 *when* questions, 60 *where* questions and 60 *why* questions. In total, 300 questions are asked in the Remedia test set. In each story, the first line is the title; the second line is the dateline. Other lines are story sentences.

### 5.2  The ChungHwa Corpus

The ChungHwa corpus comes from a bilingual book, "English Reading Comprehension in 100 days" which is published by Chung Hwa Book Co.,(H.K.) Ltd. This corpus contains 100 reading comprehension stories both in English and Chinese. The following domains are covered in the corpus: the English language, tourism, culture and society, sports, history and geography, arts, literature, economy and business, science and technology. We reserve 50 documents as the training set and the other 50 documents as the test set. The average number of sentences of each document is 9 (varies from 4 to 18). There are about 18K English words and 17K Chinese characters in the ChungHwa corpus. Each document has four questions on average. A linguistic was asked to annotate the named entities, anaphor referents and answer sentences for each document.

**Table 2.** A sample story from the ChungHwa corpus with an English story and its questions as well as a Chinese story and its questions

| |
|---|
| Imagine this: you have just won a competition, and the prize is an English language course at a famous school in Britain or the United States. You can either take a 30-week course for four hours a week, or a four-week course for 30 hours a week. Which one should you choose? ... |
| If you win a competition, what may be the prize?<br>What may be the two kinds of courses?<br>What is the advantage and disadvantage of the long course? |

Table 2 shows a sample story from the ChungHwa corpus both in English and in Chinese. Within the 50 test stories, there are 16 *who* questions, 98 *what* questions, 7 *when* questions, 17 *where* questions, 11 *why* questions, 10 *yes/no* questions, 10 *how many* questions and 25 *how* questions. In total, 194 questions are asked in the ChungHwa test set.

## 6    Experimental Results

For English stories, BOW matching, named entity filtering, pronoun resolution and verb dependency matching are applied step by step to examine the impact of each technology. The results are shown in Table 3 and Table 4 for the Remedia corpus and the ChungHwa corpus (the English part) respectively. For the purpose of comparison with previous work, the results in Table 3 and Table 4 are shown with HumSent accuracies. The abbreviations of different technologies are listed below:

- BOW: bag-of-words matching
- PR: the referents of *he, him, his, she* and *her* are resolved
- AA: all anaphors include pronouns in PR, other pronouns (e.g. *their, them, they, you, your, it,* etc.) and noun phrases anaphors are resolved
- NEF: named entity filtering
- VD: verb dependency matching

The Deep Read system achieved 29% and 36% HumSent accuracy with BOW and BOW+PR+NEF respectively [7]. With the same technologies, our results are comparable to the results of Deep Read system. The difference may caused by different stop words list and stemming process.

As our first attempt on the ChungHwa Chinese stories, BOW, PR and AA have been applied. Currently, a Chinese syntactic parser is not available in our

**Table 3.** Detail results by using different technologies on the Remedia test set

|         | BOW | BOW+PR | BOW+PR+NEF | BOW+AA+NEF | BOW+AA+NEF+VD |
|---------|-----|--------|------------|------------|---------------|
| Who     | 34% | 37%    | 49%        | 51%        | 54%           |
| What    | 31% | 33%    | 33%        | 33%        | 33%           |
| When    | 33% | 37%    | 60%        | 62%        | 62%           |
| Where   | 28% | 32%    | 28%        | 30%        | 30%           |
| Why     | 15% | 18%    | 18%        | 22%        | 23%           |
| Overall | 28% | 31%    | 38%        | 39%        | 40%           |

**Table 4.** Overall results by using different technologies on the ChungHwa test set in English

|         | BOW | BOW+PR | BOW+PR+NEF | BOW+AA+NEF | BOW+AA+NEF+VD |
|---------|-----|--------|------------|------------|---------------|
| Overall | 67% | 68%    | 68%        | 68%        | 68%           |

system. So the verb dependencies of Chinese are not used in our experiments. Our system achieved 69%, 69% and 70% HumSent accuracy with BOW, BOW+PR and BOW+AA respectively. Since the Chinese character segmentation has been annotated in the corpus, we simply use the segmentation annotation instead applying a segmentation tool in current study. PR for Chinese only focuses on

Named entity filtering technology is not used for the Chinese part because we do not have a question classification for Chinese questions. Questions can be more complicated in Chinese to ask time, location and person. Without these three classes, the corresponding named entities cannot be used to perform named entity filtering. We will study question classification and Chinese parsing in our future work.

# 7    Discussion

In Table 3, our general approach achieved 40% HumSent accuracy, which is comparable to the state of the art, 41% [5]. After applying VD, our system can improve 1%. To further analyze the result, we performed VD alone on the Remedia test set and studied the verb dependency matching situation in the true answer and its question for all 300 questions. We only found that 79 questions have matching verb dependencies against their true answers. That means the accuracy upper-bound is $79/300 = 26.3\%$ when VD is applied alone. Moreover, 74 out of 79 questions can be correctly answered by BOW+AA+NEF. Therefore, the improvement upper-bound using VD is $(79-74)/300 = 1.67\%$. The limited coverage and improvement room of VD lead to the insignificant improvement.

In Table 3, the accuracies increase for *who* and *when* questions after applying NEF. However, the accuracy drops from 32% to 28% for where questions. After manually analyzing the Remedia training set, we found that not all LOCATION tags are recognized. For example:

Question: Where do these sea animals live?

True answer: She was born in a sea animal park called Sea World.

System returned answer: (ORLANDO, FLORIDA, September, 1985) -

In this example, the system returned answer is wrong. "A sea animal park" in the true answer is not tagged as LOCATION. On the other hand, "ORLANDO, FLORIDA" is tagged as LOCATION. Even though the true answer has the maximum number of matching words against the question, the named entity filtering process gave higher priority to the sentence:

(ORLANDO, FLORIDA, September, 1985) -.

The insufficient LOCATION tags in the Remedia corpus lead to the decrease of accuracy of *where* questions.

When manually analyzing the Remedia training set, we found that inference technology and world knowledge are helpful in answering about one third of questions. For example:

Question: Who had a baby at Sea World?

True answer: Baby Shamu's mother is named Kandu.

System returned answer: The workers at Sea World will watch Baby Shamu each day and make notes.

In this example, the true answer has no matching words against the question. In order to answer the question correctly, the system must know that the one who had a baby is a mother.

For the ChungHwa English stories, the overall accuracies are greater than those of the Remedia corpus. After manually analyzing the ChungHwa training set, we found that questions in ChungHwa corpus tend to use the same words that used in their corresponding answers. That causes the baseline (BOW matching) result is higher than the result obtained from the Remedia corpus. With a higher baseline result, the improvement made by PR, AA, NEF and VD is not obvious.

## 8    Conclusion

In this paper, we describe a reading comprehension system. This system can return a sentence in a given document as the answer to a given question. This system applies BOW matching approach as the baseline and combines three technologies to improve the result. These technologies include named entity filtering, pronoun resolution and verb dependency matching. By applying these technologies, our system achieved 40% HumSent accuracy on the Remedia test set. Specifically, our system brings in verb dependencies that can be derived from syntactic parses which is not used in previous reading comprehension systems. The verb dependency matching does not lead to significant improvement because of its limited coverage and improvement room. In addition, we have developed a new bilingual corpus, ChungHwa corpus. The evaluation result on the

English corpus is 68% HumSent accuracy and on Chinese corpus is 69% HumSent accuracy. In our future work, named entity recognition approaches and pronoun resolution approaches will be studied for English reading comprehension. Moreover, Chinese question classification approaches and Chinese parsing technologies will be studied for Chinese reading comprehension.

# References

1. Allen, J.: Natural Language Understanding. The Benjamin/Cummings Publishing Company, Menlo Park, CA (1995)
2. Bloomfield, L.: An Introduction to the Study of Language. Henry Holt and Company, New York (1983)
3. Brill, E.: A Simple Rule-based Part of Speech Tagger. In Proceedings of the Third Conference on Applied Natural Language Processing (1992)
4. Buchholz, S.: Using Grammatical Relations, Answer Frequencies and the World Wide Web for TREC Question Answering. In Proceedings of the tenth Text Retrieval Conference (TREC 10) (2001) 502-509
5. Charniak, E., et al.: Reading Comprehension Programs In a Statistical-Language-Processing Class. In ANLP-NAACL 2000 Workshop: Reading Comprehension Tests as Evaluation for Computer-Based Language Understanding Systems (2000)
6. Collins, M.: Head-Driven Statistical Models for Natural Language Parsing. PhD thesis (1999)
7. Hirschman, L., Light, M., Breck, E. and Burger, J.: Deep Read: A Reading Comprehension System. In Proceedings of the 37th Annual Meeting of the Association for Computational Linguistics (1999)
8. Light, M., Mann, G. S., Riloff, E. and Breck, E.: Analyses for Elucidating Current Question Answering Technology. Journal of Natural Language Engineering (2001) Vol. 7, No. 4
9. Litkowski, K. C.: Question-answering Using Semantic Relation Triples. In Proceedings of the eighth Text Retrieval Conference (TREC 8) (1999) 349-356
10. Miller, G.: WordNet: an On-line lexical database. International Journal of Lexicography (1990)
11. Ng, H. T., Teo, L. H., Kwan, L. P.: A Machine Learning Approach to Answering Questions for Reading Comprehension Tests. In Proceedings of the 2000 Joint SIGDAT Conference on Empirical Methods in Natural Language Processing and Very Large Corpora (2000)
12. Riloff, E. and Thelen, M.: A Rule-based Question Answering System for Reading Comprehension Test. In ANLP/NAACL-2000 Workshop on Reading Comprehension Tests as Evaluation for Computer-Based Language Understanding Systems (2000)

# Sense Matrix Model and Discrete Cosine Transform

Bing Swen

Institute of Computational Linguistics, Peking University,
Beijing 100871, China
bswen@pku.edu.cn
http://icl.pku.edu.cn

**Abstract.** In this paper we first present a brief introduction of the Sense Matrix Model (SMM), which employs a word-sense matrix representation of text for information retrieval, and then a discussion about one of the document transform techniques introduced by the model, namely discrete cosine transform (DCT) on document matrices and vectors. A first system implementation along with its experimental results is discussed, which provides marginal to medium improvements and validates the potential of DCT.

## 1 Introduction

Due to the extreme variety and complexity of language use, current text information retrieval models deal with only very simple aspects of linguistic constructs, and the modeling of "deep structures" of text that are related to meaning and understanding is still a very weak (or missing) aspect. Further improvements of retrieval effectiveness necessitate such semantic aspects to be eventually taken into account. As the first step, much work has been done in the research of meaning (or concept) based retrieval, with major efforts devoted to approaches with word sense explicitly incorporated (e.g., Sussna 1993, Vooehees 1993). But such attempts have not resulted in significantly better retrieval performance (Stokeo et al 2003), partially due to the insufficiency in word sense discrimination and representation.

A well-known fundamental problem of IR models using index-term based text representations is the polysemy and synonymy issues (Kowalski and Maybury 2000), which has plagued IR research over the years. Synonymy decreases recall and polysemy decreases precision, leading to poor overall retrieval performance. For example, the vector-space model or VSM (Salton and Lest 1968, Salton 1971), so far the most widely used, may suffer from the problem of word-by-word match. Usually the synonymy issue is tackled using some "semantic" categories (word clusters) as index terms so that synonymous words are of the same category, which is a recall enhancement method to compensate the precise word-matching retrieval. Some data analysis techniques, such as LSI (Deerwester et al 1990) using "latent" semantic categories (statistical word groups) may also be incorporated into VSM, which helps the synonymy problem significantly in some cases (but may also lead to limited effectiveness for the polysemy problem), though the resulting approaches usually have nothing to do with true meaning of documents.

S. H. Myaeng et al. (Eds.): AIRS 2004, LNCS 3411, pp. 202–214, 2005.

An alternative to term-based VSM is to directly use senses (or concepts) for text indexing, sometimes termed "sense/semantic VSMs". Note that the precursor of VSM, a "notion space" introduced by Luhn (1957), was actually a VSM in a concept space. Such a text representation could be language-independent. And if documents were represented in a sense vector space, then cross-lingual IR could be carried in a straightforward manner. But a common problem for the sense/concept (or any other semantic category) based indexing and retrieval approaches is that there is not a common system of concepts for indexing and retrieval, and hence no agreement between the semantic representation of documents and queries.

In our recent research, we considered the possibility of combining the strength of term-VSM and sense-VSM in a more comprehensive text representation. We proposed a new document representation, where instead of an array of term (or sense) weights, a document is represented by a matrix of term-by-sense weights, which leads to a matrix-based model, called a "sense matrix model" (SMM) for information retrieval (Swen 2003). The model provides a framework in which many matrix properties and computation techniques can be introduced or developed for IR, and similarity measures can be properly defined by matrix norms or other attributes. The model also introduces novel techniques for document transformation using well-developed data analysis techniques such as multiway array decomposition and discrete cosine transform (DCT).

In the next section we present an outline of the major points of SMM. Section 3 introduces DCT on document vectors and matrices. Section 4 describes our DCT implementation and experimental results. A summary is presented in section 5.

## 2 A Sense Matrix Model for IR

We start from the point of view of the "semantic triangle" (Ogden and Richards 1930) as an illustration of the fundamental problem of IR (despite its known insufficiencies in semantics study). The following figure is an "augmented semantic triangle" as we apply it to IR (details of the triangle model will not be elaborated here).

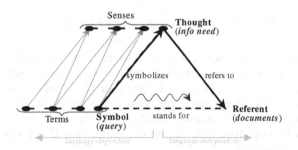

**Fig. 1.** "Augmented" Semantic Triangle applied to IR

In this model, the three ends of refrent/thing, thought/concept and symbol/word form a triangle, with the edge being the suggested relations. Note that there is no a direct association between symbols (here our info queries) and things (relevant documents). The "internal process" in IR is needs ? words ? docs, where uncertainty is inevitably introduced in the indirection. The problem may be largely attributed to the fact of irregular many-to-many mapping from words to senses. As a whole, the mapping from words and senses in any natural language is in an irregular manner.

In reality, such an irregularity must be admitted and represented in some way. Our basic idea to explicitly introduce both words and senses in the document representation then follows quite naturally, namely,

document $D$ ==> term set × sense set (with weights)

Here ==> stands for reduction. A straightforward manner to make use of such combined information in an IR formalism is that we should collect all the terms along with the senses they may or actually have in the document, and then index the document by a term-sense network for retrieval,

$$\left.\begin{array}{l} \text{doc} \Longrightarrow \text{word-sense net D} \\ \text{query} \Longrightarrow \text{word-sense net } Q \end{array}\right\} \quad \text{match}(D, Q)$$

This network relationship of words and senses may be further represented by a matrix of index-term by sense weights, resulting in a matrix representation of documents. A document collection is then represented as a term-sense-document space:

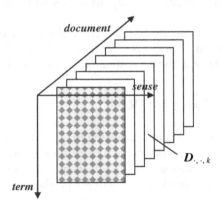

**Fig. 2.** The Term-Sense-Document Space $D_{I \times J \times K}$

In this space a sense becomes a term-by-document matrix (and hence the name *SMM*). Clearly, such a matrix-based retrieval model is a "sense expansion" of VSM: VSM's document vector of term weights is "expanded" or "split" (distributed) along the sense direction (with or without appropriate sense weighting of each term). An immediate consequence is that documents and queries sharing no common terms may still get matched if they share a few common senses and if appropriate similarity measures can be constructed.

### 2.1 Measure of Matrix Similarity

There are several possible methods to evaluate the similarity between document matrices. Since matrices are vectors with more restrictions, norms for a matrix space can be defined. For any matrix norm, we may define the matrix distance by

$$d(A, B) = \|A - B\| \ . \tag{1}$$

The concept of "angle" between matrices may also be introduced in correspondence to vector angle. First, we may introduce a "normalized distance", namely, distance between normalized matrices:

$$d_{\mathrm{norm}}(A, B) = \left\| \frac{A}{\|A\|} - \frac{B}{\|B\|} \right\| \le 2 \ . \tag{2}$$

Then a correct angle may be defined:

$$\cos \angle_{\mathrm{norm}}(A, B) =_{\mathrm{def}} 1 - \frac{1}{4} d_{\mathrm{norm}}(A, B)^2 \ . \tag{3}$$

It is easy to verify that when F-norm is used and the document matrices are vectors, this angle is proportional to the standard vector angle:

$$\cos \angle_{\mathrm{norm}}(\mathbf{q}, \mathbf{d}) = \frac{1}{2}(1 + \cos \angle_{\mathrm{VSM}}(\mathbf{q}, \mathbf{d})) \ . \tag{4}$$

Secondly, also note that we may define

$$\cos \angle(A, B) = \frac{\|AB\|}{\|A\| \cdot \|B\|} \tag{5}$$

to be the cosine of the "angle" between any two multipliable matrices $A$ and $B$, based on the *compatibility condition* of matrix norms: $\|AB\| \le \|A\| \cdot \|B\|$, which ensures that matrices may be multiplied by vectors and the resulting vectors must also preserve the defining conditions of vector norms. This definition generally holds for any multipliable matrices. However, to our document matrix, there are two different possible definitions of matrix angles:

$$\cos \angle(D_1, D_2) = \frac{\|D_1 D_2^{\mathrm{T}}\|}{\|D_1\| \cdot \|D_2\|} \ , \quad \cos \angle'(D_1, D_2) = \frac{\|D_1^{\mathrm{T}} D_2\|}{\|D_1\| \cdot \|D_2\|} \ , \tag{6}$$

where $\|D_1 D_2^{\mathrm{T}}\|$ is the term-term correlation via senses, and $\|D_1^{\mathrm{T}} D_2\|$ is the sense-sense correlation via terms. Hence there would be two models concerning the sense matrix angle similarity, namely,

$$\mathrm{sim}_{i\mathrm{SMM}}(D_k, D_{k'}) = \cos \angle(D_k, D_{k'}) = \frac{\|D_k D_{k'}^{\mathrm{T}}\|}{\|D_k\| \cdot \|D_{k'}\|} \ , \tag{7}$$

$$\mathrm{sim}_{n\mathrm{SMM}}(D_k, D_{k'}) = \cos \angle'(D_k, D_{k'}) = \frac{\|D_k^{\mathrm{T}} D_{k'}\|}{\|D_k\| \cdot \|D_{k'}\|} \ . \tag{8}$$

These are some of the options SMM provides for specific implementations. Other more possibilities of similarity measures may also be available according to specific measurability (Swen 2003). For example, if we prefer measuring the similarity between the reduced sense vectors, we would have

$$\text{sim}_{rsSMM}(D_k, D_{k'}) = \frac{|| D_k D_{k'}^{\mathrm{T}} ||_{V1}}{|| D_k ||_{V1} \cdot || D_{k'} ||_{V1}} \quad , \tag{9}$$

where $|| A ||_{V1} = \sum_{i,j} |A_{ij}|$ is a *pseudo norm*.

## 2.2  Simpler SMM Cases

SMM can be tailored in various forms, including a few simpler and straightforward applicable cases.

**Unary SMM.** The simplest SMM, where only one sense (say, "relevance degree") is defined for all index terms, and the document representation is $n$-by-1 matrix, namely a vector of single-sense weights.

**Binary SMM.** In this case, there are only two senses for all terms, with the sense set being, e.g., {+/– *relevant*} or {*positivity, negativity*}. Assume that a term $t_i$ has a probability $p_i$ to be relevant for one case, then the probability of being the other case is $q_i = 1 - p_i$. The document matrix of a binary SMM then takes the form as follows:

$$D = \begin{pmatrix} p_1 w_1 & (1 - p_1)w_1 \\ p_2 w_2 & (1 - p_2)w_2 \\ \vdots & \vdots \\ p_N w_N & (1 - p_N)w_N \end{pmatrix} , \tag{10}$$

where $w_i$ is the VSM weight of term $t_i$. Binary SMM "splits" the term weights of VSM into two components according to a further importance assignment method, which provides for additional flexibility. { $p_i$ } is the adjustable parameter set and should be determined in specific applications (e.g., a prior probability distribution).

**POS SMM.** The "top-most" senses of words are arguably the words' part-of-speeches. A "part-of-speech SMM" in which the sense dimensions are the part-of-speeches of terms may be constructed. It corresponds to splitting the VSM term weights into the weights of a term's part-of-speeches.

When the input text is POS tagged, there are 2 ways to determine the matrix elements. The simple one it to index each <Word/POS> pair as a VSM term, but record the matrix correspondence (otherwise it would result in a "POS VSM" with greatly restricted terms). The standard VSM term weightings are directly applicable to these tagged terms. The other way is to split the VSM weights with POS distributions, where the document matrix of takes the form

$$D = \begin{pmatrix} p_{1,1} w_1 & \cdots & p_{1,m} w_1 \\ p_{2,1} w_2 & \cdots & p_{2,m} w_2 \\ \vdots & \cdots & \vdots \\ p_{N,1} w_N & \cdots & p_{N,m} w_N \end{pmatrix} , \tag{11}$$

where $m$ is the number of part-of-speeches adopted and $N$ is the term number in the collection. The $p_{i,j}$ parameter may be estimated to be the frequency of the $j$th POS of word $i$ in document $D$. The advantage of this method is that for simple applications, the { $p_{i,j}$ } parameters may be set to the POS probability distribution of words in the collection be considered (instead of being computed for each document).

Since current POS tagging has succeeded considerably, the weighting of POS SMM's "senses" can be expected to be effective and robust.

When more realistic word senses beyond part-of-speeches are incorporated into SMM, an important issue is how to determine the values of the term-sense matrix elements for various senses, which we call *sense weighting*, corresponding VSM term weighting.

## 2.3  Other Applications

SMM is usable to a wide range of applications where VSM is commonly used, with additional features. First, we note that some of the similarity measures of SMM can be "neutral" to the index terms so long that document matrices with multipliable sense dimensions are used, which means that SMM has the ability to compare documents in different languages. Advantages include:

- No translation processing is needed;
- No requirement of any "parallel" (or comparable) bilingual corpora to be used as "training data";
- No restriction on the number of languages involved.

For example, if we have the matrices of two documents $A_{Ch}$ and $B_{En}$ in Chinese and English respectively, then we can immediately compute their "cosine" similarity in SMM:

$$\text{sim}(\mathbf{A}_{Ch}, \mathbf{B}_{En}) = \frac{\left\| \mathbf{A}_{Ch} \mathbf{B}_{En}^T \right\|_{V1}}{\left\| \mathbf{A}_{Ch} \right\|_{V1} \cdot \left\| \mathbf{B}_{En} \right\|_{V1}} . \tag{12}$$

There is no necessary for a preprocessing of English-to-Chinese translation (or vice versa). The correlation matrix $A_{Ch} B_{En}^T$ is sufficient for estimating the intensity of correlation (via senses) between English and Chinese words in any document pairs. The only prerequisite is that: each language involved should be indexed with the same or convertible sense set (and in the same order). Studies from multi-lingual WordNet seem to suggest the hypothesis that the cognitive semantic structures of human language are largely common is prevailing. Thus such a cross-lingual sense set is possible.

Using SMM for cross-lingual text filtering is also in the same principle, where the documents and user profiles can be in different languages but with compatible sense set. SMM is also useful for multilingual text classification (clustering or categorization). On the other hand, like VSM, SMM with some similarity measures may be limited to monolingual cases.

As with standard pattern recognition, most of the existing text classification methods are based on VSM. Many of existing methods can be adapted to use with SMM by simply modifying the similarity measures. For example, clustering using a document-document similarity matrix $\text{sim}(D_k, D_{k'})$ for non-incremental clustering is directly adaptable to SMM. An SMM version of the kNN method is also available, using the matrix distance measure to determine the $k$ nearest neighbors and to let them vote for the class of the input sample.

The SVM (support vector machine) categorization method may be modified to work with the matrix documents. To achieve an "SMM based SVM", we simply replace the decision function and the VSM-based SVM kernel

$$K(\vec{d}, \vec{d_t}) = f(\vec{d} \cdot \vec{d_t}) \; , \quad y(\vec{x}) = \vec{w} \cdot \vec{x} + b \; , \tag{13}$$

to their SMM versions

$$y(\mathbf{X}) = \mathrm{tr}\mathbf{W}^\mathrm{T}\mathbf{X} + b = \mathrm{tr}\mathbf{W}\mathbf{X}^\mathrm{T} + b \; , \quad K(\mathbf{D}, \mathbf{D}_t) = f(\mathrm{tr}\mathbf{D}\mathbf{D}_t^\mathrm{T}) \; . \tag{14}$$

In this way, we treat the matrix data as a (restricted) vector in the "flattened space".

## 2.4  Document Transformation

Documents represented as 1-way or 2-way array data can be transformed for specific purposes. In text retrieval, we normally first reduce noise from the data, e.g., document preprocessing to eliminate stop words, terms of extreme low frequency, format elements, etc. There is, however, another kind of noise due to the inter-correlation among features (indexing terms), which cannot be reduced by preprocessing. In this case, some appropriate data transforms may be applied to extract independent or less correlated elements (or factors), and to suppress elements of very low intensity. The fundamental important issue is to reduce the data samples to a simpler common space to construct the reduced representation. In this paper we introduce two kinds of document transformation, namely MAD (Multiway Array Decomposition) and DCT (Discrete Cosine Transform) for text retrieval.

We first discuss using MAD for SMM. The DCT is to be discussed in the later sections.

**SMM with MAD.** Dimensionality reduction is a common practice in multiway data analysis (Kiers 2000). The term-sense-document data set of SMM is a good candidate to perform such an analysis. To reduce the "side effects" of the sense independence simplification, truly independent components can be derived via techniques developed in *factor analysis*. A commonly used data analysis method is orthogonal decomposition of multiway arrays. It is used to reduce the dimensionality of data representations, leading to compact data space and reduced noise effects. For this purpose, a generalized E-Y decomposition of the term-sense-document "3-way array" is applicable (Sidiropoulos and Bro 2000):

$$D_{i,j,k} = \sum_f^F t_{f,i} s_{f,j} d_{f,k} \sigma_f \; , \quad \sum_{i=1}^{N_t} t_{f,i} t_{f',i} = \sum_{j=1}^{N_s} s_{f,j} s_{f',j} = \sum_{k=1}^{N_d} d_{f,k} d_{f',k} = \delta_{f,f'} \; . \tag{15}$$

$N_{t,s,d}$ are the numbers of terms, senses and documents in the collection respectively, and $\delta$ is the Kronecker symbol. These orthogonal column vectors are notated as $\mathbf{t}_i$, $\mathbf{s}_j$, $\mathbf{d}_k$ respectively. Such decomposition may be illustrated as follows:

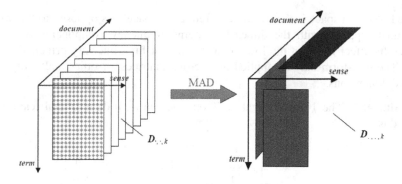

**Fig. 3.** Three-Way Array Decomposition

The largest possible $F$ is called the rank of $D$, rank($D$). For $F' \leq F$,

$$D_{i,j,k}^{(F')} = \sum_{f}^{F'} t_{f,i} s_{f,j} d_{f,k} \sigma_f$$ is commonly used as an approximation of the original 3-

way array to $D$ (under some conditions, e.g., the leas-squares).

The above result can be used to establish a dimensionality reduction model for SMM. First we introduce two quantities,

$$d_{f,k} = \sum_i \sum_j \frac{1}{\sigma_f} t_{f,i} s_{f,j} D_{i,j,k} \ , \qquad \sum_f \sigma_f d_{f,k} \cdot \sum_{f'} \sigma_{f'} d_{f',k} = \|D_k\|_F^2 \ . \qquad (16)$$

The reduced matrices over the $F$-space of an original document $D_k$ and query $Q$ are

$$\hat{D}_{f,f',k} = \sqrt{\sigma_f \sigma_{f'} d_{f,k} d_{f',k}} \quad s.t. \quad \left\|\hat{D}_{\cdot,\cdot,k}\right\|_F^2 = \left\|D_{\cdot,\cdot,k}\right\|_F^2, \quad \left\|\hat{Q}\right\|_F^2 = \|Q\|_F^2 \ . (17)$$

It is easy to see that SMM/MAD is quite different from LSI (Deerwester *et al* 1990), though the purpose looks similar. In LSI, term correlation is solely based on term co-occurrence. The "latent semantic" of a term is actually a (weighted) average of the different senses of the term, and thus the term correlation in LSI may reduce the search quality when the actual sense of a query term differs from its average meaning. In SMM/MAD, the sense dimensions add a more stable association between terms and documents. On the other hand, when using existing PARAFAC implementation of the decomposition, one could expect the computational overhead of SMM/MAD to be significantly larger than LSI. Since the document matrices are highly sparse, more efficient and/or approximate decomposition algorithms could be introduced for this specific case.

## 3 DCT on Documents

The discrete cosine transform, developed by Ahmed *et al* (Ahmed *et al* 1983), is a close relative of the discrete Fourier transform (DFT), and is widely used in image compressing coding, such as JPEG, MPEG, H.261 (video telephony), etc. In DCT, the reduced common space is the (discrete) "frequency space" (as opposed to the original "coordinate space", with the dimensions being frequency components (0, $\omega$, $2\omega$, $3\omega$, ...). Its effect is to get the low-frequency and/or high-intensity principal components. The transform may be applied on vectors, matrices, or other multi-way arrays of fixed dimensions.

**VSM with DCT.** The 1-dimensional DCT operates on an $n$-dimensional feature term vector d is

$$\hat{\mathbf{d}} = T(n)\mathbf{d}, \quad T(n) = \sqrt{\frac{2}{n}} \begin{pmatrix} \frac{1}{\sqrt{2}} & \frac{1}{\sqrt{2}} & \cdots & \frac{1}{\sqrt{2}} \\ \cos\frac{1}{2n}\pi & \cos\frac{3}{2n}\pi & \cdots & \cos\frac{2n-1}{2n}\pi \\ \vdots & \vdots & \vdots & \vdots \\ \cos\frac{n-1}{2n}\pi & \cos\frac{3(n-1)}{2n}\pi & \cdots & \cos\frac{(n-1)(2n-1)}{2n}\pi \end{pmatrix}_{n\times n} \tag{18}$$

or written in expanded form,

$$\hat{d}_i = \sum_{j=0}^{n-1} c(i)\cos\frac{i(2j+1)\pi}{2n} \cdot d_j, \quad c(i) = \sqrt{\frac{1}{n}} \text{ (if } i=0) \text{ or } \sqrt{\frac{2}{n}} \text{ (if } 1 \le i \le n), \quad i = 1..n$$

$$\tag{19}$$

The transformed vector (with components in the frequency space) has a property that elements of higher intensity occur first from lower dimension positions, including the "direct current component" (frequency = 0), components of 1, 2, ... times frequency, such that most high-frequency components will have very small values.

It is easy to verify that $T$ is an orthogonal transform:

$$T(n)T(n)^{\mathrm{T}} = T(n)^{\mathrm{T}}T(n) = 1 . \tag{20}$$

Therefore $\|\hat{\mathbf{d}}\| = \|\mathbf{d}\|$, sim($d_1$, $d_2$) = sim($\hat{d}_1$, $\hat{d}_2$). The inverse transform, inverse discrete cosine transform (IDCT), is thus

$$\mathbf{d} = T^{-1}(n)\hat{\mathbf{d}}, \quad d_i = \sum_{j=0}^{n-1} c(j)\cos\frac{(2i+1)j\pi}{2n} \cdot \hat{d}_j . \tag{21}$$

To suppress vector components of low intensity, a quantization processing is applied on the transformed vector:

$$\tilde{\mathbf{d}} = \text{round}\left(\frac{1}{q}\hat{\mathbf{d}}\right) \tag{22}$$

where $q$ is the quantization constant (usually an integer, but float numbers may be used for our system). A quantization using a single constant to divide all vector elements is called a *uniform quantization*. A *non-uniform quantization* uses different constants on different elements,

$$\tilde{\mathbf{d}} = \text{round}\left(Q^{-1}\hat{\mathbf{d}}\right), \quad Q = \begin{pmatrix} q_1 & & 0 \\ & \ddots & \\ 0 & & q_n \end{pmatrix}. \tag{23}$$

Some elements of $(q_1, q_2, \ldots, q_n)$ may be the same. We call such a quantization factor list a *quantization table*.

When documents are transformed, similarity measure is then computed using the DCT'ed and quantized vectors, $\text{sim}(\mathbf{d}_1, \mathbf{d}_2) = \cos \angle(\mathbf{d}_1, \mathbf{d}_2) \Rightarrow \cos \angle(\tilde{\mathbf{d}}_1, \tilde{\mathbf{d}}_2)$.

**SMM with DCT.** As with image compression, the 2-d document matrices of SMM are appropriate objects for DCT processing. The 2-dimensional DCT operates on a document term-sense matrix D, defined as follows:

$$\hat{\mathbf{D}} = T(I)\mathbf{D}\,T(J)^{\mathrm{T}}, \quad \tilde{\mathbf{D}}_{i,j} = \text{round}\left(\frac{1}{q_{\text{term}} \cdot q_{\text{sense}}}\hat{\mathbf{D}}_{i,j}\right), \tag{24}$$

where two quantization factors or tables for the term and sense dimensions are used.

The transformed matrix has the property that elements of higher intensity occur from the upper left corner of lower frequencies, and most high-frequency matrix element have small values. Similarity measures are then computed for the quantized document matrices: $\text{sim}(\mathbf{D}_1, \mathbf{D}_2) = \cos \angle(\mathbf{D}_1, \mathbf{D}_2) \Rightarrow \cos \angle(\tilde{\mathbf{D}}_1, \tilde{\mathbf{D}}_2)$.

## 4 Document DCT Experiments

The DCT on document vectors was one of the first SMM features tested in our research (other tests including some unary SMM cases). We have so far tried the DCT method on the standard test collections that come with SMART version 11 (SMART 1992). These test sets would reflect at least partially the actual effects introduced by document DCT. Comparison experiments were made to show the differences between the DCT extension and the original output. To keep the comparison simple, a very simple quantization table was used in all the tests:

```
q_factor                        1
q_table                         " "
dct_freq_cutoff                 -1
dct_2nd_quantization            false
```

The VSM is specified to use the simple remove_s stemming and no weighting. The same VSM term weights are then delivered to a DCT indexing procedure. The same inverted retrieval procedure is used for the output vector files of both. The results are listed as follows. In these experiments, marginal to medium improvements were observed. Such a consistent performance may be attributed to the effectiveness of DCT.

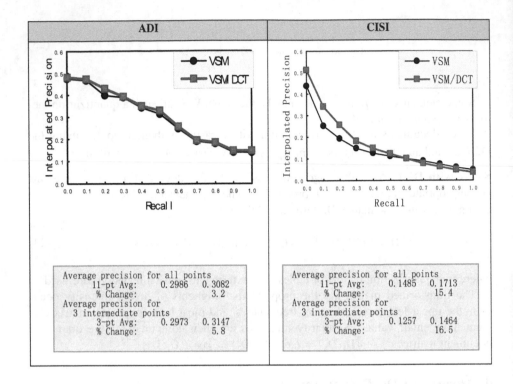

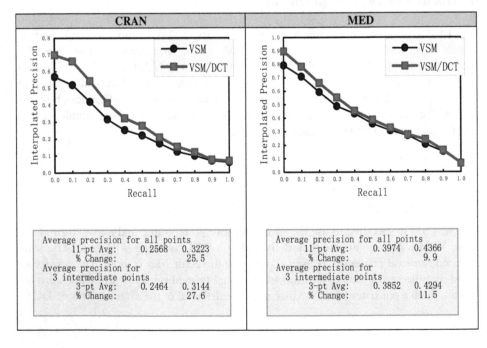

When the VSM uses optimal and robust term weighting, it will usually outperforms the DCT model using the above simple quantization configuration. We think that an optimal quantization table should play the same role for DCT as term weighting dose. If well-formed, we could expect that DCT of robust quantization may perform much better than this simplest case did.

## 5  Summary and Future Work

In this paper we present an outline of the sense matrix model SMM, on which much further work can be based. A unique feature of SMM is that it explicitly represents both words and the associated senses in the model, providing additional facilities for investigating retrieval effectiveness. The similarity judgment introduced by SMM can be regarded as a "recall device" that increases recall (compared to VSM). The problem of concern is that whether it dose so by maintaining or improving precision, hence would result in better retrieval effectiveness. We expect that SMM with effective part-of-speech tagging may provide a good example for such a study.

The document DCT experiments also indicate that DCT on documents may have the potential to lead to improved retrieval effectiveness. So far our DCT implementation has achieved marginal improvements. Many issues remain open for research. We still don't know how to construct an optimal quantization table. The closely related issue of incorporating VSM term weighting into DCT is also interesting. In general, the framework of SMM includes many aspects and subcases that each alone may need more thorough understanding and experiments. The TREC dataset that we recently applied for will allow us to conduct more experiments at large in this research. Currently we are experimenting SMM for the TREC 2004 Robust Track. The results will be discussed elsewhere.

## References

1. Ahmed, N., T.Natarajan and K.R. Rao. 1974. On image processing and a discrete cosine transform. IEEE Trans. on Computers C-23(1):90-93.
2. Deerwester, S., S. T. Dumais, T. K. Landauer, G. W. Furnas and R. A. Harshman. 1990. Indexing by latent semantic analysis. Journal of the Society for Information Science, 41(6), 391-407.
3. Greengrass, E. 2000. Information Retrieval: A Survey. Tech Report, Nov 2000.
4. Ide, N. and J. Véronis. 1998. Introduction to the Special Issue on Word Sense Disambiguation: The Start of the Art. Computational Linguistics, Vol24 No1.
5. Kiers, H. A. L. 2000. Towards a standardized notation and terminology in multiway analysis. Journal of Chemometrics, 14: 105–122.
6. Kowalski, G. and M. Maybury. 2000. Information Storage and Retrieval Systems Theory and Implementation. Kluwer.
7. Krovetz, R. and W. B. Croft. 1992. Lexical Ambiguity and Information Retrieval. ACM Transactions on Information Retrieval Systems, Vol. 10(2), 115–141.
8. Luhn, H. P. 1957. A statistical approach to mechanised encoding and searching of literary information. IBM Journal of Research and Development 1(4), 309–317, 1957.

9.  Miller, G. 1990. Wordnet: an On-line Lexical Database. In Special Issue: International Journal of Lexicography Vol. 3(4).  235 – 312.
10. Ogden, C.K. and I.A. Richards. 1930. The Meaning of Meaning. New York: Harcourt, Brace & World.
11. Salton, G. 1971. The SMART retrieval system – Experiments in automatic document processing. Prentice Hall Inc., Englewood Cliffs, NJ.
12. Salton, G. and M. E. Lesk. 1965. The SMART automatic document retrieval system – an illustration. Communication of the ACM, 8(6): 391-398, June 1965.
13. Salton, G. and M. E. Lesk. 1968. Computer evaluation of indexing and text processing. In Journal of the ACM, volume 15(1),  8–36, January.
14. Sidiropoulos, N. D. and R. Bro. 2000. On the uniqueness of multilinear decomposition of N-way arrays. Journal of Chemometrics, 14: 229–239.
15. SIGIR Forum 2003. Challenges in Information Retrieval and Language Modeling. Vol.37, No.1, 2003.
16. SMART version 11. 1992. Available via anonymous ftp from  ftp.cs.cornell.edu
17. Stokoe, C. et al. 2003. Word Sense Disambiguation in Information Retrieval Revisited. In The 26th ACM-SIGIR Conference on Research and Development in Information Retrieval (SIGIR'03).
18. Sussna, M. 1993. Word Sense Disambiguation for Free-Text Indexing Using a Massive Semantic Network. In Proceedings of the 2nd International Conference on Information and Knowledge Management (CIKM),  67 – 74, Washington, DC.
19. Swen, Bing (孙斌). 2003. Relative Information and a Sense Matrix Model for IR. Technical Report   TR-003,   ICL,   Peking   Univ.,   Nov   2003.   (available   at http://icl.pku.edu.cn/icl_tr/)
20. Voorhees, E. M. 1993. Using WordNet to Disambiguate Word Sense for Text Retrieval. In Proceedings of the 16th International ACM SIGIR Conference, 171–180, Pittsburgh, PA.

# Query Model Estimations for Relevance Feedback in Language Modeling Approach

Seung-Hoon Na[1], In-Su Kang[1], Kyonghi Moon[2], and Jong-Hyeok Lee[1]

[1] Division of Electrical and Computer Engineering, POSTECH, AITrc, Republic of Korea
{nsh1979, dbaisk, jhlee}@postech.ac.kr
[2] Div. of Computer and Information Engineering
Silla University, Republic of Korea
khmun@silla.ac.kr

**Abstract.** Recently, researchers have successfully augmented the language modeling approach with a well-founded framework in order to incorporate relevance feedback. A critical problem in this framework is to estimate a query language model that encodes detailed knowledge about a user's information need. This paper explores several methods for query model estimation, motivated by Zhai's generative model. The generative model is an estimation method that maximizes the generative likelihood of feedback documents according to the estimated query language model. Focusing on some limitations of the original generative model, we propose several estimation methods to resolve these limitations: 1) three-component mixture model, 2) re-sampling feedback documents with document language models, and 3) sampling a relevance document from a relevance document language model. In addition, several hybrid methods are also examined, which combine the query specific smoothing method and the estimated query language model. In experiments, our estimation methods outperform a simple generative model, showing a significant improvement over an initial retrieval.

## 1 Introduction

The basic idea of the language modeling approach to information retrieval, first introduced by Ponte and Croft [10], is that it does not explicitly assume relevance information, but assumes individual document models for each document and estimates them. With these models, documents are ranked by query likelihood that a document model will generate a given query. In spite of their mathematical simplicity, the language modeling approaches perform well, empirically showing comparative performance to classical probabilistic models [3], [4], [6], [12], [13], [15], [17], [18].

Relevance feedback, the well-known high performing technique is not easily interpreted in the language modeling approach, while the classical probabilistic model provides well-founded interpretations naturally. The main reason is that the language modeling approach does not explicitly assume a relevance document set [12]. Researchers have tried to incorporate relevance feedback into the language modeling approach without harming the original language modeling framework.

S. H. Myaeng et al. (Eds.): AIRS 2004, LNCS 3411, pp. 215–226, 2005.
© Springer-Verlag Berlin Heidelberg 2005

Risk minimization framework in [5], [16] incorporates the relevance feedback into its framework, which includes not only a more generalized language model, but also the classical probabilistic model as a special case. A critical problem in this framework is to estimate a 'query language model', which encodes a probabilistic model for an ideal query sample that is generated to maximally satisfy the user's information need. In this framework, relevance feedback is the process that re-estimates more elaborately the query language model that is naively approximated at initial retrieval.

Several estimation methods of a query language model were proposed: the likelihood ratio [10], the query translation model [5], the generative model over feedback documents [16] and divergence minimization over feedback documents [16]. Of these methods, the generative model over feedback documents (hereafter, the generative model) shows high precision empirically, and is based on an elegant and simple framework.

In this paper, we explore several estimation methods to improve the original generative model by focusing on two of its unsolved problems: *a variety of document topics, and incompleteness of feedback document sample.*

In addition, we examine hybrid methods that combine the above query language models with *query specific smoothing*, which optimize smoothing parameters through feedback documents. This issue is interesting because query specific smoothing is another separate criterion based on the feedback method without estimating the query language model.

The remainder of this paper is organized as follows. In section 2, we review the background of the language modeling approach and the generative model for estimating query language model. In section 3 and section 4, we describe the problems of the generative model in detail and our estimation methods to resolve them, respectively. In section 5, we describe several hybrid methods for combining query language models with query specific smoothing. In section 6, experimental results are presented. Finally, we present our conclusion in section 7.

## 2   Background

### 2.1   Query Language Model in Language Modeling Approach

Basically, the language modeling approach ranks documents in the collection with the query-likelihood (formula 1) that a given query $\mathbf{q}$ would be observed during repeated random sampling from each document model [4], [10], [12]. [1]

$$P(\mathbf{q} \mid \theta_D) = \prod_w P(w \mid \theta_D)^{c_{w,q}}. \tag{1}$$

where $c_{w,q}$ is the number of term $w$ in a given query, $D$ is a document.

To incorporate relevance feedback into the language modeling approach, [5] presented the risk minimization based language modeling framework using the 'query

---

[1]  There is some difference between authors about interpretation of a query. [10] treats a query as a set, while [4], [8], [12] interpreted a query as a sequence of words. We adopt the sequence interpretation.

language model', which is a probabilistic model about user preferences of information, the context of a query, and synonym or semantically related terms. The estimation of query language model is further explored by [16], which is adopted in this study.

Initially, a query language model is approximated by maximum likelihood estimation on a user's initial query sample. Let this initial query language model be the *original query language model* $\hat{\theta}_Q$.

At the feedback stage, *feedback query language model* $\theta_F$ is estimated through feedback documents. Then our new query language model $\theta_Q$ is

$$\theta_Q = (1-\alpha)\hat{\theta}_Q + \alpha\theta_F. \tag{2}$$

where $\alpha$ controls the influence of the feedback query language model.

After estimating new query language model $\theta_Q$ by (2), all documents are re-ranked according to the expectation of query likelihood of document $D$ on query sampled randomly from the estimated new query language model. In order words, if we want to build a new query $\mathbf{q}$ with length $|\mathbf{q}| = n$, the expected query likelihood on a document $D$ is

$$\mathbf{E}_q[\log P(\mathbf{q}|\theta_D)] = \sum_w \mathbf{E}_q[c_{w,q}]\log P(w|\theta_D). \tag{3}$$

where $\mathbf{E}_q[c_{w,q}]$ is the expected count of term $w$ in random query $\mathbf{q}$.

Because $\mathbf{E}_q[c_{w,q}]$ is $n \cdot P(w|\theta_Q)$, formula (3) is

$$\mathbf{E}_q[\log P(\mathbf{q}|\theta_D)] = n\sum_w P(w|\theta_{Q'})\log P(w|\theta_D) \tag{4}$$

$$\propto \sum_w P(w|\theta_{Q'})\log P(w|\theta_D).$$

The original query language model $\hat{\theta}_Q$ is fixed, and only the feedback query language model $\theta_F$ is varied according to several estimation methods.

## 2.2  A Generative Model of Feedback Documents

The original generative model of feedback documents is based on the assumption that each term of the feedback documents is generated by a unigram language model $\theta$. The unigram language model is the mixture model that consists of the unknown feedback query language model $\theta_F$ and a collection language model $\theta_C$. $\theta_F$ is determined to maximize the generative likelihood of feedback documents.

Given feedback documents $F$, the likelihood of them is

$$P(F|\theta) = \prod_i \prod_w P(w|\theta)^{c_{w,d_i}}. \tag{5}$$

The log likelihood of feedback documents is

$$\log P(F|\theta) = \sum_{D \in F}\sum_w c_{w,D}\log(\lambda_F P(w|\theta_F) + (1-\lambda_F)P(w|\theta_C)). \tag{6}$$

where $\lambda_F$ is constant that indicates the amount of background model when generating a document, which is different from a general smoothing parameter $\lambda$ for a document language model.

To maximize (6), we iteratively obtain $\theta_F$ by applying the EM algorithm [2].

$$t^{(k)}(w) = \frac{(1-\lambda_F)P^{(k)}(w|\theta_F)}{(1-\lambda_F)P^{(k)}(w|\theta_F) + \lambda_F P^{(k)}(w|\theta_C)}. \tag{7}$$

$$P^{(k+1)}(w|\theta_F) = \frac{\sum_j c_{w,\mathbf{d}_j} t^{(k)}(w)}{\sum_w \sum_j c_{w,\mathbf{d}_j} t^{(k)}(w)}.$$

## 3 Limitations of Simple Generative Model

As mentioned in section 1, the original generative model with mixed two-component model confronts some obstacles in successfully estimate the query language model. The first obstacle is that most documents have multiple topics: *a variety of document topics*. In the original model, terms in documents are divided into only two categories: relevant terms and general terms. Relevant terms are recognized by the difference between likelihoods generated from the current query language model and a background collection model. Because of various document topics, it is possible that a term may be considered relevant by this two mixture model but it is not truly relevant. This term may occur in non-relevant contexts of the document. To solve this problem, we use a new mixture model with three components by adding a document specific model into the original two components. As the result, we can construct a new estimation method named the *three-component mixture model*.

The second obstacle is that the likelihood function of feedback documents has *sample incompleteness*. Sample incompleteness means that feedback documents have a zero count for unseen terms. This problem is conceptually different from the smoothing issue in the language modeling approach because we want to estimate not the document language model, but the query language model from feedback documents. Feedback documents are simply un-smoothed raw samples, so the likelihood of feedback documents can be biased. If we use ideal example samples having more realistic term occurrences, then the likelihood of formula (5) can be more accurate for estimation.

We devised the two different methods to solve this sample incompleteness problem. The first method is *re-sampling feedback documents from document language models*. Re-sampling will improve the likelihood of feedback documents by substituting old counts of terms by 'expected' counts in the re-sampled document. The second method is *sampling a relevant document from a relevance document language model*. As a result, all feedback documents become samples of this single relevance document language model. In this method, the total number of language models is $N+3$ ($N$ document language models, a collection language model, a query language model and a relevance language model). It is clear that the estimation problem of a relevance document language model from 'several' samples is simpler than the estimation problem of a document language model from 'only one' sample. However, this method causes the difficult problem of estimating the relevance document language model from feedback documents. Estimation of the relevance document language model is the same as the goal of the classical probabilistic model. This problem has also been treated in the language modeling approach [5], [7]. The method of 'sampling a relevant document' is the first trial to assume both the relevance document language model and the query language model.

## 4 New Generative Models for Estimation of Query Language Model

### 4.1 Using Three-Component Mixture Model

A three-component mixture model consists of a query language model $\theta_F$, a background collection language model $\theta_C$, and a document specific topic model $\theta_D^s$.

The log likelihood of feedback documents of (6) is revised

$$\log P(F \mid \theta) = \tag{8}$$
$$\sum_{D \in F} \sum_w c_{w,D} \log(\lambda_F P(w \mid \theta_F) + \lambda_S P(w \mid \theta_D^s) + \lambda_C P(w \mid \theta_C)).$$

where $\lambda_Q + \lambda_S + \lambda_C = 1$, and $p(w \mid \theta_D^s)$ is the document specific topic model of the document $D$.

A remaining problem is to determine the document specific language models, since it highly effects estimation of the final query language model. For the document specific topic model, we use naïve approximation that applies the maximum likelihood estimated document language model. In other words,

$$\log P(F \mid \theta) = \tag{9}$$
$$\sum_{D \in F} \sum_w c_{w,D} \log(\lambda_F P(w \mid \theta_F) + \lambda_S P(w \mid \theta_D^s) + \lambda_C P(w \mid \theta_C)).$$

This approximation is still a matter of discussion. We believe that a more refined document specific topic model will solve the problem of a variety of document topics. For example, it may be revised by using a collection topic language model $\theta_{C_w}$ for term $w$.

$$\tag{10}$$
$$P(w \mid \theta_D^s) \approx \pi P(w \mid \hat{\theta}_D) + (1 - \pi) P(w \mid \theta_{C_w}).$$

where $\pi$ is the fixed constant to control the amount of document specific model over a collection topic language model.

The assumption about the three-component mixture model for a document is based on the multiple Poisson model in the classical probabilistic model [8]. [8] shows statistically that the multiple Poisson model is more adequate for most situations rather than the 2 Possion model. We believe that this issue is important not only for the initial retrieval model, but also for the relevant feedback stage.

## 4.2 Re-sampling Feedback Documents from Document Language Models

In the scheme of re-sampling feedback documents, each document language model is estimated by observing a single document, and a new document sample is generated from each smoothed document language model. If we assume that the length of each document sample is the same with $|\mathbf{d}| = l$, the expected count of term $w$ in a document sample for a document language model $\theta_D$ is

$$E_{\mathbf{d},D}[c_{w,D}] = lP(w | \theta_D). \tag{11}$$

The expected log-likelihood of feedback documents is

$$\log P(F | \theta) = \sum_{D \in F} \sum_{w} E_{\mathbf{d},D}[c_{w,D}] \log(\lambda_F P(w | \theta_F) + (1 - \lambda_F)P(w | \theta_C)) \tag{12}$$

$$= l \sum_{D \in F} \sum_{w} P(w | \theta_D) \log(\lambda_F P(w | \theta_F) + (1 - \lambda_F)P(w | \theta_C)).$$

From this log-likelihood function, we can estimate the query language model in the same manner as described in section 2.2.

## 4.3 Sampling a Relevance Document from a Relevance Document Language Model

In the method of sampling a relevant document, the relevance document language model is estimated by observing all feedback documents, and a single relevant document is sampled from this model.

Estimating the relevance document language model $\theta_R$ is an important sub-problem in this method. Here, we introduce one method used in previous works.

**Relevance Language Model.** Originally, Lavrenko's relevance language model [7] is the model that encodes the probability of observing a word $w$ in the documents relevant to the user's information need. Conceptually, the relevance language model coincides with the relevance 'document' language model to estimate in this method. We use model-1 of the two methods proposed by [7] because Model-2 belongs to the above query translation model with one step. Relevance language model is

$$P_{Rel}(w | \theta_R) \propto \sum_{D} P(\theta_D)P(w | \theta_D)\prod_{q \in \mathbf{q}} P(q | \theta_D). \tag{13}$$

# 5   Combining with Query Specific Smoothing

## 5.1 Query Specific Smoothing from Query Language Model

If an ideal query $\mathbf{q}$ is randomly generated with length $n$ from the estimated query language model $\theta_Q$, and the document language models is smoothed with Jelinek smoothing, then the log of query likelihood of given feedback documents is

$$Q = \sum_D \log P(\mathbf{q} \mid \theta_D) = \sum_q \sum_{D \in F} c_{w,q} \log P(q \mid \theta_D). \tag{14}$$

By taking expectation of $Q$ on query distribution, we obtain

$$Q = \sum_w \sum_{D \in F} \mathbf{E}_q [c_{w,q}] P(q \mid \theta_D) = n \sum_w \sum_{D \in F} P(w \mid \theta_Q) \log P(w \mid \theta_D). \tag{15}$$

With this likelihood function, the Jelinek smoothing parameter $\lambda$ can be estimated using EM algorithm. We call this re-estimation $\lambda$ *query specific smoothing* to discriminate it from 'term specific smoothing' in [3], which assumes an individual smoothing parameter for each term. Performing term specific smoothing from query language model may be dangerous because they all provide term re-weightings, so that term re-weighting is applied twice and term weights become biased. In other words, high weight terms become higher, and low weight terms become lower. Using 'query specific' smoothing instead of 'term specific' can provide more reliable estimation since it falls under 'regulation' of term weights rather than term re-weighting.

### 5.2 Mutual Estimation of Query Language Model and Smoothing Parameter

All generative models mentioned in this paper have a mixing parameter to control the amount of a query language model against a background collection language model. If we assume that this mixing parameter is the same as the smoothing parameter of the document language model, the query language model can be re-estimated further whenever the smoothing parameter (i.e. mixing parameter) is re-estimated. Thus, the query language model and smoothing parameter are complementary in the sense that changing them induces modifying the other together. Figure 4 describes the systematic view of this mutual estimation of the query language model and the query specific smoothing parameter.

Consider a query language model is estimated using a generative model of section 2 or section 4. Then, the smoothing parameters are re-estimated using the method of section 5.1. Again, the query language model is re-estimated using this smoothing parameter, and the smoothing parameter is also re-estimated. The same process is iteratively performed until the smoothing parameter converges into a specific value.

Recall that estimation of query language model and query specific smoothing is to maximize different criteria: 'likelihood of feedback documents' and 'likelihood of query'. A questionable point is whether the smoothing parameter converges or vibrates. Although theoretical issues about convergence remain, we found that the smoothing parameter always converges into a single value after 10-20 steps.

## 6 Experimentation

Our experimentations were performed on NTCIR-3 English topics and document collections. There were 32 topics, which are comprised of four fields: Title, Description, Narrative, and Concept. Table 1 summarizes the size of document set and useful information on each topic field.

**Table 1.** Collection size and the average length of query for each topic field

| # docs | Title | Desc | Conc | Narr | All |
|--------|-------|------|------|------|-----|
| 22,927 | 3.58 | 8.76 | 10.38 | 42.18 | 64.90 |

We use four different versions of each set of queries: (Title) title only, (Desc) description only, (Conc) concept only, (All) all fields. Data pre-processing is the standard. Words are tokenized and stemmed with a Poster stemmer. All stop words are removed using our own stoplist of 274 words. For smoothing, we use the Jelinek method with a mixing parameter value of $\lambda = 0.25$, and for interpolation, we set $\alpha = 0.9$ which is acquired by empirically changing a parameter. We report three performance measures averaged over queries: AvgPr (uninterpolated average precision), RPr (R-precision), Pr@10 (precision at 10 documents).

## 6.1 Evaluation of Estimation Methods

Table 2 describes the acronyms of estimation methods which evaluated in this paper. Q.s.s. indicates which estimation method is selected for query specific smoothing: (None: do not use any query specific smoothing, Normal: method of section 5.1, Mutual: method of section 5.2). Table 3 and Table 4 show the experimental results of 12 estimation methods based on four different estimation methods.

**Table 2.** Estimation methods of the query language model

| Model          Q.s.s | None | Normal | Mutual |
|----------------------|------|--------|--------|
| Two-component mixture | G2M | G2Mq | G2Mu |
| Three-component mixture | G3M | G3Mq | G3Mu |
| Re-sampling feedback doc. | SD | SDq | SDu |
| Sampling a rel doc. (*Rel*) | RR | RRq | RRu |

Here, number surrounded with '(' and ')' indicates % improvement over baseline, bold number means that the method shows significant improvement over baseline (more than 2 percent), and underlined number indicates that the method is best in our experiments.

In Table 3, we observe that the three-component mixture model (G3M) is slightly better over the baseline two-component mixture model (G2M). In the case where query specific smoothing is used, significant improvement can be easily obtained at the three-component mixture model rather than two-component mixture model. Specifically, in the measure of precision at 10 documents (Pr@10), three-component mixture model is always superior to two-component mixture model for all topic fields using the same query specific smoothing.

**Table 3.** Performance of different estimation methods of query language model

| | Title | | | Desc | | |
|---|---|---|---|---|---|---|
| | AvgPr | R-Pr | Pr@10 | AvgPr | R-Pr | Pr@10 |
| No feedback | 0.3222 | 0.3219 | 0.3531 | 0.3735 | 0.3692 | 0.4062 |
| G2M(base) | 0.4056 | 0.3744 | 0.4000 | 0.4465 | 0.4296 | 0.4344 |
| G2Mq | 0.4073 | 0.3784 | 0.4062 | 0.4481 | 0.4331 | 0.4312 |
| | (+0.42) | (+1.07) | (+1.55) | (+0.36) | (+0.81) | (-0.74) |
| G2Mu | 0.4053 | 0.3786 | **0.4094** | 0.4481 | **0.4418** | 0.4281 |
| | (-0.07) | (+1.12) | (+2.35) | (+0.36) | (+2.84) | (-1.45) |
| G3M | 0.4059 | 0.3743 | 0.4062 | 0.4461 | **0.4447** | 0.4406 |
| | (+0.07) | (-0.03) | (+1.55) | (-0.09) | (+3.51) | (1.43) |
| G3Mq | 0.4086 | 0.3813 | **0.4219** | 0.4499 | **0.4448** | 0.4344 |
| | (+0.74) | (+1.84) | (+5.47) | (+0.76) | (+3.54) | (0) |
| G3Mu | 0.4078 | 0.3759 | **0.4250** | 0.4445 | 0.4344 | 0.4344 |
| | (+0.54) | (+0.40) | (+6.25) | (-0.45) | (+1.12) | (0) |
| | Conc | | | All | | |
| | AvgPr | R-Pr | Pr@10 | AvgPr | R-Pr | Pr@10 |
| No feedback | 0.3980 | 0.4207 | 0.4406 | 0.4688 | 0.4617 | 0.4437 |
| G2M(base) | 0.4882 | 0.4856 | 0.4750 | 0.4884 | 0.4914 | 0.4656 |
| G2Mq | 0.4892 | 0.4846 | 0.4750 | 0.4887 | 0.4954 | 0.4656 |
| | (+0.20) | (-0.21) | (0) | (+0.06) | (+0.81) | (0) |
| G2Mu | 0.4919 | 0.4879 | 0.4750 | 0.4970 | 0.4975 | 0.4687 |
| | (+0.76) | (+0.47) | (0) | (+1.76) | (+1.24) | (+0.67) |
| G3M | 0.4917 | 0.4838 | 0.4812 | 0.4916 | 0.4871 | 0.4687 |
| | (+0.72) | (-0.37) | (+1.31) | (+0.66) | (-0.88) | (+0.67) |
| G3Mq | 0.4869 | 0.4863 | 0.4750 | 0.4929 | 0.4884 | **0.4781** |
| | (-0.27) | (+0.14) | (0) | (+0.92) | (-0.61) | (+2.68) |
| G3Mu | 0.4954 | **0.4981** | 0.4781 | 0.4934 | 0.4928 | **0.4750** |
| | (+1.47) | (+2.57) | (+0.65) | (+1.02) | (+0.28) | (+2.02) |

Sampling methods (SD, SDq, SDu, SR, SRq, SRu), show the different extent to which their improvement varies according to differing query fields. Methods of re-sampling feedback documents (SD, SDq, SDu) failed to improve a baseline generative model showing a worse performance for title and concept field queries, while they show a slight improvement about 1~2 percent over the baseline method for long queries. Note that methods of sampling a relevant document (RR, RRq, RRu) yield significant improvement for description and all topic field queries (All), although they show a worse performance for title queries. For title queries, the above two methods never improve the baseline method. This result implies that re-sampling may not contribute to solve the sample incompleteness problem. For this reason, in next section, we will further examine sampling methods by combining three-mixture generative model with it, and provide an evidence of potential existence of sample incompleteness problem.

**Table 4.** Performance of different estimation methods of query language model (Cond.)

| | Title | | | Desc | | |
|---|---|---|---|---|---|---|
| | AvgPr | R-Pr | Pr@10 | AvgPr | R-Pr | Pr@10 |
| No feed-back | 0.3222 | 0.3219 | 0.3531 | 0.3735 | 0.3692 | 0.4062 |
| G2M(base) | 0.4056 | 0.3744 | 0.4000 | 0.4465 | 0.4296 | 0.4344 |
| SD | 0.3836 | 0.3586 | 0.3750 | 0.4506 | 0.4318 | **0.4469** |
| | (-5.42) | (-4.22) | (-6.25) | (+0.92) | (+0.51) | (+2.88) |
| SDq | 0.3865 | 0.3625 | 0.3781 | 0.4506 | 0.4310 | 0.4406 |
| | (-4.71) | (-3.18) | (-5.48) | (+0.92) | (+0.33) | (+1.43) |
| SDu | 0.3928 | 0.3645 | 0.3813 | **0.4566** | 0.4314 | 0.4406 |
| | (-3.16) | (-2.64) | (-4.68) | (+2.26) | (+0.42) | (+1.43) |
| RR | 0.3840 | 0.3624 | 0.3844 | **0.4756** | **0.4650** | 0.4406 |
| | (-5.33) | (-3.21) | (-3.90) | (+6.52) | (+8.24) | (+1.43) |
| RRq | 0.3842 | 0.3609 | 0.3844 | **0.4751** | **0.4617** | 0.4375 |
| | (-5.28) | (-3.61) | (-3.90) | (+6.41) | (+7.41) | (+0.71) |
| RRu | 0.3841 | 0.3606 | 0.3812 | **0.4739** | **0.4487** | 0.4375 |
| | (-5.30) | (-3.69) | (-4.70) | (+6.14) | (+4.45) | (+0.71) |
| | Conc | | | All | | |
| | AvgPr | R-Pr | Pr@10 | AvgPr | R-Pr | Pr@10 |
| No feedback | 0.3980 | 0.4207 | 0.4406 | 0.4688 | 0.4617 | 0.4437 |
| G2M(base) | 0.4882 | 0.4856 | 0.4750 | 0.4884 | 0.4914 | 0.4656 |
| SD | 0.4627 | 0.4760 | 0.4781 | 0.4886 | 0.4706 | **0.4750** |
| | (-5.22) | (-1.98) | (+0.65) | (+0.04) | (-4.23) | (+2.02) |
| SDq | 0.4667 | 0.4672 | 0.4687 | 0.4896 | 0.4866 | **0.4781** |
| | (-4.40) | (-3.79) | (-1.33) | (+0.25) | (-0.98) | (+2.68) |
| SDu | 0.4660 | 0.4582 | 0.4687 | 0.4884 | 0.4769 | **0.4813** |
| | (-4.55) | (-5.64) | (-1.33) | (0) | (-2.95) | (+3.37) |
| RR | 0.4795 | 0.4885 | 0.4750 | **0.5025** | 0.4930 | 0.4625 |
| | (-1.78) | (+0.60) | (0) | (+2.89) | (+0.33) | (-0.67) |
| RRq | 0.4807 | 0.4881 | 0.4719 | **0.5026** | 0.4894 | 0.4656 |
| | (-1.54) | (+0.51) | (-0.65) | (+2.91) | (-0.41) | (0) |
| RRu | 0.4858 | 0.4899 | 0.4844 | **0.5115** | 0.4919 | 0.4719 |
| | (-0.49) | (+0.89) | (+1.98) | (+4.73) | (+0.10) | (+1.35) |

# 7  Conclusion

The estimation of query language model is an important problem for relevance feed-back in language modeling approach. Our works was motivated by Zhai's generative model, which is derived within well-founded framework and empirically well performed. We highlighted two unsolved problems in the original generative model: *a variety of document topics*, and *incompleteness of the feedback document sample*. To solve these problems, we proposed three new estimation methods and evaluated them; 1) a three-component mixture model, 2) re-sampling feedback documents with document language models, and 3) sampling a relevance document from a relevance document language model. Specifically, in our third framework, we recognized that the estimating problem of a relevance document language model [6], [7] is a sub-problem for solving relevance feedback in language modeling approach.

## Acknowledgements

This work was supported by the KOSEF through the Advanced Information Technology Research Center(AITrc) and by the BK21 Project.

## References

1. Berger, A. and Lafferty, J. Information Retrieval as Statistical Translation. In Proceedings of 22nd Annual International ACM SIGIR Conference on Research and Development in Information Retrieval, pages 222-229, 1999
2. Dempster, A. Maximum Likelihood from Incomplete Data via the EM algorithm. In Journal of Royal Statistical Society, vol. 39, no. 1, pages 1-39, 1977
3. Hiemstra, D. Term Specific Smoothing for Language Modeling Approach to Information Retrieval: The Importance of a Query Term. In Proceedings of 25th Annual International ACM SIGIR Conference on Research and Development in Information Retrieval, pages 35-41, 2002
4. Hiemstra, D. Using Language Models for Information Retrieval. In PhD Thesis, University of Twente, 2001
5. Lafferty, J. and Zhai, C. Document Language Models, Query Models, and Risk Minimization for Information Retrieval. In Proceedings of 24th Annual International ACM SIGIR Conference on Research and Development in Information Retrieval, pages 111-119, 2001
6. Lam-Adesina, A. and Jones, G. Applying Summarization Techniques for Term Selection in Relevance Feedback. In Proceedings of 24th Annual International ACM SIGIR Conference on Research and Development in Information Retrieval, pages 1-9, 2001
7. Lavrenko, V. and Croft, B. Relevance-based language models. In Proceedings of 24th Annual International ACM SIGIR Conference on Research and Development in Information Retrieval, pages 120-127, 2001
8. Miller, D., Leek, T. and Schwartz, R. A Hidden Markov Model Information Retrieval System. In Proceedings of 22nd Annual International ACM SIGIR Conference on Research and Development in Information Retrieval, pages 214-221, 1999
9. Ng, K. A Maximum Likelihood Ratio Information Retrieval Model. In TREC-8 Workshop notebook, 1999
10. Ponte, A. A Language Modeling Approach to Information Retrieval. In PhD thesis, Dept. o Computer Science, Univercity of Massachusetts, 1998
11. Robertson, S. and Hiemstra, D. Language Models and Probability of Relevance. In Proceedings of the Workshop on Language Modeling and Information Retrieval, 2001

12. Song, F. and Croft, W. A General Language Model for Information Retrieval. In Proceedings of 22nd Annual International ACM SIGIR Conference on Research and Development in Information Retrieval, pages 279-280, 1999

13. Srikanth, M. and Srihari, R. Biterm Language Models for Document Retrieval. In Proceedings of 25th Annual International ACM SIGIR Conference on Research and Development in Information Retrieval, pages 425-426, 2002

14. Xu, J. and Croft, W. Query Expansion using Local and Global Document Analysis. In Proceedings of 19th Annual International ACM SIGIR Conference on Research and Development in Information Retrieval, pages 4-11, 1996

15. Zaragoza, H. and Hiemstra, D. Bayesian Extension to the Language Model for Ad Hoc Information Retrieval. In Proceedings of 26th Annual International ACM SIGIR Conference on Research and Development in Information Retrieval, pages 4-9, 2003

16. Zhai, C. and Lafferty, J. Model-based Feedback in the Language Modeling Approach to Information Retrieval. In Proceedings of the 10th international conference on Information and knowledge management, pages 430-410, 2001

17. Zhai, C. and Lafferty, J. A Study of Smoothing Methods for Language Models Applied to Ad Hoc Information Retrieval. In Proceedings of 24th Annual International ACM SIGIR Conference on Research and Development in Information Retrieval, pages 334-342, 2001

18. Srikanth, M. and Srihari, R. Biterm Language Models for Document Retrieval. In Proceedings of 25th Annual International ACM SIGIR Conference on Research and Development in Information Retrieval, pages 425-426, 2002

19. Zaragoza, H. and Hiemstra, D. Bayesian Extension to the Language Model for Ad Hoc Information Retrieval", In Proceedings of 26th Annual International ACM SIGIR Conference on Research and Development in Information Retrieval, pages 4-9, 2003

20. Zhai, C. and Lafferty, J. Model-based Feedback in the Language Modeling Approach to Information Retrieval. In Proceedings of the 10th international conference on Information and knowledge management, pages 430-410, 2002

21. Zhai, C. and Lafferty, J. A Study of Smoothing Methods for Language Models Applied to Ad Hoc Information Retrieval. In Proceedings of 24th Annual International ACM SIGIR Conference on Research and Development in Information Retrieval, pages 334-342, 2001

# A Measure Based on Optimal Matching in Graph Theory for Document Similarity

Xiaojun Wan and Yuxin Peng

Institute of Computer Science and Technology,
Peking University, Beijing 100871, China
{wanxiaojun, pengyuxin}@icst.pku.edu.cn

**Abstract.** Measuring pairwise document similarity is critical to various text retrieval and mining tasks. The most popular measure for document similarity is the Cosine measure in Vector Space Model. In this paper, we propose a new similarity measure based on optimal matching in graph theory. The proposed measure takes into account the structural information of a document by considering the word distributions over different text segments. It first calculates the similarities for different pairs of text segments in the documents and then gets the total similarity between the documents optimally through optimal matching. We set up experiments of document similarity search to test the effectiveness of the proposed measure. The experimental results and user study demonstrate that the proposed measure outperforms the most popular Cosine measure.

## 1 Introduction

Measuring pairwise document similarity is critical to various text retrieval and mining tasks, such as clustering, filtering and similarity search. It is thus important to measure similarity as effectively as possible. To date, many similarity measures have been proposed and implemented, such as the Cosine measure [3], the Jaccard measure [8], the Dice Coefficient measure [13], the information-theoretic measure [2], the DIG-based measure [6], etc. Among these measures, the Cosine measure is the most popular one and has been successfully used in various tasks.

To our knowledge, most poplar similarity measures between documents utilize feature vectors to represent the original documents and then calculate the similarity between the feature vectors, such as the Cosine measure and the Jaccard measure. These measures do not take into account the structural information of a document. The compressed feature vectors lose the word distribution over text segments in the documents. In fact, different text segments represent different subtopics [7]. For example, one dimension in a document feature vector can be the word *"IBM"*, and its weight can be its frequency 3, but we do not know how the word *"IBM"* distributes in the document: Maybe these three instances of *"IBM"* appear in the same text segment discussing the WebSphere software, or each *"IBM"* may appear in a distinct text segment: one discussing the Lotus

S. H. Myaeng et al. (Eds.): AIRS 2004, LNCS 3411, pp. 227–238, 2005.
© Springer-Verlag Berlin Heidelberg 2005

software, one discussing WebSphere software and the other discussing the Tivoli software. The different word distributions over text segments will influence the similarity between documents: the more similar the word distributions are, the more similar the documents will be.

In this paper, we propose a new similarity measure based on optimal matching theory between documents. The proposed similarity measure takes into account the structural information of a document by considering the word distribution over different text segments. It first calculates the similarities for different pairs of text segments in the documents and then gets the total similarity between the documents optimally through optimal matching. Experiments and a user study are performed to demonstrate the effectiveness of the proposed similarity measure, which outperforms the popular Cosine measure. We also examine the approach of text segmentation and the similarity measure between segments, which are the key issues influencing the performance of the proposed measure significantly.

The rest of this paper is organized as follows: Section 2 introduces in detail the popular Cosine measure widely used for document similarity search. The optimal matching theory is explained in section 3. In Section 4, we propose a document similarity measure based on optimal matching. Experiments and results are described in Section 5. A preliminary user study is performed in Section 6. Section 7 gives our conclusions and future work.

## 2     Vector Space Model and Cosine Measure

A variety of different retrieval models have been developed to represent documents and measure the similarity between documents in the IR field, including the Boolean model, probabilistic model, Vector Space Model(VSM) [3, 15], and the more complicated Latent Semantic Analysis [5]. The Vector Space Model is one of the most effective models and has been widely used for filtering, clustering and retrieval in IR field.

The Vector Space Model creates a space in which documents are represented by vectors. For a fixed collection of documents, an m-dimensional vector is generated for each document from sets of terms with associated weights, where m is the number of unique terms in the document collection. Then, a vector similarity function, such as the inner product, can be used to compute the similarity between vectors.

In VSM, weights associated with the terms are calculated based on the following two numbers:

1. term frequency, $f_{ij}$ the number of occurrences of term $t_j$ in document $d_i$; and
2. inverse document frequency, $idf_j = \log(N/n_j)$ , where $N$ is the total number of documents in the collection and $n_j$ is the number of documents containing term $t_j$.

The similarity $sim(d_1, d_2)$, between two documents $d_1$ and $d_2$, can be defined as the normalized inner product of the two vectors $\boldsymbol{d_1}$ and $\boldsymbol{d_2}$:

$$sim_{cosine}(d_1, d_2) = \frac{\sum_{j=1}^{m} w_{1j}w_{2j}}{\sqrt{\sum_{j=1}^{m}(w_{1j})^2 \cdot \sum_{j=1}^{m}(w_{2j})^2}} \qquad (1)$$

where $m$ is the number of unique terms in the document collection. Document weight $w_{ij}$ is

$$w_{ij} = f_{ij} \cdot idf_j = f_{ij} \cdot \log(N/n_j) \ . \qquad (2)$$

Equation (1) is also known as the Cosine measure, which is the most popular similarity measure for document similarity search. The Cosine measure can capture a scale invariant understanding of similarity. An even stronger property is that the Cosine similarity does not depend on the length: $sim(\alpha d_1, d_2) = sim(d_1, d_2)$ for $\alpha > 0$. This allows documents with the same composition, but different totals to be treated identically. Also, due to this property, samples can be normalized to the unit sphere for more efficient processing.

However, in the Vector Space Model, each document is compressed into a vector, which loses the structural information about term distributions over text segments. In the extreme case, two documents with identical vectors may be composed of different sentences. When compared with a given document using the Cosine measure, they get the same similarity values. But in fact, they should have different similarity values when judged manually.

## 3   Optimal Matching in Graph Theory

Optimal matching(OM) and maximal matching(MM) are classical problems in graph theory. Let $G = \{X, Y, E\}$ be a bipartite graph, where $X = \{x_1, x_2, \ldots, x_n\}$ and $Y = \{y_1, y_2, \ldots, y_m\}$ are the partitions, $V = X \bigcup Y$ is the vertex set, and $E = \{e_{ij}\}$ is the edge set. A matching $M$ of $G$ is a subset of the edges with the property that no two edges of $M$ share the same node. Given the unweighted bipartite graph $G$, as illustrated in Figure 1, MM is to find a matching $M$ that has as many edges as possible. OM is basically an extension of MM, where a weight $w_{ij}$ is assigned to every edge $e_{ij}$ in $G$, as illustrated in Figure 1. OM is to find the matching M that has the largest total weight.

Given the weighted bipartite graph $G$, the Kuhn–Munkres algorithm [9] is employed to solve the OM problem. The algorithm is as follows:

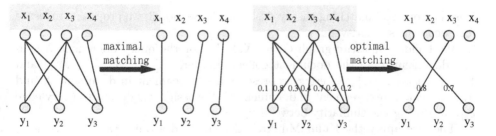

**Fig. 1.** Maximal matching and optimal matching

1. Start with initial label of $l(x_i) = max_j e_{ij}$ , $l(y_j) = 0$ ; $i, j = 1, 2, \ldots, t$ ; $t = max(n, m)$ .
2. Find edge set $E_l = \{(x_i, y_j) | l(x_i) + l(y_j) = e_{ij}\}$ , $G_l = (X, Y, E_l)$ and the matching $M$ in $G_l$.
3. If $M$ consists of all vertices in $X$, $M$ is the optimal matching of $G$ and the algorithm ends. Otherwise, go to next step.
4. Find a vertex $x_i \in X$ where $x_i$ is not inside $M$. Let $A \leftarrow \{x_i\}$ and $B \leftarrow \{\}$ . $A$ and $B$ are different sets.
5. If $N_{G_l} = B$ , then go to step 9, otherwise go to next step. $N_{G_l} \subseteq Y$ corresponds to the set of vertices that matches the vertices in set $A$.
6. Find a vertex $y_j \in N_{G_l}(A) - B$.
7. If $(y_j, z) \in M$, let $A \leftarrow A \bigcup \{z\}$ , $B \leftarrow B \bigcup \{y_j\}$ , then go to step 5. Otherwise go to next step.
8. There exists an augmenting path $P$ from $x_i$ to $y_j$, let $M \leftarrow M \bigoplus E(P)$ , and go to step 3.
9. Compute the value of $a = min_{x_i \in A, y_j \notin N_{Q_l}(A)}\{l(x_i) + l(y_j) - e_{ij}\}$ , then construct a new label $l'$ by

$$l'(v) = \begin{cases} l(v) - a & \text{if } v \in A \\ l(v) + a & \text{if } v \in B \\ l(v) & \text{otherwise} \end{cases} .$$

Then compute the value of $E_{l'}$, $G_{l'}$ according to $l'$.
10. Replace $l \leftarrow l'$ , $G_l \leftarrow Gl'$ , and go to step 6.

The computational complexity of Kuhn–Munkres algorithm is $O(t^3)$, where $t = max(n, m)$ .

To date, maximal matching and optimal matching in graph theory have been successfully employed to measure the similarity between video clips in content based video retrieval [10]. Inspired by this, we propose a novel similarity measure based on optimal matching theory to better measure the similarity between documents.

## 4     The Proposed Similarity Measure

### 4.1     The Proposed Measure

Based on optimal matching theory described in Section 3, we propose a measure for document similarity.

We build a bipartite graph $G = \{X, Y, E\}$ for the query document $X$ and any document $Y$ in the document collection. Here, $X = \{x_1, x_2, \ldots, x_n\}$ and $Y = \{y_1, y_2, \ldots, y_m\}$ , where $x_i$ represents a text segment in document $X$ and $y_j$ represents a text segment in document $Y$. We assign weight $w_{ij}$ to every edge $e_{ij}$, measuring the similarity between $x_i$ and $y_j$.

Then we apply the Kuhn–Munkres algorithm to acquire the total value of the optimal matching in the graph. In order to balance the effect of the lengths of different documents, we normalize the total value as follows:

$$sim_{optmatch}(d_1, d_2) = \frac{optmatch(d_1, d_2)}{min(length(d_1), length(d_2))} \qquad (3)$$

where $optmatch(d_1, d_2)$ represents the total value of the optimal matching for $d_1$ and $d_2$. $length(d)$ represents the count of text segments in document $d$ and $min(a, b)$ returns the minimal value of $a$ and $b$.

Lastly, we use the normalized value as the final similarity between the two documents.

The proposed similarity measure takes into account the structural information of a document by considering the term distribution over text segments, thus capturing the subtopic structure of the document.

## 4.2 Text Segmentation

As described above, $x_i$ and $y_j$ represent text segments in documents. We expect that the text segment used in the proposed measure is cohesive and coherent and may map to a distinct subtopic as well. Naturally, single sentence is a good candidate for text segment. Overlapping adjacent sentences can be used as text segment in that adjacent sentences are usually linguistically related. The number of the adjacent sentences can vary in our experiments. Noting that the paragraph structure of a document is not available for our data, we do not consider using paragraphs as text segments.

The above approaches for text segmentation are simple and heuristic–based and can not discover the subtopic structure of document. Here, we employ the algorithm of TextTiling [5] to partition texts into coherent multi-paragraph segments, which reflect the subtopic structure of the texts. For TextTiling, subtopic discussions are assumed to occur within the scope of one or more overarching main topics, which span the length of the text. Since the segments are adjacent and non-overlapping, they are called TextTiles.

The algorithm of TextTiling detects subtopic boundaries by analyzing patterns of lexical co-occurrence and distribution with the text. The main idea is that terms that describe a subtopic will co-occur locally, and a switch to a new subtopic will be signaled by the ending of co-occurrence of one set of terms and the beginning of the co-occurrence of a different set of terms. The algorithm has three parts: tokenization into terms and sentence-sized units, determination of a score for each sentence-sized unit, and detection of the subtopic boundaries, which are assumed to occur at the largest valleys in the graph resulting from plotting sentence-units against scores.

After texts are segmented with TextTiling, we can use the multi-paragraph segments as vertices in the bipartite graph.

## 4.3 The Edge Weighting Measures

In the proposed approach, the edge weighting measure is a key issue and significantly influences the final performance.

For two text segments $x_i$ and $y_j$, we can use the popular Cosine measure or the Jaccard measure to measure the similarity between them. Here, we also use $x_i$

and $y_j$ to represent the set of words in the text segments respectively. The Binary Jaccard measure measures the ratio of the number of shared attributes(words) of $x_i$ and $y_j$ to the number possessed by $x_i$ and $y_j$:

$$sim_{bi-jaccard}(x_i, y_j) = \frac{|x_i \bigcap y_j|}{|x_i \bigcup y_j|} \tag{4}$$

where $|x_i \bigcap y_j|$ denotes the number of the words shared by $x_i$ and $y_j$, and $|x_i \bigcup y_j|$ denotes the number of the words possessed by $x_i$ and $y_j$.

The Jaccard measure can be extended to continuous or discrete non-negative features as follows [14]:

$$sim_{con-jaccard}(x_i, y_j) = \frac{x_i \bullet y_j}{\|x_i\|_2^2 + \|y_j\|_2^2 - x_i \bullet y_j} \tag{5}$$

where $x_i$ and $y_j$ represent the corresponding feature vectors respectively.

Taking into account a word's $idf$ value, we have another simple extension for Jaccard measure as follows:

$$sim_{idf-jaccard}(x_i, y_j) = \frac{\sum_{t \in x_i \bigcap y_j} idf_t}{\sum_{t \in x_i \bigcup y_j} idf_t} \tag{6}$$

where $x_i \bigcap y_j$ denotes the words shared by $x_i$ and $y_j$, and $x_i \bigcup y_j$ denotes the words possessed by $x_i$ and $y_j$.

For Cosine measure and Continuous Jaccard measure, we use $tf \cdot idf$ as the weight for every word in the vector.

## 5    Experiments

### 5.1    Experimental Setup

We design a simple document retrieval system to test the effectiveness of the proposed similarity measure in comparison with the popular Cosine measure. Given a query document, the retrieval system is to find the similar documents to the given document and return a ranked document list. We implement it just by comparing each document in the document collection and the query document and then sort the documents in the collection by their similarity values.

To evaluate the proposed approach, a ground truth data set is required. To our knowledge, there is no gold standard data set for evaluation of our system. So we build the ground truth data set from the TDT-3 corpus, which has been used for evaluation of the task of topic detection and tracking [1] in 1999 and 2000. TDT-3 corpus is annotated by LDC from 8 English sources and 3 Mandarin sources for the period of October through December 1998. 120 topics are defined and about 9000 stories are annotated over these topics with an "on-topic" table presenting all stories explicitly marked as relevant to a given topic.

According to the specification of TDT, the on-topic stories within the same topic are similar and relevant. After removing the stories written in Chinese, we use 40 topics and more than 2500 stories as a test set, while the others are used as a training set. Sentence tokenization is firstly applied to all documents. The stop word list in SMART [12] is employed in order to remove stop words. Then we use Porter's stemmer [11] to remove common morphological and inflectional endings from English words. The total stories are considered as the document collection for search, and for each topic we simulate a search as follows: The first document within the topic is considered as the query document and all the other documents within the same topic are the relevant documents, while all the documents within other topics are considered irrelevant to the query document. Then the system compares this document with all documents in the document collection, returning a ranked list of 100 documents. The higher the document is in the ranked list, the more similar it is with the query document. Thus we can use classical precision($P$) at top $N$ results to measure the performance:

$$P@N = \frac{|C \bigcap R|}{|R|} \tag{7}$$

where $R$ is the set of top $N$ similar documents returned by our system, and $C$ is the set of relevant documents defined above for a given query document. In most of our experiments, we use $P@5$, $P@10$ and $p@20$ for evaluation. For each search, we can get the corresponding $P@5$, $P@10$ and $p@20$. Then we calculate the average $P@5$, $P@10$ and $p@20$ over the 40 topics as the final precisions.

Note that the number of documents within each topic is different and some topics contain less than 5 documents, so its corresponding precisions may be low. But these circumstances do not affect the comparison of the performance of different systems.

In our experiments, we use the popular Cosine measure as the baseline similarity measure for this document retrieval problem. We compare the performance of the proposed similarity measure with the baseline to show the effectiveness of the proposed approach.

The proposed approach may have different settings for text segment and similarity measure between text segments respectively. For text segment, we can use single sentence (uni-sentence or uni-), adjacent two sentences (bi-sentence or bi-), adjacent three sentences (tri-sentence or tri-), adjacent four sentences (four-sentence or four-), adjacent five sentences (five-sentence or five-) or multi-paragraph segment (TextTile). We use the JTextTile [4], which is a Java implementation of Marti Hearst's text tiling algorithm, to segment texts into coherent topic segments with the recommended parameter settings. For the similarity measure between text segments, we can use the Cosine measure, Bi-Jaccard measure, Continuous Jaccard measure, and Idf-Jaccard measure. Different settings result in different performance of the proposed approach.

## 5.2    Experimental Results

The results of $P@5$, $P@10$ and $P@20$ are shown in Figures 2-4, respectively. In Figures 2-4, "*cosine*" means the traditional Cosine measure is used to measure the similarity between documents. "*optmatch(cosine)*" means we employ the proposed measure with the Cosine measure measuring the similarity between text segments. "*optmatch(bi-jaccard)*", "*optmatch(idf-jaccard)*" and "*optmatch(con-jaccard)*" use the Bi-Jaccard measure, Idf-Jaccard measure and Continuous Jaccard measure respectively to measure the similarity between text segments. The uni-, bi-, tri-, four-, five-sentence and TextTile are various choices for text segment in the proposed measure.

As can be seen from Figures 2-4, the best performance achieved by the proposed measure is better than that of the Cosine measure. For precision at top 5 results, the proposed measures with most different settings outperform the Cosine measure. Table 1 gives the best values of $P@5$, $P@10$ and $P@20$ for the proposed measure and their corresponding settings(text segment and similarity measure for text segments). In Table 1, the upper bounds of $P@5$, $p@10$ and $p@20$ are given and these precisions are not 100% in that the number of relevant documents for a query document is not always larger than 5, 10 or 20. Seen from the figures and the table, we actually achieve a significant improvement on the $P@5$ and $P@10$ compared to the most popular Cosine measure.

The curves in above figures show that when the number of the adjacent sentences becomes larger, the performance of the proposed measure is improved, but this ascending trend stops when text segment contains adjacent five sentences. The measures using TextTile as text segment mostly get better performance of $P@10$ and $P@20$ than those using adjacent sentences as text segment. Given a similarity measure for text segments, the proposed measure with the setting of single sentence gets the worst results.

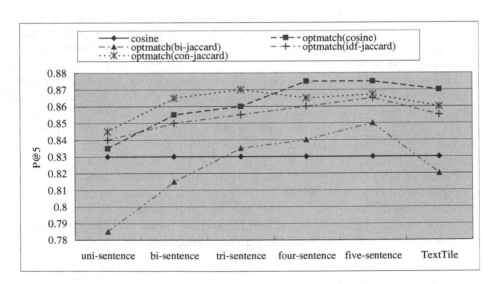

**Fig. 2.** P@5 comparison for different similarity measures

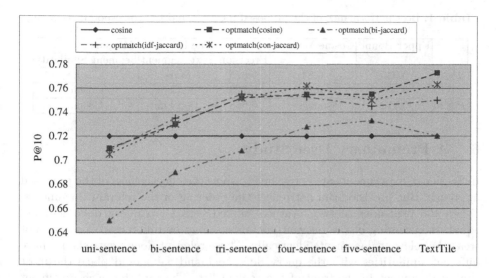

**Fig. 3.** P@10 comparison for different similarity measures

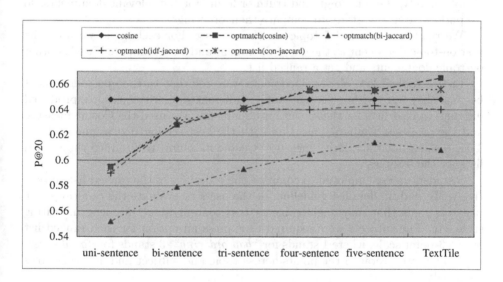

**Fig. 4.** P@20 comparison for different similarity measures

We can see from the curves that with the same text segment, the proposed measures using the Continuous Jaccard measure and the Cosine measure to measure text segment similarity get almost the same best performance. The Idf-Jaccard measure gets a little lower performance than the above two measures. The Bi-Jaccard measure always gets the worst results compared to other measures. Inverse document frequency(IDF) does improve the performance.

Overally, (TextTile+Cosine measure) is the best setting for the proposed measure.

**Table 1.** Best performance of the proposed measure and the corresponding setting

| | Upper Bound | Cosine Measure | The Proposed Measure | | |
|---|---|---|---|---|---|
| | | | Precision | Text Segment | Segment Similarity |
| $P@5$ | 0.935 | 0.83 | 0.875 | four- | Cosine |
| $P@10$ | 0.863 | 0.72 | 0.773 | TextTile | Cosine |
| $P@20$ | 0.772 | 0.648 | 0.665 | TextTile | Cosine |

## 6    A Preliminary User Study

In Section 5, experiments are performed to show the effectiveness of the proposed approach. But precision can not reflect the ordering of the relevant documents. When the precisions of two returned ranked lists for a query document are equal, the ordering of the relevant documents may be different. We should not treat the relevant documents identically. These relevant documents should have different similarities with the query document and we should place the most similar document in the front of the ranked list. The rank for each document in the ranked list represents the degree of similarity with the query document. Unfortunately, there is no ground truth ordering for the relevant documents, so we performed a user study to compare their orderings.

We randomly selected 5 topics in the test set. For each topic we used the first on-topic document as the query document and then compared it with other on-topic documents and get a ranked list.

In this ranked list, all the documents were relevant to the query document because they were within the same topic. Different similarity measures produced different ranked lists for these relevant documents. We used the Cosine measure and the proposed similarity measure with the best setting(text segment: Text-Tile; text segment similarity measure: Cosine measure) to produce the ranked lists for the 5 topics.

5 students were employed to compare the orderings of the ranked list subjectively. We had no detailed guideline for the users to evaluate the orderings. We just told them that they might consider text content, structure, length, writing style, etc. The subjects were required to express an option for each ranked list over a 3-point scale, where 1 stands for *"bad ordering"*, 3 stands for *"good ordering"*. Then we averaged the points across topic and subject and Table 2 shows the result.

**Table 2.** Subject's points for different ranked list on a scale of 1(worst) to 3(best)

| | Cosine Measure | The Proposed Measure |
|---|---|---|
| Topic 1 | 1.2 | 1.0 |
| Topic 2 | 1.6 | 2.0 |
| Topic 3 | 1.6 | 2.2 |
| Topic 4 | 2.2 | 2.2 |
| Topic 5 | 2.0 | 2.6 |
| Average | **1.72** | **2.0** |

As can be seen from Table 2, for most topics, the proposed measure produced a more satisfying ranked list than the Cosine measure. The proposed measure put the most similar documents in the front of the ranked list. In a sense, the proposed measure has a better understanding of *"similarity"* than Cosine measure.

## 7   Conclusions and Future Work

In this paper, we have proposed a new measure for document similarity. The proposed measure is based on optimal matching in graph theory. The proposed measure is able to capture the structural information of a document by considering the word distributions over text segments. Experimental results and a preliminary user study shows that the proposed measure outperforms the most popular Cosine measure. We have also discussed the two key issues in the proposed measure: text segment choice and similarity measure between text segments. Results show that taking TextTile as text segment and the Cosine measure as similarity measure for segments can get the best performance.

In future work, we will apply the proposed measure to more tasks, e.g. document clustering, duplicates detection and story link detection. Though TextTiling is a popular approach to capture the subtopic structure of a document, we will try to explore better approaches to get the discourse structure suitable to measure document similarity.

## Acknowledgements

We thank John Chen for both his great suggestions and his careful amendments for this paper. Lastly, we thank those students for the user study.

## References

1. Allan, J., Carbonell, J., Doddington, G., Yamron, J. P., and Yang, Y.: Topic Detection and Tracking Pilot Study: Final Report. Proceedings of DARPA Broadcast News Transcription and Understanding Workshop (1998) 194–218
2. Aslam, J. A. and Frost, M.: An Information-theoretic Measure for Document Similarity. Proceedings of the 26th International ACM/SIGIR Conference on Research and Development in Information Retrieval (2003)
3. Baeza-Yates, R. and Ribeiro-Neto, B.: Modern Information Retrival. (1999)
4. Choi, F.: JTextTile: A Free Platform Independent Text Segmentation Algorithm. http://www.cs.man.ac.uk/ choif
5. Deerwester, S. C., Dumais, S. T., Landauer, T. K., Furnas, G. W. and Harshman, R. A.: Indexing by Latent Semantic Analysis. Journal of the American Society of Information Science **41–6** (1990) 211–240
6. Hammouda, K. M. and Kamel, M. S.: Document Similarity Using a Phrase Indexing Graph Model. Journal of Knowledge and Information Systems **6–4** (2004)
7. Hearst, M. A.: Multi-paragraph Segmentation of Expository Text. Proceedings of the 32nd Meeting of the Association for Computational Linguistics (1994)

8. Jones, W. P. and Furnas, G. W.: Pictures of Relevance: a Geometric Analysis of Similarity Measure. Journal of the American Society for Information Science **38**–**6** (1987) 420–442
9. Lovasz, L. and Plummer, M. D.: Matching Theory. (1986)
10. Peng, Y. X., Ngo, C. W., Dong, Q. J., Guo, Z. M. and Xiao, J. G.: Video Clip Retrieval by Maximal Matching and Optimal Matching in Graph Theory. Proceedings of 2003 IEEE International Conference on Multimedia & Expo (2003)
11. Porter, M. F.: An Algorithm for Suffix Stripping. Program **14**–**3** (1980) 130-137
12. Salton, G.: The SMART Document Retrieval Project. Proceedings of the Fourteenth International ACM/SIGIR Conference on Research and Development in Information Retrieval (1991)
13. Smadja, F.: Translating Collocations for Bilingual Lexicons: a Statistical Approach. Computational Linguistics **22**–**1** (1996)
14. Strehl, A. and Ghosh, J.: Value-based Customer Grouping from Large Retail Datasets. Proceedings of the SPIE Conference on Data Mining and Knowledge Discovery (2000)
15. van Rijsbergen, C. J.: Information Retrieval. (1979)

# Estimation of Query Model from Parsimonious Translation Model

Seung-Hoon Na [1], In-Su Kang[1], Sin-Jae Kang [2], and Jong-Hyeok Lee [1]

[1] Division of Electrical and Computer Engineering, POSTECH, AITrc, Republic of Korea
{nsh1979, dbaisk, jhlee}@postech.ac.kr
[2] School of Computer and Information Technology
Daegu University, Republic of Korea
sjkang@daegu.ac.kr

**Abstract.** The KL divergence framework, the extended language modeling approach, have a critical problem with estimation of query model, which is the probabilistic model that encodes user's information need. However, at initial retrieval, it is difficult to expand query model using co-occurrence, because the two-dimensional matrix information such as term co-occurrence must be constructed in offline. Especially in large collection, constructing such large matrix of term co-occurrences prohibitively increases time and space complexity. This paper proposes an effective method to construct co-occurrence statistics by employing parsimonious translation model. Parsimonious translation model is a compact version of translation model, and it contains very small number of parameters that includes non-zero probabilities. Parsimonious translation model enables us to enormously reduce the number of remaining terms in document so that co-occurrence statistics can be calculated in tractable time. In experimentations, the results show that query model derived from parsimonious translation model significantly improves baseline language modeling performance.

## 1 Introduction

In the recent past, the language modeling approach has become popular IR model based on its sound theoretical basis and good empirical success [3], [4], [5], [8], [9], [10], [11], [14], [15], [16], [17], [19], [20], [21], [22], [23]. However, the original language modeling could have trouble with incorporation of the relevance feedback or query expansion. Relevance feedback (or pseudo relevance feedback) is the well-known technique that improves significantly initial retrieval results. In probabilistic model, relevance feedback can be well explained in its framework, while language modeling approach do not. The main reason of this limit is that language modeling does not explicitly assume a relevance document set [15].

Risk minimization framework and query model concept, suggested by Lafferty and Zhai [8], extend the language modeling approach to incorporate relevance feedback or query expansion. In risk minimization framework, language modeling is re-designed by KL(Kullback-Leiber) divergence between query model and document model. Query model is probabilistic version of user's query sample, which encodes knowledge about a user's information need.

S. H. Myaeng et al. (Eds.): AIRS 2004, LNCS 3411, pp. 239–250, 2005.

Obtaining high initial retrieval performance is very important problem, since post processing such as pseudo relevance feedback is highly dependent on initial retrieval performance. To improve initial retrieval, query expansion based on word co-occurrence can be one of good strategies. In language modeling approach, word co-occurrence is formulated into translation model [1], [8]. First translation model, suggested by Berger and Lafferty, is document-query translation model [1]. This model is expanded with Markov chain word translation model by Laffery and Zhai [8]. Both translation models, expanded language modeling approaches, showed significant improvements over baseline performance. It is highly provable that translation model is useful for query expansion problem.

However, Markov chain word translation model yields high time complexity. Especially in offline construction, its time complexity is $O(NK)$, where $N$ is number of and $K$ is average number of terms in document. Our goal is to reduce its time complexity so that the translation model can be easily used in practical situations.

To achieve this goal, we propose to use parsimonious translation model. This model conceptually belongs to Markov chain translation model, but there is difference in using document language model. In parsimonious translation model, document language model is mixture model with 'document specific topic model' and global collection language model. On the other hand, in non-parsimonious translation model, it is mixture model with 'MLE document model' and global collection language model. Document specific topic model is the model that eliminates global common portion and leaves document topic portion from MLE document model.

The paper is organized as follows. In Section 2 we briefly review KL divergence framework of the language modeling approaches and query model estimation problem. In Section 3 we examine our query model estimation method, including construction method of parsimonious translation model, in more detail. A series of experiments to evaluate our method is presented in Section 4. Finally, Section 5 concludes and points out possible directions for future work.

## 2   Kullback-Leiber Divergence Framework and Query Model Estimation

### 2.1   Kullback-Leiber Divergence Framework

Basically, the language modeling approach ranks documents in the collection with the query-likelihood (formula 1) that a given query **q** would be observed during repeated random sampling from each document model [3], [4], [13], [19], [23]. [1]

$$p(\mathbf{q} \mid \theta_D) = \prod_w p(w \mid \theta_D)^{c(w;\mathbf{q})}. \tag{1}$$

where $c(w;\mathbf{q})$ is the number of term $w$ in a given query, $D$ is a document and $p(w|\theta_D)$ is document language model for $D$.

---

[1] There is some difference between authors about interpretation of a query. [15] treats a query as a set, while [3], [4], [13], [19], [20] interpreted a query as a sequence of words. We adopt the sequence interpretation.

Laffery and Zhai [8], proposed Kullback-Leiber divergence framework for language modeling so that allows modeling of both queries and documents and incorporates relevance feedback or query expansion. The risk between documents and query is defined as follows.

$$R(\mathbf{d};\mathbf{q}) \propto -\sum_w p(w \mid \theta_Q) \log p(w \mid \theta_D).$$    (2)

where $p(w|\theta_Q)$ is query model, and documents are ranked in inverse proportion to its risk.

## 2.2 Query Model Estimation Problem

Laffery and Zhai [8] suggested Markov chain word translation model, where word translation events occur by random work processes on Markov chain [8], so that training and application costs are significantly reduced without harming performance. In translation model based on this Markov chain, model construction has high time complexity. For given term $q$, translation model on Markov chain (using one step) is calculated as follows.

$$t(w \mid q) = \sum_D p(w \mid \theta_D) p(\theta_D \mid q).$$    (3)

Translation model $t(w|q)$ means the probability to generate $w$ from document topically related to $q$. Translation model is mixture model of document models which is weighted by posterior probabilities of documents for given term $q$. Similar concepts are suggested in relevance model of Lavrenko and Croft [10].

We can rewrite formula (3).

$$t(w \mid q) = \frac{1}{p(q)} \sum_D p(w \mid \theta_D) p(q \mid \theta_D) p(\theta_D).$$    (4)

where $p(w|\theta_D)p(q|\theta_D)$ corresponds to co-occurrence probability in document between two terms $q$ and $w$. To obtain single translation probability, these co-occur probabilities must be summed across whole documents. Its time complexity is $O(N)$ for given pair $w$ and $q$, where $N$ is the number of documents.

At retrieval time, it is not practical to calculate translation probability for entire vocabulary for each query term. To make this calculation quickly, a well known strategy is to restrict extraction of term pairs within *local context*: small windows such as few words or phrase level or sentence level [7], [18], [19], [20]. However, in most application (e.g, word sense disambiguation), *topical context* and local context play different roles [7]. Therefore co-occurrence from only local context cannot completely substitute co-occurrence from global context. Especially, in query expansion problem, 'topically' related terms should be selected and expanded. Co-occurrence statistics on topical context would be primal resource in our problem, rather than those on local context.

Our method is to select highly topical terms in given document, and to construct co-occurrence statistics on only these terms, ignoring non-topical terms of document. To determine highly topical terms in document, document specific topic language model is constructed, which is a type of parsimonious language model for MLE document model [3]. Parsimonious language model enables us to build models that are significantly smaller than standard models. In this model, it is assumed that top high probable $k$ terms are topical in the document. In the parsimonious document model, there are only few terms having non-zero probabilities. By applying Markov chain on this parsimonious document model, a translation model can be constructed. We called this translation model by parsimonious translation model, discriminating from original translation model.

## 3 Estimating Query Model from Parsimonious Translation Model

In this section, we describe our method to construct parsimonious translation model and to estimate query model from it. As mentioned in Section 2, document specific topic model is constructed at first. Next, parsimonious translation model is acquired from these document specific topic models. Pseudo query model is calculated from this translation model. It is more elaborated by applying refinement process. In addition, we also argue that parsimonious translation model can be effectively used in constructing two-dimensional features such as bi-gram and tri-gram.

### 3.1 Estimating Document Specific Topic Model

As noted in Section 2, document language models are constructed by mixing MLE document language model and global collection language model. MLE for document is far from document specific model because it contains global common words. To construct document specific topic model, we assume that documents are generated from mixture model with document specific model and global collection model. For given document $D$, the likelihood of document is as follows.

$$t(w\,|\,q) = \frac{1}{p(q)} \sum_D p(w\,|\,\theta_D) p(q\,|\,\theta_D) p(\theta_D).$$ (5)

where is $p(w|\theta_D^s)$ document specific topic model for estimatation (i.e. parsimonious document language model).

To maximize the document likelihood, we apply EM algorithm [2].

E-step:

$$p[w \in D] = \frac{\lambda\ p(w\,|\,\theta_D^s)}{\lambda\ p(w\,|\,\theta_D^s) + (1-\lambda)\ p(w\,|\,\theta_C)}.$$ (6)

M-step:

$$p(w\,|\,\theta_D^s) = \frac{c(w; D) p[w \in D]}{\sum_w c(w; D) p[w \in D]}.$$ (7)

where $p[w \in D]$ is the probability such that given $w$ is document specific term. As iterations increase, global collection model is not changed and only document specific topic models are iteratively updated.

Next, selection process is performed, where only highly topical terms are selected, and non-topical terms are discarded. For non-topical terms $w$, it probability $p(w|\theta_D^s)$

becomes 0. Discarded probability is re-distributed to topical-terms, uniformly. There are two possible techniques to select topical terms. One method is *select_top(k)*, where terms are sorted by $p(w|\theta_D)$, and only top $k$ ranked terms are selected ($k$ is about between 50 and 100). Another method is *select_ratio(P)*, where top terms are selected as much as summation of probabilities of selected terms is below limit probability $P$ ($P$ is about between 0.6 and 0.9).

## 3.2 Parsimonious Translation Model

As mentioned in Section 2, translation probability $t(w|q)$ is the probability generating $w$ in the document that includes given term $q$. Since word translation model is mixture model of different document models, it is one of document language model. As substituting document language model of formula (4) into summation of document specific model and global collection model, we further derive translation model.

$$t(w|q) = \frac{1}{p(q)} \left( \begin{array}{l} \eta^2 \sum_D p(w|\theta_D^s)p(q|\theta_D^s)p(\theta_D) + \eta(1-\eta)\sum_D p(w|\theta_D^s)p(q|\theta_C)p(\theta_D) + \\ \eta(1-\eta)\sum_D p(w|\theta_C)p(q|\theta_D^s)p(\theta_D) + (1-\eta)^2 \sum_D p(w|\theta_C)p(q|\theta_C)p(\theta_D) \end{array} \right) \tag{8}$$

where $\eta$ is a smoothing parameter for mixing document specific model, and collection language model. Conceptually, although $\eta$ corresponds to the smoothing parameter $\lambda$ for initial retrieval, we treat $\eta$ differently to $\lambda$.

Translation model consists of three summation parts: Document specific co-occurrence model $\sum p(w|\theta_D^s)p(q|\theta_D^s)p(\theta_D)$, global co-occurrence model $p(w|\theta_C)p(q|\theta_C)$, and term topicality $\sum p(w|\theta_D^s)p(\theta_D)$. Parsimonious translation model $t^s(w|q)$ is defined as model which divides document specific co-occurrence model by global likelihood $p(q)$.

$$t^s(w|q) = \frac{1}{p(q)} \sum_D p(w|\theta_D^s)p(q|\theta_D^s)p(\theta_D). \tag{9}$$

At offline indexing stage, of these quantities, we need to pre-calculate only document specific co-occurrence model $\sum p(w|\theta_D^s)p(q|\theta_D^s)p(\theta_D)$. Other quantities can be calculated easily from information of basic language modeling.

When using *select_top(k)* method for document specific model, time complexity for constructing co-occurrence information is about $O(k^2N)$. Compared with $K$, the average number of unique terms in document, $k$ is very small. When $k$ is 50, $k^2$ is 2500 which is largely reduced value compared with $K^2 \approx 10,000$. In this case, reduction ratio of time complexity is about 4 times.

## 3.3 Estimation of Pseudo Query Model

Given query terms $q = q_1 q_2 \ldots q_m$, we can infer query model from translation model as following, similar to [8].

$$p(w|\theta_Q^t) = \sum_{q_i} t(w|q_i)p(q_i|\hat{\theta}_Q). \tag{10}$$

where $\theta_Q^t$ is inferred query model directly from translation model.

Final pseudo query model $\theta_R^t$ is acquired by mixing MLE query model and above inferred relevance document model using parameter $\alpha$.

$$p(w \mid \theta_R) = \alpha \ p(w \mid \theta_R^t) + (1-\alpha)p(w \mid \hat{\theta}_Q). \tag{11}$$

### 3.4 Refinement of Pseudo Query Model

Pseudo query model (relevance document model) can be more refined by using KL divergence metric between relevance document and mixture model. Relevance document model is assumed by mixture model with query model for estimation and global collection model.

$$KL(\theta_R \parallel \mu\theta_Q + (1-\mu)\theta_C) \tag{12}$$

$$\propto -\sum_w p(w \mid \theta_R) \log(\mu \ p(w \mid \theta_Q) + (1-\mu)p(w \mid \theta_C))$$

where $\mu$ is mixing parameter.

From given relevance document model $\theta_R$ and collection model $\theta_C$, query model $\theta_Q$ is selected to minimize KL divergence formula (13). To minimize (13), we apply EM algorithm, similarly to estimation of document specific model.

E-step:

$$p[w \in Q] = \frac{\mu \ p(w \mid \theta_Q)}{\mu \ p(w \mid \theta_Q) + (1-\mu) \ p(w \mid \theta_C)}. \tag{13}$$

M-step:

$$p(w \mid \theta_Q) = \frac{p(w \mid \theta_R)p[w \in Q]}{\sum_w p(w \mid \theta_R)p[w \in Q]}. \tag{14}$$

where $p[w \in Q]$ is posterior probability that $w$ belongs to query specific term.

Here, mixture parameter $\mu$ should be set in proportion to the ratio of query specific portion that included in relevance document model, against global common portion. In this sense, $\mu$ depends on smoothing parameter $\lambda$ in translation model. As smoothing parameter $\eta$ become larger, $\mu$ also should be larger. Interestingly, as we can see in next section, experimentation result shows that even if $\eta$ set to 1.0 with non-smoothing case, optimal value of $\mu$ is less than 1.0. It means that in even parsimonious translation model, there is some portion of global topic words. Our EM algorithm for query model estimation is similar to the method used in model-based feedback [21], and constructing parsimonious relevance model from feedback documents [3], except that KL divergence measure is used instead of likelihood.

## 4   Experimentation

Our evaluation database is constructed using the KT 1.0 collection and NTCIR-3 test collection of Korean. The information of each collection is summarized at Table 1: The number of documents (# Docs), the average number of unique terms in documents (UniTerms/Doc.), the number of topics (# Topics), and the number of relevant documents. KT 1.0 collection has the small number of documents in computer science domain, where each document describes abstract level of an article.

**Table 1.** Collection summaries

| Collection | # Docs | UniqTerms/Doc. | # Topics | # Rels |
|---|---|---|---|---|
| KT 1.0 | 1,000 | 125.9 | 30 | 424 |
| NTCIR-3 K | 66,147 | 176.3 | 30 | 3,868 |

For indexing, we performed preliminary experimentations using various indexing methods (Morphology, word, and bi-character). It is well known that bi-character (n-Gram) indexing units are highly reliable for Korean or other Asian Languages [12], and our experimentations show same results. Thus, the bi-character indexing unit is used in this experimentation.

In NTCIR-3 Korean collection (NTCIR-3 K), we compare four different versions of queries: (T) title only, (D) description only, (C) concept only, (N) narrative only, and (TDNC) combining all topics. Table 2 shows the average number of terms in each query topic.

**Table 2.** The average number of terms in query

| Collection | Title | Desc | Conc | Narr | All |
|---|---|---|---|---|---|
| KT 1.0 | — | 13.6 | — | — | — |
| NTCIR-3 K | 4.76 | 14.5 | 13.53 | 50.56 | 83.35 |

For baseline language modeling approach, we use Jelinek smoothing, setting the smoothing parameter $\lambda$ into 0.25. This smoothing parameter value is acquired empirically, by performing several experimentations across different parameters.

As evaluation measures, in addition to non-interpolated average precision(AvgPr), R-precision (R-Pr, the average precision at the $R$-th position, where is R is total number of relevant documents) and precision at 10 documents (Pr@10) are also considered.

## 4.1 Effectiveness of Query Model

This section describes the retrieval results using query model describes in Section 3. For query model, each parameter is selected empirically which relatively well performs compared with other parameters. In KT 1.0, $\eta$, $\alpha$ and $\mu$ are 1.0, 0.1, and 0.95, respectively. In NTCIR-3, $\eta$, $\alpha$ and $\mu$ are 1.0, 0.1, and 0.4, respectively.

To construct parsimonious translation model for two different collections, $select\_top(k)$ method in Section 3 is used. For each document, $K$, the number of effective topical terms, is set to be about 25% of the average number of unique terms in document. For KT 1.0, $K$ is 32, and 45 for NTCIR-3.

**Table 3.** Comparison of query model (QM) to baseline language model on KT and NTCIR 3 test collection

| Collection | | Baseline | QM | %chg |
|---|---|---|---|---|
| KT 1.0  (D) | AvgPr | 0.4184 | 0.4476 | +6.98% |
| | R-Pr | 0.4141 | 0.4304 | +3.93% |
| | Pr@10 | 0.4167 | 0.4467 | +7.19% |
| NTCIR-3 (T) | AvgPr | 0.2849 | 0.3204 | +12.46% |
| | R-Pr | 0.3082 | 0.3331 | +8.07% |
| | Pr@10 | 0.3567 | 0.4200 | +17.74% |
| NTCIR-3 (D) | AvgPr | 0.2307 | 0.2976 | +28.99% |
| | R-Pr | 0.2666 | 0.3149 | +18.11% |
| | Pr@10 | 0.3500 | 0.4000 | +14.28% |
| NTCIR-3 (C) | AvgPr | 0.3114 | 0.3235 | +3.88% |
| | R-Pr | 0.3265 | 0.3392 | +3.89% |
| | Pr@10 | 0.4400 | 0.4267 | -3.02% |
| NTCIR-3 (TDNC) | AvgPr | 0.3941 | 0.4052 | +2.81% |
| | R-Pr | 0.4084 | 0.4234 | +3.67% |
| | Pr@10 | 0.5267 | 0.4967 | -5.69% |

Evaluation results are shown in Table 3. QM indicates our proposed methods that uses query model from the parsimonious translation model in two different collections. For all these evaluation measures, we can see that our methods using QM yields better performances and sometimes show significant improvements over baseline method.

We observe that QM is more robust than baseline, especially when low performance query are used. For description field in NTCIR-3, the highest increasement in this experimentation is achieved, where it reaches to 29%. QM sometimes shows low performances over baseline, in Conc and All topics of NTCIR-3, at the measure of precision at 10 retrieved documents (Pr@10). In these topics, many high precise query terms are contained, so that it seems that recall improvement by expansion does not sufficiently complement the precision degradation by down-weighting high precise query terms. Although there are some exceptional cases, we can see that QM shows much better performance over baseline, in overall.

### 4.2  Effect of the Number Top Selected Terms

The important parameter is the number of top selected topical terms to construct parsimonious translation model. If we use the number as very small value, the time complexity to construct translation model can remarkably be reduced. Table 4 describes changes of effects of query model according to the number of top selected terms, in NTCIR-3 test collection. The parameter $\eta$, $\alpha$ and $\mu$ are same values to those of Section 4.1.

**Table 4.** Performance of QM across the number of top selected terms (NTCIR-3 K Title)

| K | AvgPr | R-Pr | Pr@10 | ΔPostingSize |
|---|---|---|---|---|
| Baseline | 0.2849 | 0.3082 | 0.3567 | 0 |
| K = 3 (1.7%) | 0.2469 | 0.2829 | 0.3533 | 6,721,536 |
| K = 5 (2.84%) | 0.3060 | 0.3268 | 0.4233 | 20,398,080 |
| K = 10 (5.67%) | 0.3070 | 0.3209 | 0.4333 | 82,640,896 |
| K = 45 (25.52%) | 0.3204 | 0.3331 | 0.4200 | 1,439,272,960 |

Query models constructed by using small $k$ (such as 5 or 10) increase baseline performance, showing comparative performance to query model by using large $k$ (such as 45). It is remarkable because space complexity for translation model can be highly reduced in the case of small $k$. At $k = 5$, the space complexity is only 5.74% of complexity at $k = 45$ (Time complexity also was significantly reduced). At Pr@10, some query models using small $k$ are better over query models using large $k$. From this experimentation, we have an empirical evidence that high topical terms in documents provide more important effects of query expansion rather than other non-topical terms, despite the number of topical terms is very small. In addition, we can expect that small $k$s may show better average precisions over large $k$ after post retrieval processing such as pseudo relevance feedback, because pseudo relevance feedback is highly dependent on precision of top retrieved documents.

### 4.3 Incorporating Relevance Feedback

To perform pseudo relevance feedback, we adopt Zhai' model-based feedback approach using generative model [22]. Zhai's method set final query model as mixture model combining query model acquired from feedback documents (feedback query model) and original query model as follows.

where $\theta_{Q'}$ and $\theta_F$ are final query language model and feedback query model respectively, and $\theta_Q$ is original query model. $\beta$ is mixture parameter between original query model and feedback query model.

There are two possible ways to construct a new query model. One way is to set original query model $\theta_Q$ as *non-expanded query model*, which is MLE estimation of original query sample. Another method is to use *expanded query model*, which is more expanded by using parsimonious translation model.

In this experimentation, we performed only *non-expanded query model* for evaluation. Pseudo relevance feedback results using this method are presented in Figure 1. Parameter $\beta$ is 0.9. The results of pseudo relevance feedback from baseline feedback documents (baseline + PRF), and pseudo relevance feedback from feedback documents using different selection number $k$ values, are described in Figure 4.

Interestingly, we found that as $k$ is smaller, performance result of pseudo relevance feedback is more increased. Specially, when $k$ is 5, average precision of 0.3700 is achieved, which is the best performance in our all experiments. The reason for this is to acquire high precision results in top documents (Pr@10) when initial retrieval is performed using small $k$ values. Although exhaustive experimentations are necessary in the future, from this experimentation, we can conjecture that parsimonious translation model, acquired only using very small $k$ values, is sufficient to improve significantly average precision. Table 5 reports the best performance of pseudo relevance feedback for each selection value $k$.

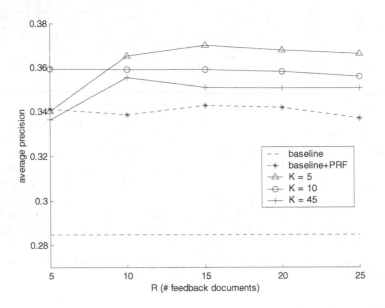

**Fig. 1.** Precision curves for pseudo relevance feedback using different feedback documents for each $k$ on NTCIR-3 (Title)

**Table 5.** The best performances (AvgPr) of pseudo relevance feedback for each selection value $K$ on NTCIR-3 (Title)

| $K$ | AvgPr | %chg | Pr@10 | %chg |
|---|---|---|---|---|
| Base | 0.2849 | — | 0.3567 | — |
| Base+PRF($R$=15) | 0.3428 | +16.89% | 0.4267 | +19.62% |
| $K = 5$ ($R$=15) | 0.3700 | +24.82% | 0.4767 | +33.64% |
| $K = 10$ ($R$=5) | 0.3594 | +21.73% | 0.4767 | +33.64% |
| $K = 45$ ($R$=10) | 0.3555 | +20.60% | 0.4500 | +26.16% |

## 5  Conclusion

Summing up, we propose effective construction method for co-occurrence statistics using parsimonious translation model. Parsimonious translation model involves an elegant method for selecting highly topical terms in documents, by document specific topic model. Basically, our idea is to use the several state of art methods in language modeling approach for information retrieval. From our experimentation on two different collections, it seems reasonable to conclude that query model based on parsimonious translation model preserves effectiveness of traditional translation models, and remarkably reduces time complexity and space complexity of traditional translation models.

## Acknowledgements

This work was supported by the KOSEF through the Advanced Information Technology Research Center(AITrc) and by the BK21 Project.

## References

1. Berger, A. and Lafferty, J. Information Retrieval as Statistical Translation. In Proceedings of 22nd Annual International ACM SIGIR Conference on Research and Development in Information Retrieval, pages 222-229, 1999
2. Dempster, A. Maximum Likelihood from Incomplete Data via the EM algorithm. In Journal of Royal Statistical Society, vol. 39, no. 1, pages 1-39, 1977
3. Hiemstra, D. Robertson, S. and Zaragoza, H. Parsimonious Language Models for Information Retrieval. In Proceedings of 27th Annual International ACM SIGIR Conference on Research and Development in Information Retrieval, pages 178-185, 2004
4. Hiemstra, D. Term Specific Smoothing for Language Modeling Approach to Information Retrieval: The Importance of a Query Term. In Proceedings of 25th Annual International ACM SIGIR Conference on Research and Development in Information Retrieval, pages 35-41, 2002
5. Hiemstra, D. Using Language Models for Information Retrieval. In PhD Thesis, University of Twente, 2001
6. Hofmann, T. Unsupervised Learning by Probabilistic Latent Semantic Analysis. Machine Learning, vol 42(1-2), 2001
7. Ide, N. and Veronis, J. Word Sense Disambiguation. Computational Linguistics, vol 24(1), 1998
8. Lafferty, J. and Zhai, C. Document Language Models, Query Models, and Risk Minimization for Information Retrieval. In Proceedings of 24th Annual International ACM SIGIR Conference on Research and Development in Information Retrieval, pages 111-119, 2001
9. Lavrenko, V., Choquette, M. and Croft, W. Cross-Lingual Relevance Model. In Proceedings of 25th Annual International ACM SIGIR Conference on Research and Development in Information Retrieval, pages 175-182, 2002
10. Lavrenko, V. and Croft, B. Relevance-based language models. In Proceedings of 24th Annual International ACM SIGIR Conference on Research and Development in Information Retrieval, pages 120-127, 2001
11. Liu, X. Cluster-Based Retrieval Using Language Models. In Proceedings of 27th Annual International ACM SIGIR Conference on Research and Development in Information Retrieval, pages 186-193, 2004
12. Lee, J., Cho, H., and Park, H. n-Gram-based indexing for Korean text retrieval. Information Processing & Management vol 35(4), 1999
13. Miller, D., Leek T. and Schwartz, R. A Hidden Markov Model Information Retrieval System. In Proceedings of 22nd Annual International ACM SIGIR Conference on Research and Development in Information Retrieval, pages 214-221, 1999

14. Nallapati, R. and Allen, J. Capturing Term Dependencies using a Language Model based on Sentence Trees. In Proceedings of the 10th international conference on Information and knowledge management, pages 383-390, 2002
15. Ponte, A. and Croft, J. A Language Modeling Approach to Information Retrieval. In Proceedings of 21st Annual International ACM SIGIR Conference on Research and Development in Information Retrieval, pages 275-281, 1998
16. Ponte, A. A Language Modeling Approach to Information Retrieval. In PhD thesis, Dept. o Computer Science, Univercity of Massachusetts, 1998
17. Robertson, S. and Hiemstra, D. Language Models and Probability of Relevance. In Proceedings of the Workshop on Language Modeling and Information Retrieval, 2001
18. Sperer, R. and Oard, D. Structured Translation for Cross-Language Information Retrieval. In Proceedings of 23th Annual International ACM SIGIR Conference on Research and Development in Information Retrieval, pages 120-127, 2000
19. Song, F. and Croft, W. A General Language Model for Information Retrieval. In Proceedings of 22nd Annual International ACM SIGIR Conference on Research and Development in Information Retrieval, pages 279-280, 1999
20. Srikanth, M. and Srihari, R. Biterm Language Models for Document Retrieval. In Proceedings of 25th Annual International ACM SIGIR Conference on Research and Development in Information Retrieval, pages 425-426, 2002
21. Zaragoza, H. and Hiemstra, D. Bayesian Extension to the Language Model for Ad Hoc Information Retrieval", In Proceedings of 26th Annual International ACM SIGIR Conference on Research and Development in Information Retrieval, pages 4-9, 2003
22. Zhai, C. and Lafferty, J. Model-based Feedback in the Language Modeling Approach to Information Retrieval. In Proceedings of the 10th international conference on Information and knowledge management, pages 430-410, 2002
23. Zhai, C. and Lafferty, J. A Study of Smoothing Methods for Language Models Applied to Ad Hoc Information Retrieval. In Proceedings of 24th Annual International ACM SIGIR Conference on Research and Development in Information Retrieval, pages 334-342, 2001

# Ranking the NTCIR Systems Based on Multigrade Relevance

Tetsuya Sakai

Toshiba Corporate R&D Center,
Kawasaki 212-8582, Japan
tetsuya.sakai@toshiba.co.jp

**Abstract.** At NTCIR-4, new retrieval effectiveness metrics called Q-measure and R-measure were proposed for evaluation based on multigrade relevance. This paper shows that Q-measure inherits both the reliability of noninterpolated Average Precision and the multigrade relevance capability of Average Weighted Precision through a theoretical analysis, and then verify the above claim through experiments by actually ranking the systems submitted to the NTCIR-3 CLIR Task. Our experiments confirm that the Q-measure ranking is very highly correlated with the Average Precision ranking and that it is more reliable than Average Weighted Precision.

## 1 Introduction

After a decade of TREC evaluations based on binary relevance assessments, the importance of Information Retrieval (IR) evaluation based on *multigrade* relevance assessments is receiving more attention than ever [1, 2, 3, 4]. NTCIR has used multigrade relevance from the very beginning, but has not fully utilised the richness of the relevance data: Following the TREC evaluation methodology, noninterpolated Average Precision (AveP) is used as the official metric for ranking the systems submitted to NTCIR. Since AveP cannot handle multigrade relevance, *two* AveP values are reported for each system using the `trec_eval` program: *Relaxed* AveP, which treats S-relevant (highly relevant), A-relevant (relevant), B-relevant (partially relevant) documents as just "relevant", and *Rigid* AveP, which ignores the B-relevant ones. Thus, Relaxed AveP *wastes* the relevance levels, while Rigid AveP wastes the partially relevant documents in addition. This situation is clearly undesirable.

At NTCIR-4, new retrieval effectiveness metrics called Q-measure and R-measure were proposed for evaluation based on multigrade relevance [6]. This paper shows that Q-measure inherits both the reliability of AveP and the multigrade relevance capability of *Average Weighted Precision* (AWP) [4, 6] through a theoretical analysis, and then verify the above claim through experiments by actually ranking the systems submitted to the NTCIR-3 CLIR Task. Our experiments confirm that the Q-measure ranking is very highly correlated with the Average Precision ranking and that it is more reliable than Average Weighted Precision.

S. H. Myaeng et al. (Eds.): AIRS 2004, LNCS 3411, pp. 251–262, 2005.

The remainder of this paper is organised as follows. Section 2 defines Q-measure and R-measure, and discusses how they are theoretically related to existing retrieval effectiveness metrics. Section 3 reports on our experiments using the systems submitted to the NTCIR-3 CLIR task for demonstrating the practicality and the reliability of Q-measure. Section 4 discusses previous work, and Section 5 concludes this paper.

## 2    Theory

### 2.1    Average Precision

Let $R$ denote the total number of known relevant documents for a particular search request (or a *topic*), and let $count(r)$ denote the number of relevant documents within the top $r$ documents of the ranked output. Clearly, the Precision at Rank $r$ is $count(r)/r$. Let $isrel(r)$ denote a binary flag, such that $isrel(r) = 1$ if the document at Rank $r$ is relevant and $isrel(r) = 0$ otherwise. Then, Average Precision (AveP) is defined as:

$$AveP = \frac{1}{R} \sum_{1 \leq r \leq L} isrel(r) \frac{count(r)}{r} \qquad (1)$$

where $L$ is the ranked output size.

Another measure often used along with AveP is *R-Precision*:

$$R\text{-}Precision = \frac{count(R)}{R} . \qquad (2)$$

### 2.2    Average Weighted Precision

We now look at Average Weighted Precision (AWP) proposed by Kando *et al.* [4], which was intended for evaluation based on multigrade relevance.

Let $X$ denote a relevance level, and let $gain(X)$ denote the *gain value* for successfully retrieving an $X$-relevant document. For the NTCIR CLIR test collections, $X \in \{S, A, B\}$ [4], and a typical gain value assignment would be $gain(S) = 3$, $gain(A) = 2$ and $gain(B) = 1$. Let $X(r)$ denote the relevance level of the document at Rank $r$ ($\leq L$). Then, the *gain at Rank $r$* is given by $g(r) = gain(X(r))$ if the document at Rank $r$ is relevant, and $g(r) = 0$ if it is nonrelevant. The *cumulative gain at Rank $r$* is given by $cg(r) = g(r) + cg(r - 1)$ for $r > 1$ and $cg(1) = g(1)$.

Let $cig(r)$ represent the cumulative gain at Rank $r$ for an *ideal* ranked output. (An ideal ranked output for NTCIR can be obtained by listing up all S-relevant documents, then all A-relevant documents, then all B-relevant documents.) Then, AWP is given by:

$$AWP = \frac{1}{R} \sum_{1 \leq r \leq L, g(r) > 0} \frac{cg(r)}{cig(r)}$$

$$= \frac{1}{R} \sum_{1 \le r \le L} isrel(r) \frac{cg(r)}{cig(r)} \ . \tag{3}$$

Kando *et al.* [4] have also proposed *R-Weighted Precision*:

$$R\text{-}WP = \frac{cg(R)}{cig(R)} \ . \tag{4}$$

Cumulative Gain was first defined by Järvelin and Kekäläinen [2], but in fact, an equivalent concept already existed in the 1960s, when Pollack proposed the *sliding ratio* measure.

If the relevance assessments are *binary* so that each relevant document gives a gain value of 1, then, by definition, both

$$cg(r) = count(r) \tag{5}$$

and

$$cig(r) = r \tag{6}$$

hold for $r \le R$. Thus, with binary relevance,

$$cg(r)/cig(r) = count(r)/r \tag{7}$$

holds for $r \le R$. Therefore, from Equations (1) and (3), if the relevance assessments are binary *and* if the system output does not have any relevant documents below Rank $R$, then

$$AveP = AWP \tag{8}$$

holds. Similarly, from Equations (2) and (4), with binary relevance,

$$R\text{-}Precision = R\text{-}WP \ . \tag{9}$$

Although AWP appears to be a natural extension of AveP, it suffers from a serious problem. Since there are no more than $R$ relevant documents,

$$cig(r) = cig(R) \tag{10}$$

holds for $r > R$. That is, after Rank $R$, $cig(r)$ becomes a *constant*, which implies, from Equation (3), that AWP cannot distinguish between System A that has a relevant document at Rank $R$ and System B that has a relevant document at Rank $L$ (i.e. at the very bottom of the ranked list). For example, suppose that $R = R(B) = 5$ for a topic, where $R(X)$ represents the number of $X$-relevant documents. Given that $gain(B) = 1$, the sequence of $cig(r)$ is clearly $(1, 2, 3, 4, 5, 5, , \ldots)$. Now, suppose that both System A and System B retrieved only one relevant document, but that System A has it at Rank 5 and that System B has it at Rank 1000. Then, for System A, the sequence of $cg(r)$ is $(0, 0, 0, 0, 1, 1, \ldots)$ and $AWP = (cg(5)/cig(5))/5 = (1/5)/5 = 0.04$. For System B, the sequence of $cg(r)$ is $(0, 0, 0, 0, 0, \ldots, 1)$ and $AWP = (cg(1000)/cig(1000))/5 = (1/5)/5 = 0.04$. Thus the two systems would be considered as identical in performance.

To sum up, AWP is not a reliable metric because its denominator $cig(r)$ "freezes" after Rank $r$. In contrast, AveP is free from this problem because its denominator $r$ is guaranteed to increase steadily. R-Precision and R-WP are also free from the problem because they only look at the top $R$ documents.

## 2.3    Q-Measure

Q-measure, proposed by Sakai [6] at NTCIR-4, is designed to solve the aforementioned problem of AWP.

First, we introduce the notion of *bonused gain at Rank $r$*, simply given by $bg(r) = g(r) + 1$ if $g(r) > 0$ and $bg(r) = 0$ if $g(r) = 0$. Then, the *cumulative bonused gain at Rank $r$* is given by $cbg(r) = bg(r) + cbg(r-1)$ for $r > 1$ and $cbg(1) = bg(1)$. That is, the system receives an *extra reward* for each retrieved relevant document. Q-measure is defined as:

$$Q\text{-}measure = \frac{1}{R} \sum_{1 \leq r \leq L, g(r) > 0} \frac{cbg(r)}{cig(r) + r}$$

$$= \frac{1}{R} \sum_{1 \leq r \leq L} isrel(r) \frac{cbg(r)}{cig(r) + r} \ . \tag{11}$$

Note that the denominator in the above equation $(cig(r) + r)$ is guaranteed not to "freeze", so that relevant documents found below Rank $R$ can be handled properly. For the example given in Section 2.2, for System A, the sequence of $cbg(r)$ is $(0, 0, 0, 0, 2, 2, \ldots)$ and $Q\text{-}measure = (cbg(5)/(cig(5)+5))/5 = (2/(5+5))/5 = 0.04$. But for System B, the sequence of $cbg(r)$ is $(0, 0, 0, 0, 0, \ldots, 2)$ and $Q\text{-}measure = (cbg(1000)/(cig(1000) + 1000))/5 = (2/(5 + 1000))/5 = 0.0004$.

Q-measure is equal to one iff a system output (s.t. $L \geq R$) is an ideal one. In contrast, both R-measure and R-WP are equal to one iff all the top $R$ documents are (at least partially) relevant: thus, for example, B-relevant documents may be ranked above the A-relevant ones. In this respect, Q-measure is superior to R-measure [6].

By definition of the cumulative bonused gain,

$$cbg(r) = cg(r) + count(r) \tag{12}$$

holds for $r \geq 1$. Therefore, Q-measure and R-measure can alternatively be expressed as:

$$Q\text{-}measure = \frac{1}{R} \sum_{1 \leq r \leq L} isrel(r) \frac{cg(r) + count(r)}{cig(r) + r} \ . \tag{13}$$

$$R\text{-}measure = \frac{cg(R) + count(R)}{cig(R) + R} \ . \tag{14}$$

By comparing Equation (13) with Equations (1) and (3), and Equation (14) with Equations (2) and (4), it can be observed that Q-measure and R-measure are "blended" metrics: Q-measure inherits the properties of both AWP and AveP, and R-measure inherits the properties of both R-WP and R-Precision. Moreover, it is clear that using large gain values would emphasise the AWP aspect of Q-measure, while using small gain values would emphasise its AveP aspect. Similarly, using large gain values would emphasize the R-WP aspect of R-measure, while using small gain values would emphasise its R-Precision aspect. For example, letting $gain(S) = 30$, $gain(A) = 20$, and $gain(B) = 10$ (or conversely $gain(S) = 0.3$, $gain(A) = 0.2$, and $gain(B) = 0.1$) instead of $gain(S) = 3$, $gain(A) = 2$, and $gain(B) = 1$ is equivalent to using the following generalised equations and letting $\beta = 10$ (or conversely $\beta = 0.1$):

$$Q\text{-}measure = \frac{1}{R} \sum_{1 \leq r \leq L} isrel(r) \frac{\beta cg(r) + count(r)}{\beta cig(r) + r} . \tag{15}$$

$$R\text{-}measure = \frac{\beta cg(R) + count(R)}{\beta cig(R) + R} . \tag{16}$$

If the relevance assessments are binary, then, from Equations (5) and (6),

$$\frac{cg(r) + count(r)}{cig(r) + r} = \frac{2count(r)}{2r} = \frac{count(r)}{r} \tag{17}$$

holds for $r \leq R$. Therefore, From Equations (1) and (13), if the relevance assessments are binary *and* if the system output does not have any relevant documents below Rank $R$, then Equation (8) can be generalised as:

$$AveP = AWP = Q\text{-}measure . \tag{18}$$

Similarly, from Equations (2) and(14), with binary relevance, Equation (9) can be generalised as:

$$R\text{-}Precision = R\text{-}WP = R\text{-}measure . \tag{19}$$

## 3   Practice

### 3.1   Data

The following files, provided by National Institute of Informatics, Japan, were used for the analyses reported in this paper.

- ntc3clir-allCruns.20040511.zip
  (45 Runs for retrieving Chinese documents)
- ntc3clir-allJruns.20040511.zip
  (33 Runs for retrieving Japanese documents)
- ntc3clir-allEruns.20040511.zip
  (24 Runs for retrieving English documents)
- ntc3clir-allKruns.20040511.zip
  (14 Runs for retrieving Korean documents)

The above files contain runs submitted by 14 different participants, and include both monolingual and cross-language runs, as well as runs using different topic fields, e.g. TITLE, DESCRIPTION etc.

## 3.2    Q-Measure Versus Other Metrics

Tables 1-3 show the Spearman and Kendall Rank Correlations for Q-measure and its related metrics based on the NTCIR-4 CLIR C-runs, J-runs and E-runs, respectively [1]. The correlation coefficients are equal to 1 when two rankings are identical, and are equal to −1 when two rankings are completely reversed. It is known that the Spearman's coefficient is usually higher than the Kendall's. Values higher than 0.99 (i.e. extremely high correlations) are indicated in italics. "Relaxed" represents Relaxed AveP, "Rigid" represents Rigid AveP. "Q-measure" and "AWP" use the *default* gain values: $gain(S) = 3$, $gain(A) = 2$ and $gain(B) = 1$. Moreover, the columns in Part (b) of the table represent Q-measure with different gain values: For example, "Q30:20:10" means Q-measure using $gain(S) = 30$, $gain(A) = 20$ and $gain(B) = 10$. Thus, "Q1:1:1" implies binary relevance, and "Q10:5:1" implies stronger emphasis on highly relevant documents. Table 4 condenses the four tables (C, J, E, K) into one by taking averages over the four sets of data.

Figures 1 and 2 visualise Tables 1 and 2, respectively, by sorting systems in decreasing order of *Relaxed AveP* and then renaming each system as System No. 1, System No. 2, and so on[2]. Thus, the Relaxed AveP curves are guaranteed to decrease monotonically, and the other curves (representing system rankings based on other metrics) would also decrease monotonically only if their rankings agree perfectly with that of Relaxed AveP. That is, an increase in a curve represents a *swop*.

From the above results regarding Q-measure, we can observe the following:

1. While it is theoretically clear that AWP is unreliable when relevant documents are retrieved below Rank $R$, our experimental results confirm this fact. The AWP curves include many swops, and some of them are represented by a very "steep" increase. This is because AWP overestimates a system's performance which rank many relevant documents below Rank $R$. For example, in Figure 1, System No. 4 outperforms System No. 3 according to AWP, even though all other metrics suggest the contrary[3].

2. Compared to AWP, the Q-measure curves are clearly more stable. Moreover, from Part (a) of each table, Q-measure is more highly correlated with Relaxed AveP than AWP is, *and* is more highly correlated with Rigid AveP than AWP is.

---

[1] He results with the K-runs can be found in the original AIRS conference paper and in [6].

[2] The graph for the E-runs can be found in the original AIRS conference paper and in [6]. That for the K-runs can be found in [6].

[3] This particular swop is discussed fully in the original AIRS conference paper.

**Table 1.** Spearman/Kendall Rank Correlations for the 45 C-runs (Q-measure etc.)

| (a) | Rigid | Q-measure | AWP |
|---|---|---|---|
| Relaxed | .9874/.9273 | *.9982/.9798* | .9802/.8990 |
| Rigid | - | .9858/.9192 | .9648/.8667 |
| Q-measure | - | - | .9851/.9152 |
| AWP | - | - | - |

| (b) | Q30:20:10 | Q0.3:0.2:0.1 | Q1:1:1 | Q10:5:1 |
|---|---|---|---|---|
| Relaxed | *.9909/.9374* | *.9997/.9960* | *.9989/.9879* | *.9947/.9556* |
| Rigid | .9788/.8970 | .9874/.9273 | .9851/.9192 | .9829/.9111 |
| Q-measure | *.9901/.9333* | *.9978/.9798* | *.9984/.9798* | *.9955/.9636* |

**Table 2.** Spearman/Kendall Rank Correlations for the 33 J-runs (Q-measure etc.)

| (a) | Rigid | Q-measure | AWP |
|---|---|---|---|
| Relaxed | .9619/.8561 | *.9947/.9583* | .9833/.9242 |
| Rigid | - | .9616/.8447 | .9505/.8182 |
| Q-measure | - | - | .9813/.9129 |
| AWP | - | - | - |

| (b) | Q30:20:10 | Q0.3:0.2:0.1 | Q1:1:1 | Q10:5:1 |
|---|---|---|---|---|
| Relaxed | .9769/.9015 | *.9980/.9811* | *.9990/.9886* | .9759/.8977 |
| Rigid | .9395/.7879 | .9592/.8447 | .9616/.8523 | .9519/.8144 |
| Q-measure | .9729/.8826 | *.9943/.9545* | *.9943/.9545* | .9706/.8864 |

**Table 3.** Spearman/Kendall Rank Correlations for the 24 E-runs (Q-measure etc.)

| (a) | Rigid | Q-measure | AWP |
|---|---|---|---|
| Relaxed | *.9922/.9565* | *.9974/.9783* | .9835/.9058 |
| Rigid | - | *.9948/.9638* | .9748/.8913 |
| Q-measure | - | - | .9843/.9130 |
| AWP | - | - | - |

| (b) | Q30:20:10 | Q0.3:0.2:0.1 | Q1:1:1 | Q10:5:1 |
|---|---|---|---|---|
| Relaxed | *.9922/.9565* | *1.000/1.000* | *.9965/.9783* | .9887/.9348 |
| Rigid | .9852/.9275 | *.9922/.9565* | *.9904/.9493* | .9887/.9348 |
| Q-measure | *.9904/.9493* | *.9974/.9783* | *.9957/.9710* | .9887/.9420 |

3. From Part (a) of each table, it can be observed that Q-measure is more highly correlated with *Relaxed* AveP than with *Rigid* AveP. (The same is true for AWP as well.) This is natural, as Rigid AveP ignores the B-relevant documents completely.

4. It can be observed that the behaviour of Q-measure is relatively stable with respect to the choice of gain values. Moreover, by comparing "Q30:20:10", "Q-measure" (i.e. Q3:2:1) and "Q0.3:0.2:0.1" in terms of correlations with "Relaxed", it can be observed that using smaller gain values implies more resemblance with Relaxed AveP (Recall Equation (15)). For example, in Ta-

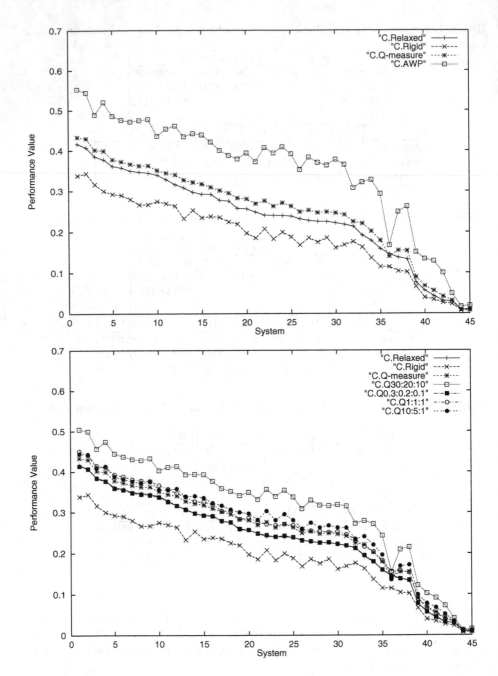

**Fig. 1.** System ranking comparisons with Relaxed Average Precision (C-runs)

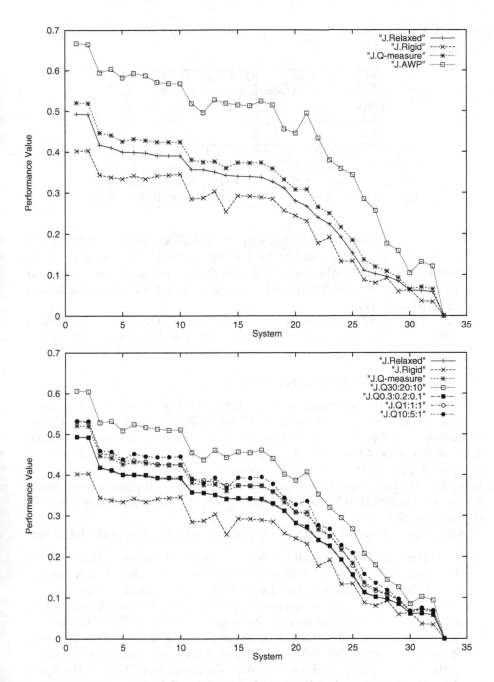

**Fig. 2.** System ranking comparisons with Relaxed Average Precision (J-runs)

**Table 4.** Spearman/Kendall Rank Correlations: Averages over C, J, E and K (Q-measure etc.)

| (a) | Rigid | Q-measure | AWP |
|---|---|---|---|
| Relaxed | .9744/.8965 | .9954/.9681 | .9846/.9213 |
| Rigid | - | .9702/.8825 | .9571/.8446 |
| Q-measure | - | - | .9877/.9353 |
| AWP | - | - | - |

| (b) | Q30:20:10 | Q0.3:0.2:0.1 | Q1:1:1 | Q10:5:1 |
|---|---|---|---|---|
| Relaxed | .9878/.9378 | .9983/.9888 | .9986/.9887 | .9876/.9360 |
| Rigid | .9605/.8537 | .9726/.8882 | .9733/.8918 | .9655/.8656 |
| Q-measure | .9884/.9413 | .9963/.9727 | .9949/.9653 | .9887/.9480 |

ble 1, the Spearman's correlation with "Relaxed" is 0.9909 for "Q30:20:10", 0.9982 for "Q-measure", and 0.9997 for "Q0.3:0.2:0.1". This property is also visible in the graphs: while each "Q30:20:10" curve resembles the corresponding AWP curve, each "Q0.3:0.2:0.1" curve is almost indistisguishable from the "Relaxed" curve.

5. From Part (b) of each table, it can observed that "Q1:1:1" (i.e. Q-measure with binary relevance) is very highly correlated with Relaxed AveP (Recall Equation (18)).

### 3.3    R-Measure Versus Other Metrics

Tables 5-7 show the Spearman and Kendall Rank Correlations for R-measure and its related metrics based on the NTCIR-4 CLIR C-runs, J-runs and E-runs, respectively[4]. Table 8 condenses the four tables into one by taking averages over the four sets of data. Again, "Q-measure", "R-measure" and "R-WP" use the default gain values, "R30:20:10" represents R-measure using $gain(S) = 30$, $gain(A) = 20$ and $gain(B) = 10$, and so on. As "R1:1:1" (R-measure with binary relevance) is identical to R-Precision (and R-WP), it is not included in the tables (Recall Equation (19)).

From the above results regarding R-measure, we can observe the following:

1. From Part (a) of each table, it can be observed that R-measure, R-WP and R-Precision are very highly correlated with one another. Moreover, R-measure is slightly more highly correlated with R-Precision than R-WP is.

2. From the tables, it can be observed that R-measure is relatively stable with respect to the choice of gain values. By comparing "R30:20:10", "R-measure" (i.e. R3:2:1) and "R0.3:0.2:0.1" in terms of correlations with R-Precision, it can be observed that using smaller gain values implies more resemblance with R-Precision (Recall Equation (16)). For example, in Table 5, the Spearman's correlation with R-Precision is 0.9939 for "R30:20:10", 0.9960 for "R-measure", and 0.9982 for "R0.3:0.2:0.1".

---

[4] The results with the K-runs can be found in the original AIRS conference paper and in [6].

**Table 5.** Spearman/Kendall Rank Correlations for the 45 C runs (R-measure etc.)

| (a) | R-Precision | R-measure | R-WP |
|---|---|---|---|
| Relaxed | .9864/.9313 | .9867/.9293 | .9863/.9293 |
| Q-measure | .9867/.9232 | .9871/.9253 | .9883/.9333 |
| R-Precision | - | .9960/.9616 | .9938/.9495 |
| R-measure | - | - | .9971/.9758 |
| R-WP | - | - | - |

| (b) | R30:20:10 | R0.3:0.2:0.1 | R10:5:1 |
|---|---|---|---|
| Relaxed | .9862/.9273 | .9870/.9333 | .9838/.9232 |
| R-Precision | .9939/.9515 | .9982/.9818 | .9845/.9152 |
| R-measure | .9972/.9778 | .9976/.9758 | .9893/.9333 |

**Table 6.** Spearman/Kendall Rank Correlations for the 33 J runs (R-measure etc.)

| (a) | R-Precision | R-measure | R-WP |
|---|---|---|---|
| Relaxed | .9886/.9356 | .9866/.9318 | .9843/.9242 |
| Q-measure | .9913/.9318 | .9903/.9356 | .9880/.9280 |
| R-Precision | - | .9923/.9583 | .9900/.9356 |
| R-measure | - | - | .9910/.9470 |
| R-WP | - | - | - |

| (b) | R30:20:10 | R0.3:0.2:0.1 | R10:5:1 |
|---|---|---|---|
| Relaxed | .9850/.9280 | .9883/.9356 | .9830/.9205 |
| R-Precision | .9920/.9470 | .9957/.9697 | .9873/.9242 |
| R-measure | .9930/.9583 | .9910/.9583 | .9883/.9356 |

**Table 7.** Spearman/Kendall Rank Correlations for the 24 E runs (R-measure etc.)

| (a) | R-Precision | R-measure | R-WP |
|---|---|---|---|
| Relaxed | .9852/.9275 | .9870/.9348 | .9870/.9348 |
| Q-measure | .9843/.9203 | .9835/.9130 | .9835/.9130 |
| R-Precision | - | .9948/.9638 | .9948/.9638 |
| R-measure | - | - | 1.000/1.000 |
| R-WP | - | - | - |

| (b) | R30:20:10 | R0.3:0.2:0.1 | R10:5:1 |
|---|---|---|---|
| Relaxed | .9870/.9348 | .9852/.9275 | .9713/.8913 |
| R-Precision | .9948/.9638 | .9983/.9855 | .9626/.8478 |
| R-measure | 1.000/1.000 | .9965/.9783 | .9591/.8551 |

## 4    Related Work

The original (Discounted) Cumulative Gain ((D)CG) proposed by Järvelin and Kekäläinen [2] has different upperbounds for different topics and does not average well. Thus, they later proposed normalisation as well as averaging across a range of document ranks for (D)CG [3]. Work is underway for comparing the above

**Table 8.** Spearman/Kendall Rank Correlations: Averages over C, J, E and K (R-measure etc.)

| (a) | R-Precision | R-measure | R-WP |
|---|---|---|---|
| Relaxed | .9868/.9376 | .9868/.9380 | .9850/.9306 |
| Q-measure | .9851/.9219 | .9847/.9215 | .9856/.9271 |
| R-Precision | - | .9958/.9709 | .9936/.9567 |
| R-measure | - | - | .9959/.9752 |
| R-WP | - | - | - |

| (b) | R30:20:10 | R0.3:0.2:0.1 | R10:5:1 |
|---|---|---|---|
| Relaxed | .9852/.9311 | .9868/.9381 | .9801/.9173 |
| R-Precision | .9941/.9601 | .9980/.9843 | .9825/.9163 |
| R-measure | .9964/.9785 | .9963/.9781 | .9831/.9255 |

metrics with Q-measure and R-measure from the viewpoint of reliability and stability. The disadvantages of Average Distance Measure (ADM) [1] compared to Q-measure and Average Precision have been discussed in [6].

## 5   Conclusions

This paper showed that Q-measure inherits both the reliability of noninterpolated AveP and the multigrade relevance capability of AWP through a theoretical analysis, and then verified the above claim through experiments by actually ranking the systems submitted to the NTCIR-3 CLIR Task. Our experiments confirm that the Q-measure ranking is very highly correlated with the AveP ranking and that it is more reliable than AWP.

## References

1. Della Mea, V. and Mizzaro, S.: Measuring Retrieval Effectiveness: A New Proposal and a First Experimental Validation. Journal of the American Society for Information Science and Technology **55–6** (2004) 530–543
2. Järvelin, K. and Kekäläinen, J.: IR Evaluation Methods for Retrieving Highly Relevant Documents. ACM SIGIR 2000 Proceedings (2000) 41–48
3. Järvelin, K. and Kekäläinen, J.: Cumulated Gain-Based Evaluation of IR Techniques. ACM Transactions on Information Systems **20–4** (2002) 422–446
4. Kando et al.: Information Retrieval System Evaluation using Multi-Grade Relevance Judgments - Discussion on Averageable Single-Numbered Measures (in Japanese). IPSJ SIG Notes **FI–63–12** (2001) 105–112
5. Sakai, T. et al.: Toshiba BRIDJE at NTCIR-4 CLIR: Monolingual/Bilingual IR and Flexible Feedback. NTCIR-4 Proceedings (2004)
6. Sakai, T.: New Performance Metrics based on Multigrade Relevance: Their Application to Question Answering. NTCIR-4 Proceedings (2004)

# X-IOTA: An Open XML Framework for IR Experimentation

## Application on Multiple Weighting Scheme Tests in a Bilingual Corpus

Jean-Pierre Chevallet

IPAL-I2R Laboratory,
21 Heng Mui Keng Terrace, Singapore 119613
viscjp@i2r.a-star.edu.sg,
http://ipal.imag.fr

**Abstract.** Carrying out experiments in Information Retrieval is a heavy activity requiring fast tools to treat collections of significant size, and at the same time, flexible tools to leave the most possible freedom during the experimentation. System X-IOTA was developed to answer the criterion of flexibility and thus to support fast installation of various experiments using automatic natural language treatments. The architecture is designed to allow a distribution of computations among distributed servers. We use this framework to test different set of weighting particularly the new Deviation from Randomness against Okapi.

## 1   Introduction

New ideas and proposals for solutions in the field of Information Retrieval (IR) must be generally validated by indexing experiments using test collections. A test collection gathers an important mass (giga byte) of documents and a significant number of queries (about hundred). These queries are solved; it means that the relevant documents for them are known. We then run our system on these queries, and compare the results. Campaigns TREC[1] are the canonical examples of this kind of evaluation. IR Experiments are specific for their uniqueness and the originality of the tests, the great amount of parameters, and the important size of the handled corpora. The developments carried out for these experiments are quite expensive (in time and energy) because prototypes are sometimes used only once. It is then very important to manage carefully the time allowed between additionnal programming, or installation of the experiment itself, and the time spent running experiments. According to the complexity of the treatments, and number of parameters that one wishes to test, the time of running experiments can be counted in weeks.

---

[1] http://trec.nist.gov

S. H. Myaeng et al. (Eds.): AIRS 2004, LNCS 3411, pp. 263–280, 2005.

Even if the field of the IR is almost as old as the development of modern computer science, there is paradoxically only a small set of IR tools easily available and quickly modifiable and adaptable to realize experiments. The most famous of these tools is certainly the SMART[2] system of Gerald Salton [1], and developed by Chris Buckley. Even if SMART is rather badly documented, it allows for good adaptation to experiments if one is willing to go through the source code to adapt it. Nevertheless, it is necessary to remain very close to the underlying IR model, and then stick to the presets schemas. Simple modifications, like changing the weighting scheme, are sometimes complicated to achieve, because changes must be made deep in the modules of the code, and lead sometimes to modifying the queries directly rather that the code of the program.

The INQUERY IR System [2] developed at the Massachusetts University Amherst's Center for Intelligent Retrieval Information, became in 1998 a system marketed by the company Sovereign Hill Software Inc. Therefore, it is no longer available for research. The famous probabilistic Okapi System [3], known for good results in test collections, is available[3] but with a financial participation. It is also an integrated system that should be used without too many modifications. HySpirit [4] comes also from a university, but became a commercial product more dedicated to applications and less to experimentations. Among the systems available for experiments, we can cite MG[4] [5] which is dedicated for the indexing of large quantities of texts, and uses compression techniques to store index files. This system is based on an architecture integrating the data. This means that indexing modules, querying modules, etc, are independent programs (Unix), connectable by data flow (ex: Unix pipe). It is then much simpler to carry out a new experiment. On the other hand, it does not allow easily access to the data (ex: weighting) without altering the code. The structures of the files on the other hand are very clearly documented.

The ECLAIR system [6] choses a different point of view: it was an attempt to set up an open and modern software framework, using programming object concepts. This system is then a set of classes, and the final system to be tested is obtained by assembling classes that are adapted to the experimentation. Unfortunately, this system does not seem available any more and is based on a strong integration at the level of the source code (in C++). The LEMUR System [7] is also an experimental modular IR system. It was conceived mainly for experimenting with IR based on a language model [8]. It also works with more traditional models. Just like ECLAIR, it is based on a strong integration at the level of the source code. The SIRE system [9] is made from a set of modules which communicate using a common text mode line based protocol: each entity is described on one line. A character indicates the type of entity (beginning of document, words, etc), then one integer describes its size, then comes the entity itself. SIRE is closed to X-IOTA, the difference is mainly in the use of XML stan-

---

[2] ftp://ftp.cs.cornell.edu/pub/smart/
[3] http://www.soi.city.ac.uk/ãndym/OKAPI-PACK
[4] http://www.cs.mu.oz.au/mg

dard for the transmission of information between modules, increasing flexibility. In that case, the description language of data is set, and not the format itself and the content. The IRTool project[5] also called TeraScale[6] aims to provide an IR experimental system under GPL license. The project does not seem enough advanced to be really usable. The Bow system[10] is a software GPL library for texts statistics and for general IR. It seems possible to build an IRS at the source code level integration (in C). It is used in Rainbow, and CrossBow which are more classifiers than IR systems. The Terrier system [11] is a set of Java classes, optimized for IR on large document collections. It is not yet available, apart from the Glasgow Research team. Lucene[7] is also in Java but is already avaliable.

There are still a very large number of IR systems available[8] like freeWAIS-sf [12] built from the Wais system, or Xapian [13] from the Muscat search engine. However, these systems are dedicated to IR applications, mainly Web sites indexing. For experimentation, they do not help that much. Finally, there are currently two choices for IR experimentation: either to use one of the existing systems, almost without modifications (or with minors modifications remaining close to the underlying IR model), or to build a new complete system from scratch for each model of IR one wants to test. We think that there is a third way: easing experiments through a framework with following characteristics:

1. Open framework: An open framework means the possibility of reorganizing elements of the system in the freest possible way. The SMART system, for example, allows certain flexibility in the construction of the experiments. One can add alternatives to the basic modules (ex: stemming, weighting scheme), but it is not possible to set up a data processing sequence drastically different from the one imposed by the underling vector space model of this system.

2. Open Data Structure: Exchange of data between modules should not be forced by any particular formats, as this would require extra work of coding for adapting the modules input and output. The XML language is a good candidate for this data exchange.

3. Distributed Architecture: A distributed architecture means the possibility of freely connecting modules developed on different sites. There is an important difference in cost between software used at its place of development, and the diffusion of this software on another site. The cost is related to the adaptations (with the system, the site, etc.) and related to the drafting of a more complete documentation. Moreover, the distribution of experimental software creates de facto two versions: the distributed version and the developed version which is inevitably more advanced, and more stable. A distributed framework enables access to an always up-to-date software version to other partners. It also reduces the problems of code property, as only

---

[5] http://sourceforge.net/projects/irtools
[6] http://ils.unc.edu/tera/
[7] http://jakarta.apache.org/lucene/docs/index.html
[8] http://www.searchtools.com

remote uses are allowed. Nevertheless, this framework is possible only if the host site agrees to provide this service: the computing time and possibly storage, and also if times of transmission of the data are not too long compared to local computing. With development of grid computing and wide band network, this solution could be a good one in the near future.

4. Programming language independence: It is unrealistic to impose a single programming language for very different and not integrated piece of software. On the contrary, independence from the programming language ensures a greater division of the software resources but on the other hand implies a connection of the modules for processing by the way of the operating system and grounded on data exchange (ex: files, pipe, network).

5. Operating System independence: For the same reason, it is preferable not to have to impose a given operating system for the connection of the modules. A simple solution can be a common network communication protocol (ex: TCP/IP) and a common application layer protocol (ex: XML, HTTP, RCP) so modules on different software platform can cooperate.

In short, we show in table 1 some characteristics of the systems currently available for research purposes with free access on the Web. The column "Integration" indicate the modules integration type; "code" means by the source code, whereas "data" indicate that only by data exchange between modules. The column "Language" is the programming language mainly used. The architecture type and the data flow diagrams for the documents and requests treatment, can be preset ("preset") and imposed by the software, or programmable ("prog") in the vase of a software library, or freer ("free") if no sequence is imposed except the compatibilities constraints of the modules. The structure of the produced data in "Data structure" can be in binary format not clearly documented ("bin"), documented binary format ("bin+doc"), or in "XML" and documented. The column "Interface" indicates if the system has a user interface other than the basic input line mode. An interface is either in text mode or imbedded in a Web site (standard "HTTP").

The constraints specified previously lead us to define a framework composed of independent modules with connections between them at the operating system level. Also, we enable module connection using network of the type socket TCP/IP with data exchanges in XML. We detail this architecture in the follow-

**Table 1.** Available IR system for experimentation

| System | Integration | Language | Architecture | Data structure | Interface |
|--------|-------------|----------|--------------|----------------|-----------|
| SMART | code | C | preset | bin | text |
| BOW | code | C++ | prog | bin | none |
| IRTool | code | C++ | prog | bin | none |
| LEMUR | code | C++ | prog | bin | none |
| MG | data | C | preset | bin+doc | none |
| X-IOTA | data | C++ | free | XML | HTTP |

ing part. Then, we give some simple examples of modules connection in Unix shell language. The last part relates the use of this framework to test basic weighting scheme, the scheme of Okapi often used and supposed to be a very good one, and a new one which is "Divergence From Randomness". We perform these tests on two collections from the CLEF conference.

## 2     An Open XML IR Framework

This IR framework is composed of two parts: a set of modules to carry out the effective experimentation, and a user interface which pilots this functional core and displays partial or final results. This user interface uses a Web technology. We mainly present the functional core and some basic modules.

The functional core proposed here is composed of basic modules that read and produces XML data to carry out a particular experiment. For example, document indexing and querying are carried out by two XML data processing module sequences. This architecture allows researcher to access the intermediate data at all times during the process. This point is important for the flexibility of this system which is oriented toward experiments. In particular the XML language ensures that at all times, data are humanly readable and can be easily checked, browsed with simple viewer tool because no particular decoding is needed. For example, a simple presentation in a web browser is possible. XML coding also brings extensibility and flexibility, while making possible parameterization of effective XML tags to be treated and while preserving flow and other additional information not understood by the module. This functionality is an important issue that turns this framework into a very flexible system.

In a practical way, treating the data in a flow using a sequence, of connected modules, will limit the risks of source code strong dependences and thus makes easier the maintenance, updating and correction of the source code. This architecture is also supported by a multiprocessors environment[9] since modules of a processing sequence will be automatically distributed by the system, on each available processors. The limitations of this choice is related to the scaling up of data size and to the coding and decoding the XML flow for all treatments, which generates some overhead. This overhead can be limited by buffering the input and output of XML lines during the process. Modules can be classified according the following characteristics:

1. Input genericity/specificity: modules with generic input accepts any XML data. They require of course some parameters for the list of tags to be treated. On the other hand, modules with specific input understand only a particular XML data structure (ex: vectors, matrix), only tags names can be changed according to the actual data.
2. Output genericity/specificity: This is the converse characteristic of the preceding one. A module having a specific output, will produce a particular data

---

[9] Ex: with any the Linux system.

structure type, as a generic output module will produces an output XML structure strictly related its input. Obviously, only input generic modules can produce generic output.

3. Transparence/opacity: this is a very interesting feature of this framework: it concerns module that left intact (transparence) or that remove information not explicitly recognized by the module (opacity). As far as possible, it is desirable that all modules remain transparent to allow the maximum of flexibility. We will speak about transparent treatments, to indicate modules which do nothing but transform input XML structure and accepting all types of added information. We will talk about opaque treatments for those modules which produce a fixed structure from their input and discard all none relevant input data from the flow.

4. Data flow/direct access processing: the data flow processing mode of treatment allows a direct output to input module connection. It is possible only when there is no need for direct access to a given XML data element into the XML tree, but rather processes are done on the fly. These modules produce some output data as soon as their inputs are fed. Direct access requires the use of a file to carry out the treatment and store XML data. It is always preferable to build modules that follow a data flow processing scheme.

The most general module type and thus the most reusable, is the one that has generic input and output, full transparency and that processes its input by flow. It is also advisable to preserve the module transparency property, even in the case specific inputs/outputs. This transparency allows variation in the XML input without disturbing the rest of the processing. For example, in the current modules of X-IOTA, the module xml2vector produces a vector, (XML type vector), composed by a set of coordinates formed by an identifier and a weight:

```
<vector id="DOC0001">
 <c id="1968" w="1"/>
 <c id="accompagn" w="1"/>
</vector>
```

The definition of this XML structure must be understood as a necessary condition, and not as a sufficient definition. That means that modules that are input specific to this type of structure, must nevertheless tolerate more information, like additional attributes which will be simply retransmitted without any modifications. In the example below, we can add part of speech information to each coordinate (pos), and continue to use modules that process vector data XML type. This added information will simply flow throughout the module.

```
<vector id="DOC0001">
 <c id="1968" w="1" pos="NUMBER"/>
 <c id="accompagn" w="1" pos="VERB"/>
</vector>
```

To illustrate operation of a real system using this framework, we develop in the following part, a canonical example of indexing and querying steps which uses very basic modules. This example is based on the vector space model.

# 3   Processing Examples

This section illustrates the possible operation of this system by presenting basic sequences for an indexing and a traditional interrogation in the vector space model.

## 3.1   Indexing

The first part is indexing which consists in producing a direct matrix file after some treatment of documents. A direct matrix file stores each document as a vector of terms and weights. The chain of treatments includes the suppression of the diacritic characters, the setting in small caps, the passage to a anti-dictionary contained in the file common_words.fr, a stemming in the French style, then the production of the document vector. Each XML entity identified by the tag -docTag will be regarded as a document. This process is described as set of piped shell commands interpreted by a system shell script. We are using here, the pipe mechanism (symbol |) of Unix system. Any other script language can be used instead of shell script (make, PHP, Perl, etc). The only constraint is the connection of the input and output of the modules in the process flow. Also, any kind of transparent treatment can be included into this chain to supplement it, and adapt it to a particular experiment.

```
cat OD1 | xmldeldia | xmlcase | xmlAntiDico -dico common_words.fr
         | xmlStemFr | xml2vector -docTag div -id id > OD1.vector
```

In this sequence, the module xml2vector builds the document vector by extraction of the text contained between the tags div. In this example, each document has an identification in the parameter id of tag div which is preserved. All the modules are generic in input and output and transparent. Only the module xml2vector produces a particular type of data: a weighted vector of terms. It is then generic in input and specific at output. It can preserve a certain degree of transparency[10]. This data processing sequence produces, for example, the following vector for the first document of the french test collection OFIL of the Amaryllis part of the CLEF test campaign:

```
<vector id="2271448" size="188">
 <c id="1968" w="1"/>
 <c id="accompagn" w="1"/>
```

---

[10] In the current version, only comments can cross this module freely. Each module leaves a trace of its treatment in an XML comment: the final file then contains the trace of all the treatments carried out in the chain from the document source.

```
<c id="achev" w="1"/>
<c id="an" w="2"/>
<c id="analys" w="1"/>
...
</vector>
```

The data structure `vector` produced by the opaque treatment `xml2vector` is a simple list of coordinates identified with a weight. The `size` value of each vector, indicates the number of coordinates that follows. Vector identifier is the one which was taken from the document in the tag indicated in the parameter `-docTag` and `-id`. The weighting scheme associated with `w` attribute is in this example only the number of occurrence of the terms in the document (i.e. term frequency).

A direct document/term matrix is the simple concatenation of all produced vectors. The identifier corresponds here directly to the document one, found in the collection itself. The final stage of indexing consists of reversing this direct matrix and calculating weights. The current version of X-IOTA, proposes a inversion module working in memory. That enables high speed inversion times, but limits the size of the handled matrices to that of main memory. For OFIL which is about 34Mo, approximately 50Mo of memory is necessary. The type of weighting is coded in this inversion module. It is possible to carry out an inversion without weighting and to build separately a module which calculates all weightings of the matrix. See the next section for more details on the weight scheme we have used.

```
cat OD1.vector | xmlInverseMatrix -w ltc > OD1.inverse.ltc
```

The inverse matrix has same XML format as the direct matrix, except that the vectors identifiers are terms and those of the coordinates correspond to the documents id. A generic module (`xmlIdxSelect`) with direct access selects and extracts an XML tree, knowing an identifiers value. This functionality is sufficient to query the matrix. For obvious efficiency reasons, the file must be first indexed (the module `xmlindex`) to builds a direct access index to the file XML knowing the tag to be indexed and the identifier attribute name. This module can be also used to access the document corpus using the document id, as it works on any XML file.

```
xmlIndex OD1.inverse.ltc vector id
```

This command means that one must index each vector `vector` by its attribute id of the file `OD1.inverse.ltc`. The matrix is then ready for querying. At this stage it is possible to extract any term vector, knowing its id term using the direct access module `xmlIdxSelect`:

```
xmlIdxSelect OD1.inverse.ltc baudoin
```

This command extracts one document vector identified by the indexing term `baudoin`. This example illustrates the simplicity of the implementation of an indexing chain using this framework, and also proves it is possible to intervene at any moment in the indexing process thanks to all data remaining coded in XML. We now will examine how to run queries.

## 3.2    Querying

Querying within this framework consists in launching all the queries of the test collection. Queries must undergo the same type of treatment, so that dimensions of the vectors are compatible. The querying procces is in fact a matrix product between the direct matrix of the queries (queries_id x term_id) and the inverse document matrix (term_id x doc_id) done by the `xmlVectorQuery` module.

```
cat OT1.vector | xmlVectorQuery OD1.inverse.ltc
               | xmlVector2trec > trec_top_file
```

Again, this module uses the index file of the inverse matrix for an efficient direct access to term vectors. The module `xmlvector2trec` is charged to convert the answer matrix (query x document), into a compatible format with the `trec_eval` tool which produces the recall and precision standardized curve. We note that the installation of such an experimental indexing chain and querying breaks up into just few modules. To vary the parameters or to introduce a particular treatment, it is easy to insert a special treatment into the XML flow of the original chain. For example, it is enough to insert the generic module `xmldelsub` to suppress some fields of the XML query tree in order to choose the fields of the requests to be used:

```
cat OT1 | xmldelsub ccept | xmlAntiDico -dico antiDico.txt
        | xml2vector -docTag record -idTag num > OT1.vector
```

In that case, we simply removed from the flow, the fields "concept" before transforming them into vectors. In the same way, modification of the weighting of the terms of the request can be carried out either in an integrated way by modifying the code of construction of the vectors (i.e. `xml2vector`), or by adding a specialized module in to the flow. The code of `xml2vector` is of a small size (like the majority of the modules), so the modification is very localized and thus involves less risks than the modification of the source code of a bigger integrated system like SMART.

Indeed, a programming error will affect only small and isolated modified module, hence not affecting other modules. It is not the case, when system integration is carried out by the source code: a writing error can affect any part of the system, because no memory protection exists between functions of the same process, but strong memory protection is carried out by the system on separate running programs.

## 3.3    Distributed Operation

A modular nature at the level of the operating system enables one to quickly set up a distribution of the modules on other computers and operating systems.

Modules can easily be transformed into server that receive and re-emit data in XML format on a particular internet port. Any input/output redirection technique can be used to access a server. The complexity to set up such a distributed mechanism depends on the operating system used. For example, Unix systems propose a very simple method[11]. Users must only provide the module that implements the protocol layer on standard input/output. The rest is provided by the OS. Every module can then become a server. The only adaptation to be realized is related to the parameter exchanges. One possible format can be the RCP protocol for XML and HTTP transactions. The French parser of the IOTA system is already installed in this way on one of the machines of the Grenoble IR team. The problem of the rights for the access and safety of such distributed architecture, can be solved with technologies of SSH protocol coupled with the mechanism of tunneling for a connection of modules through fire-walls.

## 3.4     Efficiency

In this section we compare the XIOTA framework with the system SMART using English collections of the CLEF2003 campaign. We just want to test the speed of both systems, and the index size so we have chosen the simple indexation: case changing on words (SMART full option). Tables 2 and 3 present results.

**Table 2.** Comparison with SMART on GH95

| Operation | SMART | XIOTA | ratio |
|---|---|---|---|
| Conversion time | 16 s | 0 s | |
| indexing time | 25 s | 3m | 7.2 |
| Index Size | 100 Mo | 377 Mo | 3.8 |

**Table 3.** Comparison with SMART on LAT94

| Operation | SMART | XIOTA | ratio |
|---|---|---|---|
| Conversion time | 1m 14 s | 0 s | |
| indexing time | 1m 8s | 8m 26s | 7.4 |
| Index size | 468 Mo | 887 Mo | 1.9 |

The table 4 give some features of the collections used in this test. XIOTA is compatible with SGML format and can handle these collections, without any conversion. The SMART system need a conversion or need a new module to be programmed and integrated into the source code. We have used an external conversion written in Perl. For the XIOTA system, the index size is the sum of the inverted matrix, and the hash code index of this matrix. For the SMART system, the index size is the sum of all dictionary files, with the inverse file.

---

[11] It is xinet, a system tool that centralized all server processes.

**Table 4.** Collections

| Collection | LAT94 | GH95 |
|---|---|---|
| Name | Los Angeles Times | Glasgow Herald |
| Original format | SGML | SGML |
| Original size | 423 Mo | 153 Mo |
| Size in SMART | 361 Mo | 138 Mo |
| Nb of documents | 113005 | 56472 |

**Table 5.** Sum up comparizon

| System | Flexibility | Performance | Speed |
|---|---|---|---|
| SMART | average | same | good |
| XIOTA | good | same | average |

The use of XML and separate modules, and also the fact that the original collection is stored in several files, creates an overhead of about 7 times more to compute the same index. It is then the "price" to pay for having more flexibility in the treatments. There is also an overheard in the size used to store the index because of the use of XML and also because all terms and document references are stored in the actual inverse file. In fact, SMART is not able to manage document ID that has to be extracted and stored separately.

All these tests are performed on the same machine, a Lunix Fedora core 1 system running on a P4 3.0 Gz with 2Go of main memory and 120 Go of disk of the I2R lab. The table 5 suggest that an integrated IR system like SMART, is a good choice when very few modification are needed. The XIOTA system is a good choice when experimenting new ideas, because in this framework, a change is counted in minutes or hours of programming time (ex: adding a XML line filter in perl), as for SMART it can take days if the modification is possible.

## 4 Testing Weighting Scheme

In this part, we present how we have used this system on a multilingual collection of the CLEF conference. Our purpose is the studies of classic and statistical weighting scheme in order to choose the best one for the CLEF experiments. We have work on the Finnish and French collections. We will present at first, the weighting scheme we would like to test. It has been shown that for textual IR purposes, one of the best weighting schemes is the one used in Okapi system [3]. This weighting is based on a probabilistic IR modeling. We will compare this weighting with a new challenger that is based on a deviation from random behavior [14].

## 4.1    The Underlying IR Model

These experiments are grounded on the classic vector space model, because it includes both models we want to test. The goal of the experiment is to compare the statistical Okapi model with Deviation From Randomness model, versus more classical weightings. This comparison will be done on two different languages. Basically, the final matching process is achieved by a product between query and document vectors, which computes the Relevant Status Value (RSV) for all documents against one query. For a query vector $Q = (q_1 \ldots q_t)$ of $t$ indexing terms, and document vector $D_j = (d_{1j} \ldots d_{tj})$, the RSV is computed by:

$$RSV(Q, D_j) = \sum_{i \in [1..t]} q_i * d_{ij} \; . \tag{1}$$

We keep this matching process for all tests, the changes are in the documents and query processing to select indexing terms, and in the weighting scheme. We recall here the scheme that is inspired by the SMART system. We suppose the previous processing steps have produced a matrix $(d_{ij})$. Initially the value $d_{ij}$ is only the result of indexing term $i$ counting in the document $j$, called term frequency $tf_{ij}$. Each weighting scheme can be decomposed in three steps: a local, a global and a normalization step. The local is related to only one vector. All these transformations are listed in table 6. For all measure we use the following symbols:

$n$       number of document in the corpus
$t$       number of unique indexing terms in the corpus
$tf_{ij}$   frequency of term $i$ in document $j$
$f_i$      frequency of term $i$ in the corpus: $f_i = \sum_{j \in [1..n]} tf_{ij}$
$S$       the corpus size: $S = \sum_{i \in [1..t]} f_i$
$d_{ij}$    current value in the matrix (initialy $tf_{ij}$)
$w_{ij}$   new value in the matrix
$d_{ij}^*$   a normalization of $d_{ij}$ (see below)
$\lambda_i$    the fraction $f_i/S$
$df_i$     number of document indexed by term $i$ (document frequency)
$c, k_1, b$ constants for DFR and Okapi
$L_j$      the length of document $j$: $L_j = \sum_{i \in [1..t]} d_{ij}$
$awrL$   mean document length: $awrL = (\sum_{k \in [1..n]} L_k)/n$
$q_i$      weight of term $i$ of query $q$

The global weighting is related to the matrix: it is a weighing which takes into account the relative importance of a term regarding the whole document collection. The most famous is the Inverse Document Frequency Idf. The table 7 lists the global weighting we have tested. The Okapi measure described in [15, 3], uses the length of the document and also a normalization by the average length of all documents in the corpus. This length is related to the number of indexing terms in a document. The Okapi measure uses 2 constants values called $k_1$ and

**Table 6.** Local weighting

| Letter | Formula | Meaning |
|--------|---------|---------|
| n | $w_{ij} = d_{ij}$ | none, no change |
| b | $w_{ij} = 1$ | binary |
| a | $w_{ij} = \frac{0.5+0.5*d_{ij}}{max_i(d_{ij})}$ | local max |
| l | $w_{ij} = ln(d_{ij} + 1)$ | natural log |
| d | $w_{ij} = ln(ln(d_{ij} + 1) + 1)$ | double natural log |

*b*. Finally we tested a new measure describe in [14] called "Divergence from Randomness" (DFR). We just give here the final formula we have implemented in our system:

$$w_{ij} = (\log_2(1 + \lambda_i) + d^*_{ij} * \log_2 \frac{1 + \lambda_i}{\lambda_i}) * \frac{f_i + 1}{df_i * (d^*_{ij} + 1)} \ . \tag{2}$$

The value $d^*_{ij}$ is a normalization by the length $L_j$ of the document $j$ regarding the average size of all document in the corpus : $awrL$. A constant value $c$ adjusts the effect of the document length in the weight:

$$d^*_{i,j} = d_{ij} * \log_2(1 + c * \frac{awrL}{L_j}) \ . \tag{3}$$

Finally, the last treatment is the normalization of the final vector. A weighting scheme is composed by the combination of the local, global and final weighting. We represent a weighting scheme by 3 letters. For example, **nnn** is only the raw term frequency. The scheme **bnn** for both documents and queries leads to a sort of Boolean model where every term in the query is considered connected by a conjunction. In that case the RVS counts the terms intersection between documents and queries.

The **c** normalization applied to both document and query vector leads to the computation of the cosine between these two vectors. This is the classical vector space model if we use the **ltc** scheme for document and queries. The scheme **nOn** for the documents, and **npn** with the queries, is the Okapi model, and the

**Table 7.** Global weighting

| Letter | Formula | Meaning |
|--------|---------|---------|
| n | $w_{ij} = d_{ij}$ | none, no global change |
| t | $w_{ij} = d_{ij} * log\frac{n}{df_i}$ | Idf |
| p | $w_{ij} = d_{ij} * log\frac{n-df_i}{df_i}$ | Idf variant for Okapi |
| O | $w_{ij} = \frac{(k_1+1)*d_{ij}}{k_1*[(1-b)+b*\frac{L_j}{awrL}]+d_{ij}}$ | Okapi |
| R | (see above) | DFR |

**Table 8.** Final normalization

| Letter | Formula | Meaning |
|--------|---------|---------|
| n | $w_{ij} = d_{ij}$ | none, no normalization |
| c | $w_{ij} = \dfrac{d_{ij}}{\sqrt{\sum_i d_{ij}^2}}$ | cosine |

use of **nRn** for document and **nnn** for the queries is the DFR model. For these two models, constants has to be defined.

Notice that the c normalization of the queries, leads to divide the RSV for this query by $\sqrt{\sum_i q_i^2}$. For each query this is a constant value which does not influence the relative order of answered document list. It follows that this normalization is useless for queries and we will not use it.

## 4.2    Finnish IR

We have used this framework to choose the best weighting scheme for this language in order to compare classic IR approached with IR based on Natural language techniques. We present in this paper only this first part and not the NPL techniques and results. We have used the 2003 collection of CLEF documents in Finnish and in French. This section presents the Finnish results. This collection is composed of 55344 documents. We have used 2003 topics. Documents and queries follow the same treatment path[12]. The first step is filtering the relevant tags from documents or queries. Then we transform XML special characters to their ISO counterpart. We delete all diacritic characters, and change to lower case. At this stage we still have special Finnish characters and accents. We eliminate common words using a list provided by Savoy[13] and then suppress all accents from characters. We apply a Finnish stemmer also proposed by Savoy and modifies to accept XML input/output to produce the final vector.

```
xmlFilterTag | xml2Latin1 | xmldeldia | xmlcase
 | xmlAntiDico -dico common_word.fi | xmlcase -noAcc | xmlStemFi
```

For the queries, we have used the following fields: FI-title FI-desc FI-narr. For documents only the text field has been used. Table 10 sum up the average precision results for all combination of weighting. We only keep a maximum of 1000 documents per query. For the DFR we have fixed the constant c to 0.83. The table 9 shows the results of the constant variation for the nRn nnn weighting scheme. The second column, we show the number of relevant and effectively retrieved document.

We discover that the maximum of relevant documents is in the range 2 of twice the c constant. This value is the one promoted by Amati in [16]. The

---

[12] For more readability, we do not keep modules parameters
[13] http://www.unine.ch/info/clef

**Table 9.** Variation of $c$ constant (nRn nnn)

| $c$ constant | precision | ret_rel | $c$ constant | precision | ret_rel |
|---|---|---|---|---|---|
| 0.00 | 4.89 | 286 | 0.85 | 41.02 | 449 |
| 0.10 | 30.24 | 436 | 0.86 | 41.01 | 449 |
| 0.50 | 39.63 | 448 | 0.87 | 41.02 | 450 |
| 0.70 | 40.40 | 448 | 0.90 | 40.16 | 450 |
| 0.75 | 40.90 | 449 | 0.95 | 39.98 | 450 |
| 0.80 | 40.97 | 449 | 1.00 | 39.86 | 450 |
| 0.81 | 41.04 | 449 | 1.50 | 39.41 | 451 |
| 0.82 | 41.06 | 449 | 2.00 | 39.26 | **452** |
| 0.83 | **41.07** | 449 | 5.00 | 39.03 | 449 |
| 0.84 | **41.07** | 449 | 10.0 | 37.96 | 447 |

maximum average precision is obtained by the value 0.83, so we keep this value for all experiments. For the Okapi weighting, we have use the same value as in [16], that is $k_1 = 1.2$, and $b = 0.75$. In table 12, we have also tested some other value for the French collection: it seems these values are on average good ones.

Table 10 sum up all experiments we have conducted. We have tested all interesting documents and queries weighting schemes. Results show clearly that the DFR weighting is very stable under every weighting scheme except for the binary (bnn). Also this measure is better than Okapi and all others, but Okapi is still very close. Initially, the DFR measure is supposed to be used with only the query term frequency (nnn). We discover one little improvement using ltn and dtn for the queries.

## 4.3   French IR

We have used the French corpus of CLEF 2003. We have used our own stemmer, and our own list for removal of common French terms. In this collection, there

**Table 10.** Finnish average precision

| Doc weight | Query weight | | | | | | | |
|---|---|---|---|---|---|---|---|---|
| | nnn | bnn | lnn | ntn | ltn | atn | dtn | npn |
| nnn | 13.16 | 9.80 | 12.22 | 19.54 | 19.55 | 19.44 | 19.16 | 19.82 |
| bnn | 28.64 | 16.61 | 25.54 | 34.30 | 33.67 | 33.94 | 32.50 | 34.41 |
| atn | 26.77 | 22.65 | 25.87 | 28.35 | 28.02 | 28.11 | 27.85 | 28.31 |
| ntc | 25.72 | 26.38 | 25.95 | 29.26 | 29.39 | 29.60 | 29.57 | 29.25 |
| lnc | 29.57 | 23.88 | 29.75 | 34.06 | 35.35 | 35.38 | 25.44 | 33.99 |
| ltc | 32.22 | 27.84 | 32.22 | 32.63 | 33.00 | 32.90 | 32.44 | 32.63 |
| ltn | 37.71 | 32.37 | 37.91 | 35.99 | 37.85 | 37.86 | 37.65 | 36.01 |
| nRn | **41.07** | **36.99** | **40.08** | 40.02 | **41.29** | **41.05** | **41.92** | 40.00 |
| nOn | 37.16 | 29.35 | 35.95 | **40.39** | 40.12 | 40.32 | 40.68 | **40.12** |

**Table 11.** French average precision

| | | | | Query weight | | | | |
|---|---|---|---|---|---|---|---|---|
| Doc weight | nnn | bnn | lnn | ntn | ltn | atn | dtn | npn |
| nnn | 7.72 | 2.78 | 5.71 | 16.71 | 15.86 | 15.53 | 14.47 | 17.49 |
| bnn | 16.01 | 4.25 | 13.19 | 29.73 | 25.13 | 24.97 | 23.30 | 29.15 |
| atn | 31.02 | 27.03 | 31.16 | 29.91 | 29.76 | 30.28 | 29.47 | 29.95 |
| ntc | 33.53 | 34.68 | 35.86 | 32.09 | 33.89 | 33.99 | 33.08 | 31.98 |
| lnc | 36.20 | 32.22 | 36.74 | 39.06 | 40.69 | 40.82 | 39.37 | 38.77 |
| ltc | 35.39 | 35.37 | 37.40 | 34.38 | 34.17 | 34.29 | 34.73 | 33.40 |
| ltn | 35.65 | 22.36 | 32.68 | 37.87 | 36.64 | 36.99 | 35.44 | 37.89 |
| nRn | **46.98** | **38.15** | **45.01** | **49.06** | **48.16** | **48.76** | **47.03** | **48.78** |
| nOn | 42.25 | 33.02 | 40.39 | 49.01 | 47.07 | 47.36 | 45.65 | 48.38 |

are 3 sets of documents. For each collection we have selected the following fields: lemonde94 TITLE TEXT, and TI KW LD TX ST for sda 94 and 95. For the queries, we have selected the fields FR-title FR-desc FR-narr. We have tested the same combination of weighting schemes as the one tested in the Finnish collection. The results are in the table 11.

Finally, we have taken the best weighting query scheme for the Okapi model (nOn) and we have computed some variation of the two constant $k_1$ and $b$. The results are in the table 12. The best values are obtained with the couple $(1, 0.75)$ which confirm the choice usually taken for this measure. In this language, we also note that the stability of the DFR measure (nRn) is better than other query weightings, except with binary queries (bnn).

**Table 12.** $k_1$ and $b$ variation for nOn ntn

| | $b$ values | | | | |
|---|---|---|---|---|---|
| $k_1$ values | 0.25 | 0.5 | 0.75 | 1 | 1.25 |
| 0.5 | 42.83 | 45.83 | 47.04 | 46.95 | 46.43 |
| 1 | 46.01 | 47.96 | **49.48** | 47.86 | 44.67 |
| 1.5 | 46.95 | 48.69 | 49.36 | 45.08 | 41.92 |
| 2 | 46.97 | 48.56 | 49.01 | 43.98 | 39.04 |
| 2.5 | 46.76 | 48.19 | 46.31 | 43.18 | 11.81 |

We obtain the best average precision with the inverse document frequency (ntn). We have not performed any special treatments for the queries, like removing terms that are not related to the theme (ex: document, retrieved, etc). The results show that a natural language analysis of the query to remove these empty words should improve the results.

# 5 Conclusion

The goal of this article is solve some of the difficulties in setting up experiments in IR, by providing a flexible and open architecture based on recent technologies. This architecture allows fast integration of new modules for natural language treatments in IR. We think that to sacrifice the speed of treatment in order to gain flexibility in the construction of experiments, is a worthwhile bet. It is true that one can argue against the choice of XML coding, for its lack of compactness and also the overhead generated by coding and decoding of information. The question is also raised in the XML community and leads to the definition of a possible "XML Binary" format which represents original data in a more compact way. System X-IOTA was already used for the test campaign CLEF 2003/04 [17]. Indexing and query time performances are reduced compared to the integrated system SMART. Nevertheless, the effectiveness is still sufficient for this type of test collections. The current version of the system with a sufficient number of modules to carry out experiments of bases in IR is freely available at `http://xiota.imag.fr`. Because of its simplicity, this system could also be adapted to teaching IR experiments. Concerning the experiment of the weighting scheme, we recommend the use of the DFR which its very stable weighting that gives very good results almost independently of the query weighting.

I thanks all those which took part in the coding of system X-IOTA, like François Paradis. The experience of his PIF experimental IR system, has inspired orientation of initial IOTA system [18] towards current system X-IOTA. I also thank Lonce LaMare Wyse and Leong Mun Kew from the Institute for Infocomm Research for their useful comments on this paper.

# References

1. Salton, G.: The SMART retrieval system. Prentice Hall, Englewood Cliffs (1971)
2. Callan, J.P., Croft, W.B., Harding, S.M.: The INQUERY retrieval system. In: Proceedings of DEXA-92, 3rd International Conference on Database and Expert Systems Applications. (1992) 78–83
3. Robertson, S.E.: Overview of the okapi projects. Journal of Documentation **53** (1997) 3–7
4. Lbeck, R., Rölleke, T.: Hyspirit (hypermedia system with probabilistic inference for the retrieval of information) a probabilistic deductive system. http://www.hyspirit.com/go.html (1999)
5. Witten, I.H., Moffat, A., Bell, T.C.: Managing Gigabytes: Compressing and Indexing Documents and Images. Morgan Kaufmann Publishing, San Francisco (1999)
6. Harper, D.J., Walker, A.D.M.: Eclair: an extensible class library for information retrieval. The Computer Journal : Special issue on information retrieval **35** (1992) 256 – 267
7. Languate Technologies Institute, C.M.U., for Intelligent Information Retrieval University of Massachusetts, C.: The lemur toolkit for language modeling and information retrieval. www.cs.cmu.edu/ lemur (2004)

8. Zhai, C., Lafferty, J.: A study of smoothing methods for language models applied to ad hoc information retrieval. In: Proceedings of the 24th annual international ACM SIGIR conference on Research and development in information retrieval, ACM Press (2001) 334–342

9. Sanderson, M.: System for information retrieval experiments (sire). Technical report, Technical report, Department of Computing Science, Glasgow University, Glasgow G12 8QQ, Scotland, UK (1996)

10. McCallum, A.K.: Bow: A toolkit for statistical language modeling, text retrieval, classification and clustering. http://www.cs.cmu.edu/ mccallum/bow (1996)

11. Ounis, I., Amati, G., van Rijsbergen, C.., Plachouras, V., He, B., Cacheda, F., Macdonald, C., Johnson, D., Lo, R.T.W., Chan, S.K.: Terrier toolkit in java for the rapid development of information retrieval applications. http://ir.dcs.gla.ac.uk/terrier/index.html (2004)

12. Pfeifer, U., Gövert, N.: Freewais-sf is an extension of the freewais software. http://ls6-www.cs.uni-dortmund.de/ir/projects/freeWAIS-sf Information Retrieval Group in the Computer Science Department of University of Dortmund, Germany (1995)

13. Limited, O.: Xapian - the open source search engine. Xapian is an Open Source Probabilistic Information Retrieval library, released under the GPL http://www.xapian.org/ (2003)

14. Amati, G., van Rijsbergen, C.J.: Probabilistic models of information retrieval based on measuring the divergence from randomness. ACM Transaction on Information Systems **20** (2002) 357–389

15. Robertson, S.E., Walker, S., Baulieu, M.: Okapi at trec-7: Automatic ad hoc, filtering, vlc and interactive track. In: Preceedings of TREC-7. (1998)

16. Amati, G., Carpineto, C., Romano, G.: Comparing weighting models for monolingual information retrieval. In: Proceedings of CLEF 2003, Trondheim, Norway (2003)

17. Sérasset, G., Chevallet, J.P.: Simple translations of monolingual queries expanded through an association thesaurus. x-iota ir system used for clips bilingual experiments. In: CLEF : Cross-Language Evaluation Forum. (2003)

18. Chiaramella, Y., Defude, B., Bruandet, M., Kerkouba, D.: Iota: a full text information retrieval system. In: ACM conference on research and development in information retrieval., Pisa, Italy. (1986) 207–213

# Recognition-Based Digitalization of Korean Historical Archives

Min Soo Kim, Sungho Ryu, Kyu Tae Cho, Taik Heon Rhee,
Hyun Il Choi, and Jin Hyung Kim

AIPR Lab., CS Div., Korea Advanced Institute of Science and Technology,
373-1 Guseong-dong, Yuseong-gu,Daejeon 305-701, Republic of Korea
{mskim, shryu, ktcho, three, hichoi, jkim}@ai.kaist.ac.kr

**Abstract.** We present a recognition-based digitization method for building digital library of large amount of historical archives. Because the most of archives are manually transcribed in ancient Chinese characters, their digitization present unique academic and pragmatic challenges. By integrating the layout analysis and the recognition into single probabilistic framework, our system achieved 95.1% character recognition rates on test data set, despite the obsolete characters and unique variants used in the archives. Compared with intuitive verification and correction interface, the system freed the operators from repetitive typing tasks and improved the overall throughput significantly.

## 1   Introduction

Recently, Korean national agencies launched an ambitious project of building a digital library of historical archives that have been kept by various institutions. Leveraging the Internet infrastructure of Korea, the project aims to provide instant access to the archives for the researchers and the public, who had been endowed with limited chance due to maintenance reasons.

The first phase of the project, which began in 2000, has been a pilot experiment that explores the possibility. During this phase, the chronicles of King's secretaries have been digitalized, which is shown in figure 1. A number of unique and challenging technical issues have been identified as a result.

The main challenge came from the fact that the most of archives are composed of handwritten Chinese characters. Basically, they are similar to *traditional* Chinese characters, but there are a lot of obsolete characters that are hardly used in contemporary texts. The archives also contains significant number of unique variants, which are depicted in figure 2. The unique layout of documents, in which the characters are written in vertical order, also provide additional challenges.

These prevented applying the conventional technologies directly. Dedicated methods had to be developed for almost everything, including the character code set and the keyboard layout. As a consequence, the initial system had to rely on humans for analyzing and annotating the scanned document images. Initially, operators who were trained with new input methods typed in the content of

S. H. Myaeng et al. (Eds.): AIRS 2004, LNCS 3411, pp. 281–288, 2005.

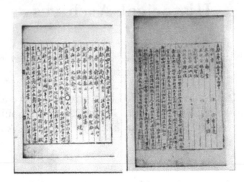

**Fig. 1.** Images from chronicles of King's Secretaries

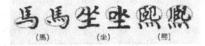

**Fig. 2.** Variants found in historical archives

the document. These annotations were later verified by experts who could interpret and understand the content of archives. This approach was inevitably time-consuming and showed only limited performance. During four years, documents with about one billion characters were digitized, which accounts for about 1.25 % of entire archives. This suggests that it will take centuries to complete the digitization project in current pace. In order to improve the overall throughput, we developed a dedicated recognition system for handwritten Chinese characters in historical archives. Our system does not intend to completely replace the operators. The recognition of handwritten Chinese characters has been one of the most challenging pattern recognition problem, and current recognition performance are far from perfect yet [1][2][3][4][5]. Rather, our system aims to augment human operators and experts by freeing them from repetitive input task. Our system provides the most likely recognition candidates by analyzing the entire layout of document, the shape of each character, and the linguistic contextual likelihood in a single probabilistic framework. The results can be regarded as sensible initial hypotheses. As a consequence, the human operators only need to verify the recognition results in most cases instead of typing in every character. Combined with intuitive verification interface, our system achieved considerable increase in overall throughput.

## 2    Recognition of Handwritten Chinese Characters

### 2.1    Overview

The problem of recognizing a document can be regarded as identifying its layout $L$ and the content $S$, given the image $X$. In our system, the relationships between these elements are modeled as a single probabilistic model. The system yields the recognition result by choosing the one with maximum posterior probability $p(L, S|X)$ among hypotheses.

$$S, L = \arg\max_{S,L} p(S, L|\mathbf{X}). \tag{1}$$

It is difficult to model the posterior $p(S, L|\mathbf{X})$ directly. However, we can transform the posterior into product of likelihoods of the image given the layout and the content by using Bayes' rule. Assuming the likelihoods of the layout and the content $S$ to be conditionally independent of each other, equation 1 can be represented as follows:

$$S, L = \arg\max_{S,L} p(X|S, L)p(S)p(L). \tag{2}$$

Equation 2 indicates that there are three major components in our system - the layout model $p(L)$, the language model $p(S)$, and the character model $p(X|S, L)$. The layout model evaluates the likelihood of the segmentation of document image. The character model represents the likelihood of the image being generated by given character. The language model evaluates the linguistic likelihood of recognition result.

In following sections, we describe each model in detail.

## 2.2   Layout Analysis

The objective of the layout analysis is to segment the entire document image into individual character images. The layout model evaluates the likelihood of each segmentation hypothesis by inspecting simple geometric characteristics of character image candidates.

As depicted in figure 1, the documents in the chronicles of King's secretaries have relatively simple layout. However, it is almost impossible to get correct segmentation using simple geometric features only. As a consequence, we employ over-segmentation approach: all possible character boundaries are first identified using a dedicated algorithm, and then a lattice that indicates possible character image hypotheses are created.

In the chronicles of King's secretaries, the segmentation lattice is built for each line, which is identified through projection profile analysis. In order to deal with complex character boundaries, we used a Tseng's nonlinear segmentation algorithm [6]. As figure 3 shows, the algorithm can effectively identify complex boundaries between overlapped and touched characters.

These possible character boundaries form the nodes in the segmentation lattice as figure 4 shows. Each node in lattice represents a possible character boundary, and each arc a candidate for a character image which is enclosed between a pair of boundaries. Because it is computationally infeasible to investigate all possible boundary pairs, we only add character image candidates that have significant layout likelihoods. The layout likelihood $p(L)$ utilizes three geometric features of character image: the height $h(\mathbf{x})$, the squareness $sq(\mathbf{x})$ and the internal gaps $gap(\mathbf{x})$. The squareness measures the similarity of bounding box to

**Fig. 3.** Pre-segmentation by nonlinear segmentation path

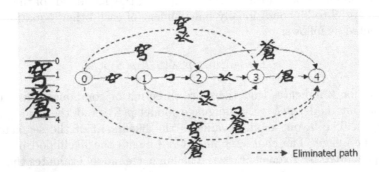

**Fig. 4.** Possible character boundaries from segmentation lattice

the square. The internal gap indicates blank lines contained in character image. These features are assumed to be independent of each other.

$$p_L(\mathbf{x}) = p(h(\mathbf{x}))p(sq(\mathbf{x}))p(gap(\mathbf{x})). \tag{3}$$

The probability distribution of each feature was estimated using Parzen windows [7].

Once the lattice is compiled, $n$ best segmentation candidates within it is identified using Viterbi algorithm as figure 4. These segmentation candidates are used for evaluating the posterior in later stages.

## 2.3   Character Model

The character model evaluates the geometric likelihood of each individual character image, which is provided by a layout hypothesis $L$. The geometric variations for individual character shapes are assumed to be independent of each other. Therefore, the geometric likelihood of entire sentence/document can be factorized into that of individual characters as shown in equation 4.

$$p(\mathbf{X} = \mathbf{x}_1\cdots\mathbf{x}_n | S = s_i\cdots s_n, L) = \prod_i p(\mathbf{x}_i | s_i, L). \tag{4}$$

The geometric model for each character image $p(\mathbf{x}_i | s_i, L)$ uses a contour direction feature, which has been preferred for handwritten Chinese character recognition. The contour direction features are extracted from respective images

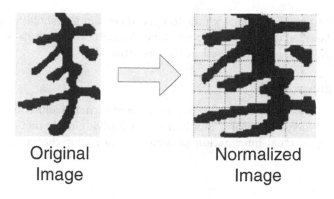

Original
Image

Normalized
Image

**Fig. 5.** Normalized images with $8 \times 8$ blocks

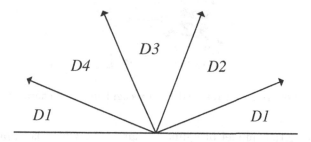

**Fig. 6.** Four groups of contour directions

as follows. First, the image is normalized into 64 by 64 pixels by a nonlinear normalization algorithm and smoothed using Gaussian filter in order to minimize the effect of geometric variations, aliasing and noises[8]. Then, it is divided into 8 by 8 independent blocks. Within each block, the distribution of contour direction is estimated using the gradients of the image. The directions of contours are quantized into four groups as shown in figure 6. Consequently, the dimension of the feature vector is 256.

The geometric model of each character $p(\mathbf{x}|s_i, L)$ is represented as a multivariate Gaussian with mean vector $\mu_i$ for each character and common covariance $\boldsymbol{\Sigma}$.

$$p(\mathbf{x}|s_i, L) = (2\pi)^{-d/2}|\boldsymbol{\Sigma}|^{-1/2} \exp\left(-\frac{1}{2}(\mathbf{x} - \mu_i)^T \boldsymbol{\Sigma}^{-1}(\mathbf{x} - \mu_i)\right). \quad (5)$$

For the sake of computational efficiency, $n$ best character candidates of each image are identified using the geometric likelihoods before evaluating the full posterior $p(S, L|\mathbf{X})$. The candidates can be identified by measuring the Mahalanobis distance from the mean vector of each character $\mu_i$.

$$d_i(\mathbf{x}) = \sqrt{(\mathbf{x} - \mu_i)^T \boldsymbol{\Sigma}^{-1}(\mathbf{x} - \mu_i)}. \quad (6)$$

If the distances of an image **x** fails to exceed certain threshold for every character, it is classified as an out-of-vocabulary character. These out-of-vocabulary characters are manually processed in the last stage.

## 2.4    Language Model

The language model assesses the linguistic contextual likelihoods of recognition results. It helps to resolve ambiguities in individual character level by incorporating contextual information provided by adjacent character recognition result.

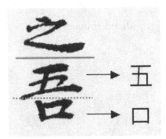

**Fig. 7.** Example of error that resulted from recognizing each character individually

We used a bigram model of Chinese characters, which is implemented using Katz's back-off method[9]. The back-off model hierarchically cascades the maximum likelihood estimates of high order conditional probability models with lower order ones in order to provide more robust estimate, as shown below.

$$p(w) = p_{bo}(s_i|s_{i-1}^{i-d}) = \begin{cases} d_{s_i^{i-d}}p_{ML}(s_i|s_{i-1}^{i-d}) & \text{if } p_{ML}(s_i|s_{i-1}^{i-d}) > 0 \\ \alpha_{s_{i-1}^{i-d}}p_{bo}(s_i|s_{i-1}^{i-d+1}) & \text{otherwise} \end{cases} \tag{7}$$

where $d_{s_i^{i-d}}$ is a discounting coefficient $\alpha_{s_{i-1}^{i-d}}$ a back-off coefficient for given $s_i^{i-d}$.

## 2.5    Verification

After evaluating the posterior, the system creates groups of character images with same recognition results and create the indices into their original positions in the respective images. This character groups are presented to operator, sorted by their posterior likelihoods as shown in figure 8. The operator then visually identifies the recognition errors by looking at the character in the original image. After removing all errors, the recognition result is assigned to all images in the group with confirmation of the operator. As a consequence, the operators only have to type in the identities of misrecognized character images and out-of-vocabulary character image.

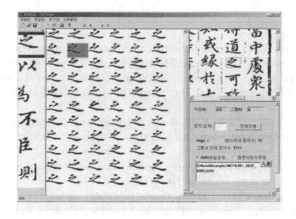

**Fig. 8.** Verification interface

# 3   Experimental Results

We evaluated the effectiveness of the proposed methods using the manually digi-tized chronicles of King's secretaries. For building character model, 100 example images for each character were used. The character models were built for most frequently used 2,568 characters that constitute 99% of frequency of usage in training data. The remaining characters were considered as out-of-vocabulary characters. For testing the recognition performance, we used other 1,000 doc-ument images. As shown in table 1, the proposed system achieved 95.1% of recognition rates. Compared to the recognition rates of baseline system that is based on Euclidian distance template matching, the proposed system reduces error rates by about 8%.

**Table 1.** Recognition rates

| Classifier | Baseline | Proposed |
|---|---|---|
| Recognition Rates | 87.5 | 95.1 |

# 4   Conclusion

We proposed a recognition-based digitization system for building digital library of Korean historical archives. By integrating layout analysis and recognition into single probabilistic framework, the proposed system could automate significant amount of annotation process. Experiments showed that the character recogni-tion rates of the propsed system is 95.1%, which means only 5% of characters demand time-consuming manual annotation process. As a result, the overall throughput of the system and the reliability of final results was significantly improved.

# References

1. S. Hara : OCR for CJK classical texts preliminary examination. Proc. Pacific Neighborhood Consortium(PNC) Annual Meeting,Taipei, Taiwan. (2000) 11–17
2. Z. Lixin and D. Ruwei : Off-line handwritten Chinese characterrecognition with nonlinear pre-classification. Proc. Inc. Conf. On Multimodal Interfaces(ICMI 2000). (2000) 473–479
3. C. H. Tung, H. J. Lee and J. Y. Tsai : Multi-stage precandidate selection in handwritten Chinese character recognition system. Pattern Recognition. **27(8)** (1994) 1093–1102
4. L.C.Tong and S.L.Tan : Speeding up Chinese character recognition in an automatic document reading system. Pattern Recognition. **31(11)** (1998) 1601–1612
5. Y.Mizukami. A handwritten Chinese character recognition system using hierachical displacement extraction based on directional features. Pattern Recognition Letters. **19(7)** (1998) 595–604
6. Y.H.Tseng and H.J.Lee : Recognition-based handwritten Chinese character segmentation using a probabilistic Viterbi algorithm. Pattern Recognition Letters. **20** (1999) 791–806
7. R.O.Duda , P.E.Hart and D.G.Stork, *Pattern Classification, 2nd Eds.* Wiley-Interscience, 2001
8. S.W.Lee and J.S.Park :Nonlinear shape normalization methods for the recognition of large-set handwritten characters. Pattern Recognition.**27(7)** (1994) 895–902
9. D.Jurafsky and J.H.Martin, *Speech and Language Processing: An Introduction to Natural Language Processing, Computational Linguistics and Speech Recognition* Prentice-Hall, 2000

# On Bit-Parallel Processing of Multi-byte Text

Heikki Hyyrö[1], Jun Takaba[2], Ayumi Shinohara[1,2], and Masayuki Takeda[2,3]

[1] PRESTO, Japan Science and Technology Agency (JST)
helmu@cs.uta.fi
[2] Department of Informatics, Kyushu University 33, Fukuoka 812-8581, Japan
{j-takaba, ayumi, takeda}@i.kyushu-u.ac.jp
[3] SORST, Japan Science and Technology Agency (JST)

**Abstract.** There exist practical bit-parallel algorithms for several types of pair-wise string processing, such as longest common subsequence computation or approximate string matching. The bit-parallel algorithms typically use a size-$\sigma$ table of match bit-vectors, where the bits in the vector for a character $\lambda$ identify the positions where the character $\lambda$ occurs in one of the processed strings, and $\sigma$ is the alphabet size. The time or space cost of computing the match table is not prohibitive with reasonably small alphabets such as ASCII text. However, for example in the case of general Unicode text the possible numerical code range of the characters is roughly one million. This makes using a simple table impractical. In this paper we evaluate three different schemes for overcoming this problem. First we propose to replace the character code table by a character code automaton. Then we compare this method with two other schemes: using a hash table, and the binary-search based solution proposed by Wu, Manber and Myers [25]. We find that the best choice is to use either the automaton-based method or a hash table.

## 1 Introduction

Different types of pair-wise string processing algorithms are fundamental in many information retrieval and processing tasks. Let the two processed strings be $P$ and $T$, of length $m$ and $n$, respectively. The most basic task is exact string matching, where $P$ is a pattern string and $T$ a text string, and one searches for occurrences of $P$ inside $T$. Other typical examples include case-insensitive search, regular expression matching, and approximate string comparison. So-called *bit-parallel* algorithms have emerged as practical solutions for several of such string processing tasks. Let $w$ denote the computer word size. We list here some examples of practical bit-parallel algorithms. Each of these has a run time $O(\lceil m/w \rceil n)$. Baeza-Yates and Gonnet [3] proposed an algorithm for exact string matching, and that algorithm can handle also for example case-insensitive search. In [17] Navarro presents methods for allowing repeatable or optional characters in the pattern. Allison and Dix [2], Crochemore et al. [5], and Hyyrö [11] have presented algorithms for computing the length of a *longest common subsequence* between $P$ and $T$. Myers [15] presented an $O(\lceil m/w \rceil n)$ algorithm for finding

S. H. Myaeng et al. (Eds.): AIRS 2004, LNCS 3411, pp. 289–300, 2005.
© Springer-Verlag Berlin Heidelberg 2005

approximate occurrences of $P$ from $T$, when Levenshtein edit distance is used as the measure of similarity. This algorithm can be modified to compute Levenshtein edit distance between $P$ and $T$ [12] as well as to use Damerau edit distance [10].

The above-mentioned, as well as numerous other, bit-parallel algorithms typically use a size-$\sigma$ table of match bit-vectors, where $\sigma$ is the size of the alphabet $\Sigma$, and the characters in $\Sigma$ are mapped into the interval $[0 \dots \sigma - 1]$. Let us call the match table $PM$. For each character $\lambda \in \Sigma$, the bit values in the corresponding match vector $PM_\lambda$ mark the positions in $P$ where the character $\lambda$ occurs. The cost of preprocessing and storing $PM$ is reasonable with small alphabets, such as the 7- or 8-bit ASCII character set. But in case of more general alphabets, perhaps most importantly Unicode text, the range of possible numerical character codes is much larger. To be specific, Unicode character codes fall into the range $[0 \dots 1114111]$. This makes using a naively stored $PM$ table impractical. A basic observation is that the value $PM_\lambda$ needs to be explicitly computed only for those $O(m)$ characters $\lambda$ that occur in $P$. All other characters share an identical "empty" match vector. One quite straightforward solution is then to store only the non-empty vectors $PM_\lambda$ into a hash table whose size is $O(m)$ instead of $\sigma$. Another solution, similar to the one proposed by Wu, Manber and Myers [25], is to sort into one size-$O(m)$ table the character codes of those $\lambda$ that occur in $P$, and store the match vectors $PM_\lambda$, in corresponding order, into another size-$O(m)$ table. The value $PM_\lambda$ is then determined by doing an $O(\log m)$ binary search in the code table (the bound assumes that two character codes can be compared in constant time). If the code of $\lambda$ is found at the $i$th position, then the vector $PM_\lambda$ is in the $i$th position of the match vector table, and otherwise $PM_\lambda$ is an empty match vector.

In this paper we propose another approach for storing and locating the $PM_\lambda$ vectors. The idea is to build an automaton that recognizes the alphabet character codes and whose accepting states identify the corresponding match vectors. The automaton reads the characters in byte-wise manner. We compare this method with the above described alternatives on UTF-8 encoded Unicode text and find that using an automaton is competitive. The results show that the choice of how to handle the match bit-vectors can have a significant effect in terms of the overall processing time: using the binary-search based method of Wu, Manber and Myers [25] may result in the overall processing time being almost three times as much as with the other two methods. We also include a basic direct table lookup in the comparison. This is done by using a restricted multi-byte character set that allows us to use a small table in storing the match vectors. The comparison provides a characterization about the feasibility of using bit-parallel algorithms with Unicode text. This is important as Unicode is becoming more and more widely used. This is true especially on the Internet, which allows people with very different cultures (and character sets) share textual information. XML (eXtensible Markup Language), which is an increasingly popular format for storing data for example on www-pages, uses by default UTF-8 encoded Unicode text. To our best knowledge, the current paper provides the first study about using bit-parallel algorithms in processing multi-byte encoded strings.

## 2    Unicode

The 7-bit ASCII (American Standard Code for Information Interchange) is a fundamental computer character encoding. It, or some 8-bit extended ASCII form of it, is used with variations of the Latin (or Roman) alphabet. Many common computer systems/programs, such as the UNIX operating systems variants as well as the C programming language, are inherently designed to use such a single-byte ASCII code. For example a zero-byte is typically interpreted to mark an end of file.

In many languages, such as Japanese or Chinese, the commonly used alphabets require a multi-byte character encoding. There are several specialized encodings. For example Japanese Extended Unix Code (EUC), BIG5 (Taiwanese), shift-JIS (Japanese), EUC-KR (Korean), and so on. For compatibility with ASCII-oriented systems, such multi-byte encodings usually reserve the code range 0...127 for the single-byte 7-bit ASCII characters, and the multi-byte characters consist of byte values in the range 128...255. In terms of being able to recognize the characters, an important property in practice is also that the multi-byte code should be a *prefix code*. This means that no character code should be a continuation of another, or conversely, that no character code should be a prefix of another character code.

In order to avoid compatibility problems when processing texts with different languages and alphabets, the Unicode Consortium has created a common international standard character code, Unicode, that can express every character in every language in the world [24, 23]. In its present form, Unicode can express 1114112 different characters. Out of these, currently more than 96000 are actually mapped into some character. Unicode defines a numeric code for each character, but it does not specify how that code is actually encoded. The following three are common alternatives:

UTF-32: A simple fixed-length encoding, where each character is encoded with 4 bytes. The main advantage over UTF-16 and UTF-8 is that processing the text is simple. Downsides are the large space consumption and that different computer architectures may represent multi-byte sequences in different orders ("endianness"), which results in compatibility issues.

UTF-16: A variable-length encoding: each character is encoded by 2 or 4 bytes. The main advantage over UTF-32 and UTF-8 is that the method uses typically only 2 bytes for example for Chinese, Korean or Japanese characters. A downside is that also UTF-16 is affected by the endianness of the used computer architecture.

UTF-8: A variable-length encoding: each character is encoded by 1, 2, 3 or 4 bytes. The main advantage over UTF-32 and UTF-16 is the high level of compatibility: UTF-8 is directly compatible with the ASCII code, and it is not affected by the endianness of the hardware. UTF-8 is the default encoding for the XML format. In terms of space, an advantage is that ASCII characters take only a single byte. A downside in comparison to UTF-16 is that for example Chinese, Korean or Japanese characters take typically 3 bytes.

In this paper we concentrate on UTF-8 as it is the most compatible of these three choices and also serves as an example of a general variable-byte encoding.

In the following we describe the basic structure of UTF-8 encoding. We show the structure of each byte as an 8-bit sequence, where the bit significance grows from right to left, and a value 'x' means that the corresponding bit value is used in storing the actual numeric value of the encoded character. Below each 8-bit sequence we also show the corresponding possible range of numerical (base-10) values for the byte.

$$
\begin{array}{ll}
\text{1 byte: 0xxxxxxx} \\
\qquad\ \ 0 \ldots 127 \\
\text{2 bytes: 110xxxxx\ 10xxxxxx} \\
\qquad\quad 192 \ldots 223\ 128 \ldots 191 \\
\text{3 bytes: 1110xxxx\ 10xxxxxx\ 10xxxxxx} \\
\qquad\quad 224 \ldots 239\ 128 \ldots 191\ 128 \ldots 191 \\
\text{4 bytes: 11110xxx\ 10xxxxxx\ 10xxxxxx\ 10xxxxxx} \\
\qquad\quad 240 \ldots 247\ 128 \ldots 191\ 128 \ldots 191\ 128 \ldots 191
\end{array}
$$

The length of a UTF-8 code can be inferred from the most significant (here leftmost) bits of its first byte. If the first bit is zero, the code has a single byte. Otherwise the code has as many bytes as there are consecutive one bits when counting from the most significant bit towards the least significant bit. A byte is a continuation byte of a multi-byte UTF-8 code if and only if its value is in the range $128 \ldots 191$. UTF-8 is clearly a prefix code. The number of available bits ('x') for encoding a character code is 7 for a single-byte code, 11 for a 2-byte code, 16 for a 3-byte code, and 21 for a 4-byte code. Hence UTF-8 encoding can in principle express $2^7 + 2^{11} + 2^{16} + 2^{21} = 2164864$ distinct characters.

## 3    Basic Variants of String Processing

In this section we review three fundamental and much studied forms of string processing. They were chosen as typical representatives of string processing that can be solved by efficient bit-parallel algorithms. The motivation is to lay basic background: The discussed tasks are the ones we will concentrate on in the tests with multi-byte encoded text in Section 5. But let us first introduce some further basic notation. The length of a string $A$ is denoted by $|A|$, $A_i$ is the $i$th character of $A$, and $A_{i..j}$ denotes the *substring* of $A$ that begins from its $i$th character and ends at the $j$th character. If $j < i$, we interpret $A_{i..j}$ to be the empty string $\epsilon$. If $A$ is nonempty, the first character of $A$ is $A_1$ and $A = A_{1..|A|}$. The substring $A_{1..j}$ is a *prefix* and the substring $A_{j..|A|}$ is a *suffix* of $A$. The string $C$ is a *subsequence* of the string $A$ if $C$ can be derived by deleting zero or more characters from $A$. Thus $C$ is a subsequence of $A$ if the characters $C_1 \ldots C_{|C|}$ appear in the same order, but not necessarily consecutively, in $A$.

**Exact String Matching.** Exact string matching is one of the most fundamental string processing tasks. When one is given a length-$m$ pattern string $P$

and a length-$n$ text string $T$, the task is to find all text indices $j$ for which $P = P_{1..m} = T_{j-1+m..j}$. A common variant of this, and also the following two other tasks, is *case insensitive matching*, where no distinction is made between lower- and uppercase characters.

**Longest Common Subsequence.** The string $C$ is a *longest common subsequence* of the strings $P$ and $T$, if $C$ is a subsequence of both $P$ and $T$, and no longer string with this property exists. We denote a longest common subsequence between the strings $P$ and $T$ by $\text{LCS}(P,T)$, and $\text{LLCS}(P,T)$ denotes the length of $\text{LCS}(P,T)$. Both $\text{LCS}(P,T)$ and $\text{LLCS}(P,T)$ convey information about the similarity between $P$ and $T$. This may be used for example in molecular biology (see e.g. [21]), file comparison (e.g. the Unix "diff" utility), or assessing how closely related two words are to each other (e.g. [20]).

**Edit Distance and Approximate String Matching.** Edit distance is another measure of similarity between two strings. The edit distance $ed(P,T)$ between the strings $P$ and $T$ is in general defined as the minimum number of edit operations that are needed in transforming $P$ into $T$ or vice versa.

The task of *approximate string matching* is to find all text locations where a text substring is within a given edit distance from the pattern. A more formal definition is that the task is to find all text indices $j$ for which $ed(P, T_{h..j}) \leq k$, where $h \leq j$ and $k$ is the given error threshold.

Above we did not specify the type of the edit distance. The following distances are typical. We denote by $ed_s(P,T)$ a simple edit distance that allows one edit operation to delete or insert a single character. The values $ed_s(P,T)$ and $\text{LLCS}(P,T)$ are connected by the equality $2\times\text{LLCS}(A,B) = n + m - ed_{id}(A,B)$ (e.g. [6]). Probably the most common form of edit distance is Levenshtein edit distance [14], which extends the simple edit distance by allowing also the operation of substituting a single character with another. We denote Levenshtein edit distance between $P$ and $T$ by $ed_L(P,T)$. Damerau distance [8], which we denote by $ed_D(P,T)$, is used especially in spelling correction related applications. It extends Levenshtein distance by allowing also a fourth edit operation: transposing (swapping) two adjacent characters.

## 3.1    Bit-Parallel Algorithms

During the last two decades, so-called *bit-parallel algorithms* have emerged as practical choices for several string processing tasks. The principle of such algorithms is in general to take advantage of the fact that computers process data in chunks of $w$ bits, where $w$ is the computer word size (in effect the number of bits in a single register within the processor). Currently most computers have a word size $w = 32$, but also the word size $w = 64$ is becoming increasingly common. In addition, most current personal computers support specialized instruction extension sets, such as MMX or SSE, that allow one to use $w = 64$ or even $w = 128$. Bit-parallel algorithms store several data-items into a single computer word, and then update them in parallel during a single computer operation.

```
ComputePM(P)
   For λ ∈ Σ Do PM_λ ← 0^m
      For i = 1...m Do PM_{P_i} ← PM_{P_i} | 0^{m-i}10^{i-1}

Bit-ParallelProcessing(P, T)
   ComputePM(P)
   InitializeVectors
   For j = 1...n Do
      ProcessVectors(T_j)
   ProcessResult (if required)
```

**Fig. 1.** Preprocessing the $PM$-table and a basic skeleton for the discussed bit-parallel algorithms

We use the following notation with bit-vectors: '&' denotes bitwise "and", '|' denotes bitwise "or", '^' denotes bitwise "xor", '~' denotes bit complementation, and '<<' and '>>' denote shifting the bit-vector left and right, respectively, using zero filling in both directions. Bit positions are assumed to grow from right to left, and we use superscripts to denote repetition. As an example let $V = 1110010$ be a bit vector. Then $V[1] = V[3] = V[4] = 0$, $V[2] = V[5] = V[6] = V[7] = 1$, and we could also write $V = 1^3 0^2 10$.

The general high-level scheme for bit-parallel string processing algorithms is as follows. First a size-$\sigma$ match table $PM$ is computed for the length-$m$ string $P$. $PM$ holds a length-$m$ match bit-vector $PM_\lambda$ for each character $\lambda \in \Sigma$. The bit-vector $PM_\lambda$ identifies the positions in the string $P$ where the character $\lambda$ occurs: the $i$th bit of $PM_\lambda$ is set if and only if $P_i = \lambda$. For simplicity, we will assume throughout this paper that $m \leq w$. The case $m > w$ can be handled by simulating a length-$m$ bit-vector by concatenating $\lceil m/w \rceil$ length-$w$ bit-vectors, and thus the table $PM$ occupies in general $\sigma \lceil m/w \rceil$ bits. Once $PM$ is preprocessed and the data bit-vectors used by the algorithm have been initialized, the bit-parallel algorithm processes the string $T$ sequentially. At each character $T_j$ the algorithm updates the data bit-vectors by using bit-operations. Depending on the task, the algorithm may at this point also update some score value and/or check whether a match was found at position $j$. Fig. 1 shows pseudocode for preprocessing $PM$ and a skeleton for the actual processing phase. The sub-procedure "ProcessVectors" encloses all steps that a particular algorithm conducts at character $T_j$. In what follows we will show some specific choices for the sub-procedures. Each bit-parallel algorithm that we discuss runs in $O(n)$ time when $m \leq w$ and in general in $O(\lceil m/w \rceil n)$ time. But a detailed discussion of any of these algorithms is outside the scope of this paper; the reader should look into the given references for more information about them. The algorithms are shown as examples of different types of bit-parallel algorithms, and they are the ones we use in testing.

Due to the nature of the match table $PM$, it is a well-known fact that bit-parallel algorithms can be easily modified to be case-insensitive. This is usually said more broadly: the algorithms can use *classes of characters*. For each character $\lambda$ we may define a set of characters that are deemed to match with $\lambda$. This can be done simply by setting the $i$th bit of $PM_{\lambda'}$ for all such $\lambda'$ for which we wish to define $P_i = \lambda = \lambda'$.

InitializeVectors-SA
  $R \leftarrow 0^m$

ProcessVectors-SA$(T_j)$
  $R \leftarrow ((R << 1) \mid 0^{m-1}1)$ & $PM_{T_j}$
  If $R$ & $10^{m-1} \neq 0^m$ Then
    Report an occurrence of
    $P$ ending at $T_j$.

InitializeVectors-LLCS
  $P \leftarrow 1^m$

ProcessVectors-LLCS$(T_j)$
  $X \leftarrow P$ & $PM_{T_j}$
  $P \leftarrow (P + X) \mid (P - X)$

ProcessResult-LLCS
  LLCS$(P, T)$ is the number of
  zero bits in $P$

InitializeVectors-ASM
  $currDist \leftarrow m$
  $VN \leftarrow 0^m$
  $VP \leftarrow 1^m$

ProcessVectors-ASM$(T_j)$
  $D0 \leftarrow (((PM_{T_j}$ & $VP) + VP)$ ^ $VP) \mid PM_{T_j} \mid VN$
  $HP \leftarrow VN \mid \sim (D0 \mid VP)$
  $HN \leftarrow D0$ & $VP$
  If $HP$ & $10^{m-1} = 10^{m-1}$ Then
    $currDist \leftarrow currDist + 1$
  Else If $HN$ & $10^{m-1} = 10^{m-1}$ Then
    $currDist \leftarrow currDist - 1$
  If $currDist \leq k$ Then
    Report an approximate occurrence of $P$ at $T_j$
  $VP \leftarrow (HN << 1) \mid \sim (D0 \mid (HP << 1))$
  $VN \leftarrow D0$ & $(HP << 1)$

**Fig. 2.** Bit-parallel procedures for exact string matching (*upper left*), LLCS computation (*lower left*), and approximate string matching (*right*)

Baeza-Yates and Gonnet proposed the bit-parallel shift-and algorithm [3] for exact string matching. When $m \leq w$, its behaviour is similar, although much faster, than that of the well-known linear-time string matching algorithm of Knuth, Morris and Pratt [13]. Shift-and processes all text characters in sequential order, and thus it is typically somewhat slower than algorithms that try to skip quickly over such text areas that are seen not to contain a match (e.g. [4, 7, 18]). The latter approach is, however, more difficult in the case of variable-length encoded text. The pseudocode for the bit-parallel processing of the shift-and algorithm at the character $T_j$ is shown in the upper left part of Fig. 2.

Allison and Dix [2] proposed the first bit-parallel algorithm for the longest common subsequence problem. To our best knowledge, this was also the first bit-parallel approximate string processing algorithm. Later Crochemore et al. [5] and Hyyrö [11] have proposed similar variants. The lower left part of Fig. 2 shows the pseudocode for the bit-parallel LLCS processing of [11] that makes four operations per character of $T$. As discussed in [11], these bit-parallel algorithms are very practical for LLCS-computation.

Myers [15] presented an efficient bit-parallel algorithm for approximate string matching under Levenshtein edit distance. The tests in [16] show that this algorithm is in many cases the fastest in practice. Here we refer to so-called "verification capable" algorithms that are based on actually computing edit distance. It is easy to transform Myers' algorithm to compute edit distance [12], and it has also been modified to use Damerau distance [10]. The right side of Fig. 2 shows the pseudocode for the slightly simpler variant of Hyyrö [9, 19].

## 4   Storing the Match Vectors

As discussed in Section 1, storing the match vectors $PM_\lambda$ into a size-$\sigma$ table is not practical in the case of Unicode encoded text or similar large alphabets. In this section we first propose an approach that uses a code automaton to overcome this problem. Then we also discuss two other options.

**Code Automaton.** Our proposal is to build a minimized code automaton that uniquely recognizes the encoding of each character that appears in $P$, and in addition accepts the encodings of all those characters that do not appear in the pattern. Let $u$ be the number of different characters that appear in $P$. Then the code automaton has $u + 1$ accepting states: one for each different character in $P$, and one that represents all other characters in $\Sigma$. If the character $\lambda$ appears in $P$, we associate $PM_\lambda$ with the state that accepts the encoding of $\lambda$. The state that represents those characters that do not appear in $P$ will be associated with a zero match vector $0^m$. In our case of multi-byte character encoding, we will read the text $T$ with the code automaton one byte at a time. Whenever the automaton recognizes a character, a bit-parallel algorithm can process the currently read text character $T_j$ by using the match vector that is associated with the current accepting state.

Such a code automaton can conceptually be built by first composing a trie over the encodings of all characters in the alphabet, and then minimizing it so that all leaves that correspond to characters that do not appear in the pattern $P$ are merged into a single leaf. The leaves corresponding to the characters that appear in the pattern are not merged. When $u$ has the same meaning as above, the resulting DAG (Directed Acyclic Graph) has $u + 1$ leaf nodes, which are the accepting states of the corresponding automaton. The final automaton is then composed by augmenting the DAG with Aho-Corasick *failure links* [1] and associating the match vectors with the accepting states. The process (except for the match vectors) is similar to how the pattern matching automaton used in [22] is built. The main difference is that here the "set of patterns" of the pattern matching automaton is formed by those character encodings that appear in $P$. Fig. 3 shows an example.

**Hash Table.** The second approach is to use a hash table, which is a standard text-book procedure for storing keys. In this scheme the range of numerical values of the character encodings (for example $1 \ldots 1114112$ in the case of the full range of Unicode encodings) is mapped onto a relatively small integer range $1 \ldots x$. Let $code(\lambda)$ denote the numerical value of the encoding of the character $\lambda$, and let the function $hash(code(\lambda))$ give the mapping onto the range $1 \ldots x$. For each $\lambda$ that occurs in $P$, the value $code(\lambda)$ is stored into the position $hash(code(\lambda))$ of the match vector table. If two non-equal characters in $P$ have the same mapping, different mechanisms can be used. We describe here a simple linear hashing scheme. If the position $hash(code(\lambda))$ in the table is already used when we are attempting to store $code(\lambda)$ into it, we continue probing the next positions one-by-one until an empty position is found and store the value there. If the end of the table is reached, we continue from the first position of the table. This works as long as the table is not yet full, but the process takes $h$ steps in the worst case, where $h$ is the number of items currently in the table. But the scheme works well if the number of stored items is small in comparison to $x$. The match vectors are

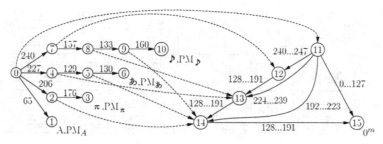

**Fig. 3.** An example of a code automaton for UTF-8. Here the pattern has characters
with bytewise decimal UTF-8 encodings 65 (= 'A'), 206 176 (= 'π'), 227 129 130 ('o'
in Japanese hiragana), and 240 157 133 160 (= a note symbol). The corresponding
accepting states are numbered 1, 3, 6 and 10, respectively. State 0 is the initial state,
and state 15 is the accepting state for those characters that do not appear in the pattern.
The matched pattern character and the corresponding match vector are shown next to
each accepting state. The dashed arrows correspond to failure links that are followed
when the current state does not have an outgoing solid arrow for the current byte
value. After reaching an accepting state, the automaton resets itself into state 0 (this
is an empty transition)

associated with the corresponding character encodings in the table. Finding the
encoding value of a text character $T_j$ from the table works in similar fashion: first
the mapping value $hash(code(T_j))$ is computed, and then the table is checked
from the corresponding position onwards until either the value $code(T_j)$ or an
empty position is found. In the former case we use the associated match vector.
In the latter case the table does not contain $code(T_j)$, and we use an empty
match vector $0^m$.

In our case we know in advance that the table will hold exactly $u$ values,
where $u$ is again the number of distinct characters in $P$. For efficiency we use an
extended table of size $x + u$ so that we do not need to worry about reaching the
end of the table. With multi-byte text we have tested a very simple mapping.
It maps a multi-byte code onto the range $0 \ldots 255$ (corresponds to $x = 256$) by
using the value of the last byte in the code. As far as the encodings are random
enough not to share too many identical last bytes, this works very efficiently.

**Binary Search.** The third approach is derived from the proposition of Wu,
Manber and Myers [25]. In it the numerical values of the character encodings of
the $u$ distinct pattern characters are stored into a size-$u$ table, and the values
are sorted. The match vectors are associated with the corresponding values. The
value $code(T_j)$ is looked up from the table by doing an $O(log_2 u)$ binary search.
Again we use the corresponding match vector if the value $code(T_j)$ was found
from the table, and otherwise an empty match vector $0^m$.

| AMD Athlon64 | | | | | | | | | | | |
|---|---|---|---|---|---|---|---|---|---|---|---|
| | Shift-and | | | | LCS | | | | ASM | | | |
| $m$ | 4 | 8 | 16 | 32 | 4 | 8 | 16 | 32 | 4 | 8 | 16 | 32 |
| AUT | 128 | 127 | 127 | 128 | 129 | 128 | 129 | 130 | 114 | 113 | 112 | 113 |
| BIN | 209 | 248 | 293 | 380 | 199 | 232 | 270 | 351 | 167 | 192 | 219 | 278 |
| HASH | 120 | 122 | 130 | 150 | 122 | 126 | 133 | 154 | 109 | 110 | 114 | 129 |
| SIMP | 100 | 100 | 100 | 100 | 100 | 100 | 100 | 100 | 100 | 100 | 100 | 100 |

| Intel Pentium 4 | | | | | | | | | | | |
|---|---|---|---|---|---|---|---|---|---|---|---|
| | Shift-and | | | | LCS | | | | ASM | | | |
| $m$ | 4 | 8 | 16 | 32 | 4 | 8 | 16 | 32 | 4 | 8 | 16 | 32 |
| AUT | 84 | 84 | 86 | 87 | 99 | 89 | 90 | 90 | 97 | 95 | 95 | 95 |
| BIN | 185 | 221 | 269 | 323 | 205 | 223 | 269 | 322 | 192 | 218 | 252 | 302 |
| HASH | 109 | 112 | 121 | 138 | 104 | 96 | 104 | 122 | 101 | 100 | 105 | 121 |
| SIMP | 100 | 100 | 100 | 100 | 100 | 100 | 100 | 100 | 100 | 100 | 100 | 100 |

| AlphaStation XP1000 | | | | | | | | | | | |
|---|---|---|---|---|---|---|---|---|---|---|---|
| | Shift-and | | | | LCS | | | | ASM | | | |
| $m$ | 4 | 8 | 16 | 32 | 4 | 8 | 16 | 32 | 4 | 8 | 16 | 32 |
| AUT | 140 | 141 | 140 | 141 | 131 | 132 | 132 | 133 | 131 | 132 | 132 | 132 |
| BIN | 220 | 256 | 299 | 354 | 196 | 224 | 251 | 293 | 182 | 211 | 246 | 285 |
| HASH | 114 | 118 | 125 | 145 | 113 | 115 | 118 | 138 | 111 | 114 | 120 | 135 |
| SIMP | 100 | 100 | 100 | 100 | 100 | 100 | 100 | 100 | 100 | 100 | 100 | 100 |

| Sparc Ultra 2 | | | | | | | | | | | |
|---|---|---|---|---|---|---|---|---|---|---|---|
| | Shift-and | | | | LCS | | | | ASM | | | |
| $m$ | 4 | 8 | 16 | 32 | 4 | 8 | 16 | 32 | 4 | 8 | 16 | 32 |
| AUT | 103 | 103 | 106 | 107 | 108 | 110 | 113 | 115 | 103 | 104 | 105 | 106 |
| BIN | 218 | 245 | 277 | 310 | 202 | 229 | 258 | 289 | 168 | 186 | 206 | 230 |
| HASH | 103 | 104 | 108 | 119 | 106 | 107 | 109 | 119 | 100 | 102 | 104 | 111 |
| SIMP | 100 | 100 | 100 | 100 | 100 | 100 | 100 | 100 | 100 | 100 | 100 | 100 |

**Fig. 4.** The results for the three tested string processing tasks on four different computer architectures. $AUT$: code automaton, $BIN$: binary search, $HASH$: hash table, $SIMP$: direct table lookup (the multibyte characters allowed in the patterns were restricted to a small subset, thus allowing to use a simple table)

## 5    Test Results

We implemented and tested the three match table handling schemes from the previous section. In order to characterize their performance in conjunction with bit-parallel algorithms of various complexity, we did separate tests with each of the three bit-parallel algorithms discussed in Section 3.1. In order to evaluate hardware-dependency, we tested on four different computers: AMD Athlon64, Intel Pentium 4, AlphaStation XP1000 and Sparc Ultra 2. The code was exactly the same with all computers, and the different bit-parallel methods used the same file-handling framework. On the AlphaStation we used CC compiler, and on the other computers we used GCC. All code was compiled with the "-O9" optimization switch. The tested strings were UTF-8 encoded, and they were generated randomly. The lengths of $P$ were $m = 4, 8, 16$ and $32$. The length of $T$ was at least one million characters in the case of searching, and $P$ and $T$ were of equal length in the case of computing $LLCS(P, T)$. Each test included 100 different choices for $P$. In searching we used a single text $T$, and in computing the value $LLCS(P, T)$ we used as many $T$ as was necessary for their combined length to be at least one million characters. In order to estimate the overhead of these match vector handling methods in comparison to the simple lookup from a size-$\sigma$ table, we included also a test where the strings contain only UTF-8 characters that have distinct last byte values. This way our simple hash table method could be turned into a direct table lookup. Fig. 4 shows the results as a percentage of the running time of the direct table lookup.

In each case, using binary search was clearly the worst method on all computers. In some cases the overall processing time was almost three times longer than with the code automaton. The relative performance of the hash table and the code automaton varied depending on $m$ and the computer. On Pentium 4 the code automaton was always the fastest scheme, in fact even faster than the direct table lookup. We re-checked this with another compiler, and the situation remained the same. This is perhaps due to some pipelining effect etc. We note that this does not depend on the fact that the direct table lookup used restricted character encodings: we tested also the other schemes on the specially encoded strings, and their running times were practically the same as with the regular random strings. On Sparc and AMD the automaton and hash table performed fairly equally. With small $m$ the hash table tended to be often a little faster (always less than 10%), and with larger $m$ the code automaton became the better of the two. On AlphaStation the hash table was up to roughly 20% faster than the code automaton, but still a little bit slower with $m = 32$.

One conclusion is that the code automaton is typically very competitive against the other methods. We also note that the overall penalty for not being able to use a direct table lookup is reasonably small: never more than roughly 40%. Since the advantage of the bit-parallel methods over other kinds of algorithms is often much larger than this, they seem to be practical also with multi-byte encoded text. In addition, also the other types of string processing algorithms will have to pay some penalty for having to deal with multi-byte encoding. We also point out that the automaton is quite insensitive to the value of $m$ or the properties of the strings. Hence it is a feasible option for use with bit-parallel multi-byte string processing.

# 6   Conclusions

In this paper we proposed a scheme that uses a code automaton for looking up match vectors of multi-byte encoded characters. We also discussed two other schemes for the same task, and compared the three quite extensively. The test results showed that using the automaton is often the fastest choice, and never more than roughly 25% slower than the next best of these schemes. The binary search based method proposed by Wu, Manber and Myers in [25] was found to perform very slow. Using it resulted always in the longest processing time, in one case almost three times longer than when using the code automaton. Overall the test results give an idea about the feasibility of processing multi-byte encoded text with bit-parallel algorithms. As the test indicated the penalty to be at most roughly 40%, bit-parallel algorithms are a viable option with multi-byte text.

# References

1. Aho, A., Corasick, M.: Efficient string matching: an aid to bibliographic search. *Communications of the ACM*, 18(6):333–340, 1975.
2. Allison, A., Dix, T.L.: A bit-string longest common subsequence algorithm. *Information Processing Letters*, 23:305–310, 1986.
3. Baeza-Yates, R., Gonnet, G.: A new approach to text searching. *Communications of the ACM*, 35(10):74–82, 1992.
4. Boyer, R. S., Moore, J. S.: A fast string searching algorithm. *Communications of the ACM*, 20(10):762–772, 1977.
5. Crochemore, M., Iliopoulos, C. S., Pinzon, Y. J., Reid, J.F.: A fast and practical bit-vector algorithm for the longest common subsequence problem. *Information Processing Letters*, 80:279–285, 2001.

6. Crochemore, M., Rytter, W.: *Text Algorithms.* Oxford University Press, Oxford, UK, 1994.

7. Czumaj, A., Crochemore, M., Gasieniec, L., Jarominek, S., Lecroq, T., Plandowski, W., Rytter, W.: Speeding up two string-matching algorithms. *Algorithmica*, 12:247–267, 1994.

8. Damerau, F.: A technique for computer detection and correction of spelling errors. *Communications of the ACM*, 7(3):171–176, 1964.

9. Hyyrö, H.: Explaining and extending the bit-parallel approximate string matching algorithm of Myers. Technical Report A-2001-10, Dept. of Computer and Information Sciences, University of Tampere, Tampere, Finland, 2001.

10. Hyyrö, H.: Bit-parallel approximate string matching with transposition. In *Proc. 10th International Symposium on String Processing and Information Retrieval (SPIRE'2003)*, LNCS 2857, 2003.

11. Hyyrö, H.: Bit-parallel LCS-length computation revisited. In *Proc. 15th Australasian Workshop on Combinatorial Algorithms (AWOCA 2004)*, 2004.

12. Hyyrö, H., Navarro, G.: Faster bit-parallel approximate string matching. In *Proc. 13th Combinatorial Pattern Matching (CPM'2002)*, LNCS 2373, 2002.

13. Knuth, D. E., Morris, J. H. Jr, Pratt, V. R.: Fast pattern matching in strings. *SIAM Journal on Computing*, 6(1):323–350, 1977.

14. Levenshtein, V.: Binary codes capable of correcting deletions, insertions and reversals. *Soviet Physics Doklady*, 10(8):707–710, 1966.

15. Myers, G.: A fast bit-vector algorithm for approximate string matching based on dynamic progamming. *Journal of the ACM*, 46(3):395–415, 1999.

16. Navarro, G.: A guided tour to approximate string matching. *ACM Computing Surveys*, 33(1):31–88, 2001.

17. Navarro, G.: NR-grep: a fast and flexible pattern matching tool. *Software Practice and Experience*, 31:1265–1312, 2001.

18. Navarro, G., Raffinot, M.: Fast and flexible string matching by combining bit-parallelism and suffix automata. *ACM Journal of Experimental Algorithms*, 5(4), 2000.

19. Navarro, G., Raffinot, M.: *Flexible Pattern Matching in Strings – Practical on-line search algorithms for texts and biological sequences.* Cambridge University Press, 2002.

20. Robertson, A. M., Willett, P.: A comparison of spelling-correction methods for the identification of word forms in historical text databases. *Literary and Linguistic Computing*, 8(3):143–152, 1993.

21. Sankoff, D., Kruskal, J. (eds.): *Time Warps, String Edits, and Macromolecules: The Theory and Practice of Sequence Comparison.* Addison-Wesley, 1983.

22. Takeda, M., Miyamoto, S., Kida, T., Shinohara, A., Fukumachi, S., Shinohara, T., Arikawa, S.: Processing text files as is: Pattern matching over compressed tests, multi-byte character texts, and semi-structured tests. In *Proc. 9th International Symposium on String Processing and Information Retrieval (SPIRE'2002)*, LNCS 2476, 2002.

23. Unicode Consortium.: Unicode Home Page, http://www.unicode.org/.

24. Unicode Consortium.: *The Unicode Standard 4.0.* Addison-Wesley, 2003.

25. Wu, S., Manber, U., Myers, E.: A sub-quadratic algorithm for approximate limited expression matching. *Algorithmica*, 15(1):50–67, 1996.

# Retrieving Regional Information from Web by Contents Localness and User Location*

Qiang Ma[1] and Katsumi Tanaka[1,2]

[1] National Institute of Information and Communications Technology,
3-5 Hikaridai, Seika-cho, Soraku-gun, Kyoto 619-0289, Japan
qiang@nict.go.jp
[2] Graduate School of Informatics, Kyoto University,
Yoshida Honmachi, Sakyo, Kyoto, 606-8501, Japan
tanaka@dl.kuis.kyoto-u.ac.jp

**Abstract.** As use of the World Wide Web (WWW) and mobile devices spreads and becomes more advanced, people are increasingly able to use mobile devices, such as cell phones, to access various kinds of local information through the Internet for use in daily activities. Conventional mobile information services focus on providing location information related to the user's current location. In contrast, we propose an information retrieval method that searches for 'local information' pertaining to a wider geographic region. In this paper, we consider local information to be the kind of information that is of interest only to users within certain regions or organizations. From this perspective, 'local information' differs from location information. Moreover, we are interested in searching for local information pertaining to a wider region (an area) that extends beyond a specific location (a place). Our local information retrieval method includes three stages: 1) searching for information using a query generated with user location data and keywords, 2) mapping the content coverage of searched Web pages onto the user's current region, and 3) ranking the regional information based on the notion of localness degree.

## 1   Introduction

As Internet use has spread and evolved, more and more information related to our daily life and residential region has become available. In other words, there is an increasing variety of Web pages whose content is 'local'. In this paper, we

* This research is partly supported by the Special Research Area's Grant In Aid For Scientific Research(2) for the year 2004 under the project title "Research for New Search Service Methods Based on the Web's Semantic Structure" (Project No: 16016247, Representative: Katsumi Tanaka) and The Scientific Research Fund Foundation (A)(2) for the year 2004 under the project title Multi-modal Searches, Views on Mobile Content,and the Generation of Broadcasting Contenth(Project No: 14208036, Representative: Katsumi Tanaka), the Japanese Ministry of Education, Culture, Sports, Science and Technology.

S. H. Myaeng et al. (Eds.): AIRS 2004, LNCS 3411, pp. 301–312, 2005.

consider a Web page to be local if it is of interests to only the residents of a certain region or organization. From this perspective, local information differs from location information. Conventional information retrieval systems and search engines, such as Google [5], are very useful for helping users to find interesting information. However, users cannot easily find or exclude 'local' information about their daily lives and residential regions since it is difficult to specify keywords to form an appropriate query. Some portal Web sites [11, 16] provide directory-type search services for regional information. In such sites, the regional information is organized manually. The resources available to do this may be limited, though, and some valuable information may be missed.

At the same time, because of the advances of mobile computing, more and more information systems have begun providing information related to each user's current location, which is obtained from users' PHS or GPS devices. For example, they may provide information concerning the area near the user's current location, such as information on nearby restaurants, public transportation, or local weather. Most of these conventional mobile information systems[3, 7, 9, 14] thus search for information in a "location-oriented" manner.

As an alternative that will let users search for local information about daily activities and their residential regions, we earlier proposed a concept called the localness degree [10, 12]. We say a Web page is local when it is of interest to only the residents (users) of a certain region or organization. From this viewpoint, we compute the localness degree of a Web page in two ways: a) by estimating its region dependence - the occurrence frequency of geographic words and the area of its content coverage, and b) by estimating how ubiquitous its topic is - in other words, we estimate whether the information on the page is relevant everywhere and everyday in our daily life. Information retrieval based on localness degree differs from a conventional "location-oriented" information system in that it searches for information in a "localness-oriented" manner.

In this paper, we propose a method to search for local-information of a particular region based on a coverage mapping of Web page and user's location data. With this method, information is searched for from both the "localness-oriented" and the "location-oriented" perspectives. In other words, we propose a hybrid information retrieval method that considers the localness degree of a Web page and the location data for a user.

Note that the localness degree is a relative concept. For example, from the standpoint of "Kyoto Prefecture", a Web page may be local. However, it may not be local from the standpoint of "Kyoto City". In this paper, we modify our previous computation of the localness degree by considering a hierarchical region structure when searching for local information. Moreover, we compare the content coverage of a Web page and a geographic region to select the local information regarding a certain region.

In contrast to conventional location-oriented methods, we search for information regarding a wider region that extends beyond the current location of a user. In many cases, user will be familiar with their current locations, but will want more information about other nearby places that they will visit or are considering visiting. In short, we search for "region" (area) information not just "place" (location point) information.

The reminder of this paper is organized as follows. In Section 2, we introduce related work. In Section 3, we describe the mapping of content-coverage and the user's region of interest. The notion of localness degree is described in Section 4. We show some preliminary experiment results in Section 5. We conclude and look at future research in Section 6.

## 2   Related Work

The KOKONONET system [13], which was developed as part of Mobile Info Search (MIS) project[14], exchanges information bi-directionally between the Web and the real world based on location data. KOKONONET uses the location data, such as an address string in a document (Web page), to identify the geographic position of the document. Actually, KOKONONET is a typical location-oriented information system. Although we also use location data, our research differs in that we automatically discover 'local' information pertaining to a wider region (an area) that extends beyond a specific location (a place).

Buyukkokten et al. [1] discussed how to map a Web site to a geographic location, and studied the usage of several geographic keys for assigning site-level geographic context. By analyzing "whois" records, they built a database that correlates IP addresses and hostnames to approximate physical locations. By combining this information with the hyperlink structure of the Web, they were able to make inferences about the geography of the Web at the granularity level of a site.

Geographic Search [4] adds the ability to search for Web pages within a particular geographic locale to traditional keyword searching. To accomplish this, Egnor converted street addresses found within a large corpus of documents to latitude-longitude-based coordinates using the freely available TIGER and FIPS data sources, and built a two dimensional index of these coordinates. Egnor's system provides an interface that allows the user to augment a keyword search with the ability to restrict matches to within a certain radius of a specified address. This differs from our method with respect to how strongly a page is related to an area. Moreover, our work focuses on obtaining local information from the Web through a contextual approach based on page content and the correlation between pages and places.

Cityguide[3] is a major location information service which provides information about specific cities in USA. In Japan, the Digital City Kyoto project [7] developed systems and technologies to provide information on Kyoto city. These services provide high quality information through collecting and indexing Web pages by human. In Digital City Kyoto project, a system called Kyoto-SEARCH [9] has been developed to add the ability to collect and index Web pages automatically. KyotoSEARCH helps users navigate among Web information, maps, and Web-based geographic knowledge in an integrated way. In other words, KyotoSEARCH indexes Web pages by mapping each page with its relevant locations on a map and provides a visual interface to help users navigate regional information.

Lee et al. [8] and Yamada et al. [17] proposed some methods to optimize the MBR of a given Web page. They used the tag information of HTML to identify the roles of geographical words appearing in the Web page. However, their modified methods are limited to the Web pages describing about Kyoto city, which is a well-known tourist city in Japan. In addition, they did not show any experiment results. The performances of their methods are not clear.

## 3    Relationship Between Web Page and Geographic Region

### 3.1    Content Coverage of a Web Page

We convert geographic words (place and organization names) contained in a Web page into pairs of location data such as *(latitude, longitude)* and estimate the content coverage of the Web page based on the MBR method.

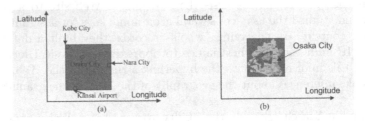

**Fig. 1.** Examples of Content Coverage

Each geographic word is converted to a two-dimensional point (latitude, longitude). We plot all of these points on a map where the y- and x-axes are latitude and longitude, respectively. The content coverage of a Web page can be approximately computed using the MBR which contains these points (**Fig.** 1(a)).

If a Web page contains only one geographic word, we roughly construct the MBR based on the geographic boundaries. For example, to construct an MBR on a map using only the geographic word in "Osaka City", we could use the maximum and minimum values of the longitude and latitude data for all places in Osaka city (**Fig.** 1(b)). In addition, if the geographic word is part of an organization name, we can analyze its address and pick out the place name to construct the MBR.

Some detailed place names have no match in the location information database, such as [15]. In such cases, we upgrade the administrative level for such place names to find approximate location data. For example, if the location data of "C street, B city, A state" is not found, we can use the location data of "B city, A state" to obtain an approximate matching.

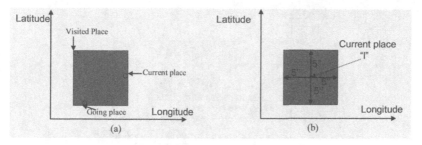

**Fig. 2.** Examples of Geographic Regions

Different places may share the same name. To avoid a mismatch, we analyze the page's context to clearly specify its location data. For example, to match up "Futyu City" to its correct location data, we can examine the context. If "Hiroshima" appears near "Futyu City", we can use the location data of "Futyu City, Hiroshima Prefecture" to find the location data of "Futyu City" in this page. On the other hand, if "Tokyo" appears, we should use "Futyu City, Tokyo metropolitan" to get the proper location data.

### 3.2    Geographic Region

In the same way as we compute the content coverage of a Web page, we can plot the user's location data on a map and construct an MBR as the user's current geographic region.

As mentioned, users may be interested in information about a region wider than their current locations. For example, suppose you are staying at a hotel and want to go to dinner - you may be interested in information about restaurants in a wider area, rather than just about those in the hotel with which you are already familiar. In other words, you need regional information more than location information.

Let the places the user is visiting, has visited, and will visit be $L=\{l_1,l_2,...,l_m\}$. First, we convert place $l_i \in L$ to a pair of location data $(latitude_i, longitude_i)$. We then plot these points on a map and construct a MBR containing them. **Fig.** 2 shows some examples of constructing a geographic region.

We cannot construct the MBR, though, if $L$ contains just one place $l$ ($|L| = 1$). In such a special case, we assume the MBR is a square whose center point is $l$. Figure 2(b) shows an example of such a region. The distance from the current place to each side is 5″ (longitude or latitude) which ranges from 5000 to 10000 meters in Japan. That is, we assume that users are most likely to be interested in regional information from around their current positions and the distance from the current position to a target place should be within 10 km. In practice, this distance can be modified according to user intentions and application systems.

### 3.3    Coverage Mapping of Web Page and Geographic Region

We compare the MBRs constructed from a Web page and the user's location data. If they are the same, we conclude that the Web page provides information

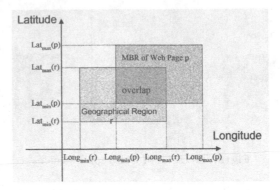

**Fig. 3.** Example of Overlap

about the user's current geographic region. This is extract mapping. However, this is a rare case in the real world. In most cases, these two kinds of MBR differ from each other. Therefore, we compute the overlap between the MBRs to estimate the strength of the relationship between a Web page and a geographic region (**Fig. 3**). In other words, if there is a large overlap between the content coverage of a Web page and the user's geographic region, we say that the Web page is strongly related to the user's interest.

The relationship between Web page $p$ and the user's geographic region $r$ is computed as follows.

$$Relation(p, r) = \frac{OverLap(p, r)}{MBR(p) + MBR(r) - OverLap(p, r)}$$
$$OverLap(p, r) = (min(max_{lat}(p), max_{lat}(r)) - max(min_{lat}(p), min_{lat}(r)))$$
$$\times (min(max_{log}(p), max_{log}(r)) - max(min_{log}(p), min_{log}(r)))$$
$$MBR(p) = (max_{lat}(p) - min_{lat}(p)) \times (max_{log}(p) - min_{log}(p))$$
$$MBR(r) = (max_{lat}(r) - min_{lat}(r)) \times (max_{log}(r) - min_{log}(r)) \tag{1}$$

where $max_{lat}(x)$ and $min_{lat}(x)$ are, respectively, the maximum and minimum latitude data used to construct the MBR of $x$. $max_{log}(x)$ and $min_{log}(x)$ are, respectively, the maximum and minimum longitude data used to construct the MBR of $x$. Obviously, if $long_{min}(p) > long_{max}(t)$, or $long_{max}(p) < long_{min}(t)$, or $lati_{max}(p) < lati_{min}(t)$, or $lati_{min}(p) > lati_{max}(t)$, the $OverLap(p, r) = 0$. $MBR(x)$ is the area of the MBR of $x$. Here, the standard unit of latitude and longitude used to compute $MBR(x)$ is seconds.

If Web page $p$ is strongly related to the user's geographic region $r$, the possibility of $p$ containing local information regarding region $r$ is high. Of course, we also need to compute the localness degree of $p$ as described in Section 4.

## 4    Localness Degree

We say a Web page is local when it is of interest to only users from a certain region or organization. From this viewpoint, we compute the localness degree of

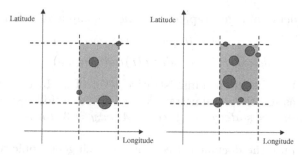

**Fig. 4.** Example of Region Dependence Based on Content Coverage

a Web page in two ways: a) estimating its region dependence (the frequency of geographic words and the area of its content coverage), and b) estimating the ubiquitousness of its topic (i.e., we estimate whether the information is general in that it applies everywhere and everyday in our daily life).

## 4.1    Region Dependence

If a Web page provides information about a narrow region, its region dependence may be high. In other words, if the content coverage of a Web page is very narrow, we say that it is local.

We have also found that Web pages that are highly dependent on a region include a lot of geographic information, such as the names of the country, state (prefecture), city, and so on. Therefore, we compute the frequency of these geographic words occurring in a Web page to estimate its region dependence. In other words, the number and sizes of points plotted on the MBR (content coverage) are also used to estimate the localness. Here, the point size represents the level of detail of a geographic word. An example is shown in **Fig.** 4. Four points are plotted on the left MBR, and eight points are plotted on the right MBR. Even if the areas of the two MBRs are equal, their localness degrees should differ because the right one probably provides information about the region that is more detailed.

That is to say, a Web page may contain many detailed geographic words, and thus be strongly related to particular regions. Here, 'detailed geographic word' means that a geographic word identifies a detailed location. Usually, we can use the administrative level of each geographic word (region name) as its level of detail. In our current work, we just consider the level of detail of geographic words based on administrative levels. For example, "Kyoto" is more detailed than "Japan". As mentioned, local information is a relative concept: a Web page may be local within a wider region but not local within a narrow region. From this perspective, for a given region $(r)$ or a given administrative level $(l)$, if the administrative level of a geographic word $gw$ is less than $l$, its level of detail is high. The given level or given administrative level relates to the user region and should be specified by the user manually or by the user location data automatically.

The level of detail of a geographic word $gw$ is computed as follows.

$$detail(gw, l) = l/admin(gw) \qquad (2)$$
$$detail(gw, r) = admin(r)/admin(gw) \qquad (3)$$

where $admin(x)$ is the given administrative level of $x$. In our current work, we are able to assign the administrative level value as follows: $country = 5$, $organization\ name = state(prefecture) = 4$, $city = 3$, $town = 2$, and $street$ $(road) = 1$.

We also compute the document frequency of each geographic word within a Web page's corpus (i.e., Web pages from the same Web site as the one whose region dependence is being estimated) when estimating the region dependence of a Web page. If the document frequency of a geographic word contained in the Web page is high, such a geographic word is a common one and has less effect on the localness of that page.

In short, if the content coverage is narrow and the content contains many detailed geographic words, the Web page is local. As mentioned, it is also important to consider the document frequency of each geographic word to prevent common geographic words affecting the localness degree. Therefore, we compute the localness degree based on the region dependence as follows.

$$local_d(p) = \frac{\sum_{i=1}^{k} detail(gw_i) \cdot tf(gw_i)/df(gw_i)}{MBR(p) \cdot words(p)} \qquad (4)$$

where $local_d(p)$ is the localness degree, $detail(gw_i)$ is the level of detail of geographic words, $tf(gw_i)$ is the frequency of $gw_i$ within $p$, $df(gw_i)$ is the document frequency of $gw_i$ within a page corpus, $MBR(p)$ is the content coverage of page $p$, and $words(p)$ is the number of words contained in $p$.

## 4.2   Topic Ubiquitousness

A ubiquitous occurrence may have a high localness degree. For example, summer festivals, athletic meets, and weekend sales are ubiquitous events that frequently occur. A ubiquitous occurrence may be a normal part of our daily life. Therefore, if a page describes a ubiquitous event, it contains local information of interest to users in a certain region.

We estimate the topic ubiquitousness of a Web page using two factors: the similarity between pages and the locations where events are held. If the pages are similar in content, but differ in terms of location or time, their topics are general. Here, the similarity between pages $a$ and $b$ is calculated based on the vector space model.

In short, the localness degree $local_u$ based on the ubiquitousness of the topic of Web page $p$ is computed as

$$local_u(p) = m/n \qquad (5)$$

where $m$ is the number of pages whose similarities (with page $p$, excluding geographic words and proper nouns) are greater than threshold $\theta$. $n$ is the number of pages used for the similarity comparison.

The ubiquitousness of a Web page is a relative concept. A Web page may be local for a wide region, but not considered local for a very narrow region. For example, although a particular event rarely occurs within Kyoto city, it may occur often within the Kansai area (which includes Kyoto city). A Web page describing such an event is general within the Kansai area and it is local. However, it is neither general nor local within Kyoto city. Similarly, an event occurring often during summer will be judged local within the time scope of summer, but may not be within the time scope of one year. Thus, we modify the computation of the localness degree based on ubiquitousness. Within time duration $\tau$ and region scope $\Omega$, the localness $(local'_u(p, l, \Omega, \tau))$ of Web page $p$ based on the ubiquitousness at region level $l$ is computed as

$$local'_u(p, l, \Omega, \tau) = m'/n' = \frac{\sum\limits_{x \in area(\Omega, l)} (countif(x, \theta))}{count(\Omega, l)} \qquad (6)$$

where $m'$ is the number of regions containing many similar pages $p$ at the region level $l$ within region scope $\Omega$ and time duration $\tau$. $n'$ is the number of regions at level $l$ within $\Omega$. $area(\Omega, l)$ is the set of regions at level $l$ within $\Omega$. If the ratio of similar pages (of $p$) among all pages of region $x$ is bigger than threshold $\theta$, $countif(x, \theta)$ returns 1; else, it returns 0. $count(\Omega, l)$ is the number of regions at region level $l$ within $\Omega$.

### 4.3    Integrated Localness Degree

We also can define an integrated localness degree based on the region dependence and ubiquitousness, such as the sum of $local_d(p)$ and $local'_u(p)$. The integrated localness $local_i(p)$ of page $p$ is computed as

$$local_i(p) = f(local_d(p), local'_u(p)) \qquad (7)$$

where $f$ stands for an integration function of $local_d(p)$ and $local'_u(p)$, such as sum or logical OR.

## 5    Preliminary Experiment

In this section, we describe the preliminary experiment of our local-information retrieval method. We used Google[5] to search for candidate Web pages and then ranked them based on their localness degrees and relationships between them and geographic region in which the user is interested. If a Web page has high localness degree (region dependency or ubiquitousness) and related to the specified geographic region[1], we consider that it is local-information.

---

[1] The geographic region is constructed by using the location names contained in the query.

**Table 1.** Search Keywords (location names)

| | |
|---|---|
| $Q_1$ | "Daito Shijyonawate" |
| $Q_2$ | "Daito Higasiosaka" |
| $Q_3$ | "Daito Tunumi" |
| $Q_4$ | "Kobe Osaka Kyoto" |
| $Q_5$ | "Kobe Osaka Nara" |
| $Q_6$ | "Himeji Kobe Osaka" |

**Table 2.** Evaluation Results of Preliminary Experiment

| | Recall ratio | Precision Ratio |
|---|---|---|
| Region Dependency | 0.482 | 0.762 |
| Ubiquitousness | 0.789 | 0.735 |
| Converage Mapping | 0.689 | 0.790 |
| Search (A) | 0.472 | 0.641 |
| Search (B) | 0.5 | 0.518 |

We have issued 6 queries (in Japanese, see **Table 1**) to Google through GoogleAPI[5] and then selected the top 50 results as the candidate Web pages. Initially, we excluded all of the structural information (e.g., HTML tags) and the advertisement content from the HTML sources of these candidate Web pages. We used ChaSen[2] for Japanese morphology analysis. To exclude stop words, we built a stop word list containing 593 terms in English and 347 terms in Japanese.

The comparison targets to compute the ubiquitousness of a Web page are collected from ASAHI.COM, a well known news Web site in Japan, during 3 month period and within 47 regions at the region level of prefecture. In our experiment, we had collected 1918 Web pages. Thus, in this experiment, at the prefecture level, $count(\Omega, l) = 47$. We used a location database[15] containing the longitude and latitude data of 3252 places in Japan.

The experiment results are shown in **Table 2**. In Table 2, "Search (A)" means that we search for Web pages which is local and related to our specified geographic region. On the other hand, "Search (B)" means that we search for Web pages which is related to our specified geographic region but not local.

The main considerable failure reasons of region dependency computation and coverage mapping are described as follows.

– The location database used in our experiment just contained the longitude and latitude data of main places in Japan, such as seats of municipal governments. Thus, we could not construct well the MBR to compute the content coverage sometimes. Using a database containing more detailed data is necessary.

- We failed in extracting pure body-text of some Web pages. In other words, we had gotten many noise terms. For example, a Web page may have a frame containing many region names which are anchor text. Although such frame is not related to the subject of that page, we had extracted these unrelated anchor text (region names) as its geographical words. It is necessary to analyze the HTML source and find the pure body text of a Web page.

On the other hand, we have failed in computing the ubiquitousness by the following causes.

- We just used news articles (Web pages) during 3-month period and could not find enough similar Web pages to compute the ubiquitousness sometimes. Actually, as we mentioned before, "ubiquitousness" is a relative concept. It is necessary to organize well the comparison collection from time, region and domain (category) perspectives.
- Some Web pages contain a small number of words. If we excluded proper nouns from them, we may fail in similarity computation.

Since our experiment was limited, there are further works needed to do. Nevertheless, from the experiment results, at least, we could assume that our method is useful for discovering or excluding local information concerning in a certain region. Moreover, comparing to our previous experiment results[10], we could say that we have succeed in improving the computation of localness degree.

## 6  Conclusion

We have developed an information retrieval method from localness-oriented and region-oriented perspectives. We generate queries by using user location data and use these to search for the local Web pages. For each searched Web page, we compute its localness degree and estimate its relationship with the user's region. That is, we rank the searched Web page by considering whether it contains local information pertaining to a certain region and return the pages having a high probability of containing such information. We compute the localness degree of a Web page from two perspectives: region dependence and ubiquitousness of its topic. In addition, in this paper, we have extended the localness degree to a relative concept by considering a hierarchical region structure. We have also proposed a method to compute the relationship between a Web page and a geographic region based on the MBR method. Based on our method, we can search for regional information extending beyond the location information, which the user may know in advance.

We plan to develop and evaluate a prototype search system based on our proposed methods. In our current work, we generate the query by simply using the location data and user specified keywords. We need to further refine the query generation. For instance, we could modify the query by more precisely considering the relationships between the current place, previous place, and future place of the user. We will carry out further study on the method to compute the geographic region and content coverage of a Web page. Specially, we need carry out further natural language processing to improve the localness degree computation and coverage mapping. In addition, we hope to study the integration function applied to the region dependence and ubiquitousness for various application systems.

# References

1. Buyukkokten O., Cho J. Garcia-Molina H., Gravano L., and Shivakumar N. Exploiting geographical location information of web pages. In *WebDB (Informal Proceedings)*, pages 91–96, 1999.
2. ChaSen. http://chasen.aist-nara.ac.jp/index.html.en.
3. Cityguide. http://www.digitalcity.com.
4. Egnor D. Google programing contest, 2002.
5. Google. http://www.google.com.
6. Guttman A. R-trees: A dynamic index structure for spatial searching. *Proc. ACM SIGMOD Conference on Management of Data*, 14(2):47–57, 1984.
7. Ishida T. Digital city kyoto: Social information infrastructure for everyday life. In *Communications of the ACM, Vol.45, No.7*, pages 76–81, 2002.
8. Lee R., SHINA H., Takakura H., and Kambayashi Y. Two-dimensional rang query processing for geographical web search (in Japanese). In *IPSJ SIG Technical Report, 2003-DBS-131*, pages 413–420, 2003.
9. Lee R., Takakura H., and Kambayashi Y. Virtual query processing for gis with web contents. In *proceedings of the 6th IFIP working conference on visual database systems*, 2002.
10. MA Q., Matsumoto C., and Tanaka K. A localness-filter for searched web pages. In *Lecture Notes in Computer Science (Proc. of APWeb03), LNCS 2642*, pages 525–536, 2003.
11. MACHIgoo. http://machi.goo.ne.jp.
12. Matsumoto C., Ma Q., and Tanaka K. Web information retrieval based on the localness degree. In *Lecture Note of Computer Science (Proc. of DEXA02), LNCS 2453*, pages 172–181, 2002.
13. MIS2. http://www.kokono.net/.
14. Miura N., Takahashi K., Yokoji S., and Shima K. Location oriented information integration -mobile info search 2 experiment - (in Japanese). *The 57th National Convention of IPSJ*, 3:637–638, 10 1998.
15. Takeda T. The latitude / longitude position database of all-prefectures cities, towns and villages in japan, 2000.
16. Yahoo!regional. http://local.yahoo.co.jp.
17. Yamada N., Lee R., Takakura H., and Kambayashi Y. Classification of web pages with geographic scope and level of details for mobile computing (in Japanese). In *IPSJ SIG Technical Report, 2003-DBS-131*, pages 509–514, 2003.

# Towards Understanding the Functions of Web Element

Xinyi Yin[1] and Wee Sun Lee[2]

[1] Department of Computer Science
National University of Singapore, Singapore 117543
yinxinyi@comp.nus.edu.sg
[2] Department of Computer Science and Singapore-MIT Alliance,
National University of Singapore, Singapore 117543
leews@comp.nus.edu.sg

**Abstract.** A web page is a collection of basic elements, and the role of each element in a page is different. For example, an image element can be part of the main content, advertisement, or banner of the site. This paper describes ongoing work using a machine learning approach to classify each element in a web page into six functional categories: Content (C), Related Link (R), Navigation (N), Advertisement (A), Form (F) and Other (O). This allows the extraction of only certain categories of content in a webpage to be delivered to a mobile device to fit user's specific needs, or to facilitate web information processes like web mining or mobile search. We manually labeled 18,864 elements from 150 websites. For each element we extracted both local features (such as the text length, URL, tag name etc) and global features (such as the text match with the other elements) to construct a feature vector. We trained the training set 10,650 elements with a decision tree learning algorithm J48, and it achieved 82% accuracy for stratified cross-validation, and an average F value 0.78 for the six different categories. Testing on 3,043 elements from pages that are not included in the training set gives 58% accuracy rate. Although this is not satisfactory overall, the F value for content category reaches 0.795, indicating that the method could be useful for less demanding applications. We are working on improving the results in order to make automatic functional classification of web elements feasible and to provide new opportunities to push the state of art in the mobile internet and mobile search.

## 1  Introduction

A web page is a collection of basic elements, and the role of each element in a page is different. Without accurate understanding of the role of each element, it will be very difficult to further process the web page for different purposes. For example, mobile devices have very small screen sizes and memory capacities, converting web pages for mobile device requires the selection of only subset of all elements based on the needs of the user. If the user wants to read only the main content and the related articles, understanding of the role of each element will help to deliver only the desired material. Another example is in web page classification, where the performance is often affected by the fact that the main content is hidden among a lot of unrelated content. If the classification is based on only the main content, the performance can

S. H. Myaeng et al. (Eds.): AIRS 2004, LNCS 3411, pp. 313–324, 2005.
© Springer-Verlag Berlin Heidelberg 2005

be improved. We believe that most areas of web processing can be improved if we get a better understanding of the role of the elements in the web page.

Researchers have spent a lot of effort in solving the problem of partitioning the web page and understanding the importance of each part. For example, in [9], the web page is partitioned into blocks, and each block is assigned a value (1 to 4) to indicate its importance. In [15], a ranking algorithm similar to Google's PageRank algorithm is used to rank the content objects within a web page to allow the extraction of only important part.

This paper presents a machine learning approach to classify each element in a web page into six functional categories: Content(C), Related Link (R), Navigation (N), Advertisement (A), Form (F) and Other (O). This allows the extraction of only certain categories of content in a webpage to be delivered to a mobile device to fit the user's specific needs, or to facilitate web information processes like web mining or mobile search. The method that we propose first divides a web page into basic elements. For each element we extracted both local features and global features. The local features are obtained from the element itself without checking other elements in the same page, such as the text length, URL or tag name etc., the global features is obtained through comparing the elements with the rest of the elements in the page, such as the text match with the other elements. We construct a feature vector and use machine learning to classify each element.

We organize the paper in the following way. In section 2, we discuss our criteria of functional category selection. In section 3, we will present the design. In section 4, we proposed a new interface for the mobile Internet system based on the element classification algorithm. In section 5 we will discuss the dataset, and describe the evaluation of the system. Section 6 is about related works. In section 7, we will give our conclusion and our direction for future research.

## 2   Functional Category Selection

One web page has many different types of content. Each of them is designed to affect the reader's behavior in a different manner. For example, the reader usually reads the main content, ignores the advertisement, and clicks on related links to further explore a topic. In our research work we aim at classify web content into meaningful functional categories to facilitate the surfing experience, especially on a mobile devices. We have two goals to achieve in selecting the functional category. First, the functional category needs to have a meaningful and useful theme that is related to the user's daily surfing experience, so content like copyright statement, company banner etc, will not be considered as a category. Second, each functional category should have a distinct and coherent meaning which has a predictable implication on the user behavior and minimize the confusion from the reader. For example, it would be proper to divide links into Navigation (N) and Related Link ®, because Navigation (N) is leading to another section of the site which may or may not have the similar topic with current page, while the Related Link (R) is leading to another page that tells a related story. At same time, further dividing an article into categories like author, article, date, title may not be so necessary. In this research we will classify the content into six categories

- Content (C): Main content that is the centre topic of a web page, which includes the main article, title, author, date, supporting picture or any other material used to describe the topic. The main topic is the most important content in a web page.
- Related Link (R): A web page normally contains links to web pages that that tells the related story, which helps a reader to further explore a topic.
- Navigation (N): A website is normally designed in a well-structured form and divided into different sections, on each web page there are fixed links to those sections, which helps a reader to navigate in the site.
- Form (F): Forms can be used to enable the reader to interact with the webpage. For example, a user can search with keyword, or submit certain request through a form.
- Advertisement (A): Today's web page is normally filled with internal or external advertisements. They may or may not be useful to the reader. It is ideal if we can separate them and give reader a choice on how to deal with them, especially on mobile devices.
- Other (O) All the other contents, for example, the banner, copyright information and the logo, are not informative, are normally invisible and do not have much influence on user's behavior.

**Fig. 1.** The top CNN banner is manually labeled as Other (O), and the title, author, date, article and the pictures are labeled as Content (C). In the page there are three Forms (F), two Navigations (N), and three Advertisements (A). Notice that the text advertisement is actually embedded in the main article, which provide a great challenge to normal classification methods

## 3   Dividing a Web Page into Basic Elements

### 3.1   Basic Elements

To understand the functional category of each element in the web page, we need to develop a consistent method to divide a web page into basic elements. There are two basic requirements for the algorithm: firstly, the performance of the method should not be affected by the source of the web page and the different ways of implementing the layout. Secondly, the granularity of the partition should be independent from the author, the reader and the content in the webpage, so that the data will present a consistent pattern for the machine learning algorithm to learn from.

Many different methods have been proposed to partition an HTML page into blocks. For example, [14] proposed a visual based method to analyze the structure of a web page, and [17] provides a method to automatically understand the semantic structure of HTML pages based on detecting visual similarities of content objects. However, these methods are not suitable for our application because the granularity is subjective and affected by the semantic meaning of the content.

We develop an algorithm similar to [15] that uses the DOM interface provided by the web browser to divide the web page into non-overlapping visible elements in an HTML page from the bottom up. The algorithm identifies the elements using two rules:

- A visible object like an image, link or text paragraph will be a basic element if it is not overlapping with another child or its parent node.
- For overlapping objects, the minimal container of the two objects will be a potential element to be verified by the rule 1. The algorithm will seek from bottom up to locate the nearest common container, and the container will be treated as one element.

For example, a web page may contain many links that are not overlapping with each other. Each of the links will be treated as basic element. Another web page may have a text paragraph with a link. Here we have two overlapping objects. The bigger one, the text paragraph, will be chosen to be checked by rule 1. If the text paragraph is not overlapping with other elements at a higher level, it will be chosen; otherwise we will recursively search upward. In this manner, all the visible objects in a web page will be allocated to the basic elements.

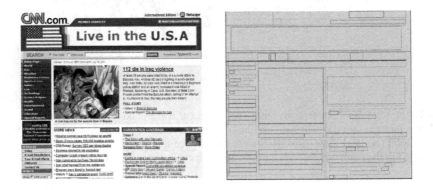

**Fig. 2.** The result of division a webpage from www.cnn.com using this method

### 3.2  Feature Vector for Element

After converting a web page into basic elements, and we will use a machine learning algorithm to classify each element into different functional categories. For each element a feature vector is constructed. The feature vector is made up of the internal features and external features. The internal features are those features which are presented by the element itself, and we can observe them without checking other elements in the same web page. For example, the element type, the URL or the anchor text is typical internal feature. Second, the external features are those features that are available through checking a group of elements in a same web page, for example, the semantic match of the text of one element with the other elements in the web page. For our experiment we collect 22 internal features and 6 external features for each of the element to construct the feature vector.

Internal features:

1. Width/height ratio (*Pr*): The shape of an element reflects its category. For example, for an image, an irregular image is usually more likely to be banner or navigation bar. We use the following formula to calculate the value for images:

$$Pr(i) = Width\ (i)\ /\ Height\ (i)\ .\tag{1}$$

2. Element (Sz): The size of an element reflects its category. For example, for content normally have a larger size, a Navigation (N) element normally have a smaller size. We use the following formula to calculate the value for images:

$$Sz(i) = Width(i) * Height(i)\ .\tag{2}$$

3. Text length: The length of visible plain text in an element is often a strong indication of the content element. The content elements normally have more text than advertisement element. We use the length as one feature.
4. Anchor Text length: The length of anchor text can differentiate Navigation (N) element from Related Link (R) element, because normally Navigation (N) element has shorter anchor text length than related element. At the same time this feature differentiates the link elements from those non-link elements like the content.
5. Physical offset: Physical position calculated by pixels is actually a very important feature. Element closer to the center point is more likely to be content then those in the edge, which tends to be Navigation (N) element. So we use four features LeftEdge, RightEdge, TopEdge and BottomEdege, and use a binary value to represent whether the element's border touches one of the four edges of the web page,

$$LeftEdge\ (i) = \begin{cases} True & if\ leftBoard\ (i) < 0.1 * ScreenWidt\ h \\ False & else \end{cases}\tag{3}$$

$$RightEdge(i) = \begin{cases} True & if\ rightBoard(i) > 0.9 * ScreenWidh \\ False & else \end{cases}\tag{4}$$

$$TopEdge\ (i) = \begin{cases} True & if\ topBoard\ (i) < 100 \\ False & else \end{cases}\tag{5}$$

$$BottomEdge\ (i) = \begin{cases} True & if\ bottomBoar\ d(i) < ScreenHeig\ ht - 100 \quad\quad (6) \\ \\ False & else \end{cases}$$

As shown in the formula, if an element's left boundary falls in left 10% of the screen, its LeftEdge is true, otherwise is False. In the same way if its top boundary is located in a position higher than 100 pixels, its TopEdge is True.

1. Tag Category: The web designer tends to uses different type of element to present different content, typically a Flash is object normally used for advertisements and plain text is more likely to be used for main content. In our context we select four category

   - Image: Any embedded image. <img>
   - Flash: element with a Macromedia Flash object
   - Control: element with tag <input> <select> <textarea>
   - Link: element with the tag <a>
   - Text: element with the tag <P>
   - Mix: the element has more than one of above category

6. Link Property: For link object we can capture additional features which could help the machine learning algorithm to separate many confusing cases. For example, a picture which links to external site is more likely to be an Advertisement (A) than an internal one. The features we selected from a link object includes:

   - The number of components: we decompose the url into component for example, "www.nus.edu.sg/soc/" will be decomposed into five component "www", "nus", "edu", "sg", "soc", same type of element may have different link but the number of the URL component is often the same.
   - External URL: whether the site is linked to an external site.
   - Multiple site: if two sites are inside the link, it is normally an advertisement reference program

7. Image Property: for image object we extract the following features.

   - Animated GIF: whether the image is animated GIF
   - Logo image: if the image file name contains word like "logo" or "banner" etc., which indicate that the image is likely to be the banner or logo for the site
   - Space image: if the image file name contains word like " blank". "corner", "space" etc, which indicate the image is not related to the content.

8. Key word: For each element category we select a group of typical words that can represent that category. For example, we will check if an element has word like "ads", "advertise", "sponsor" or "click" to check for Advertisement (A) content. If the element contains the typical word from each category becomes a feature.

External feature:

1. Content Match: For each element we calculate the sum of the cosine similarity with every other element in the same pages. If we assume that there is one center topic in the web page, this calculation will give us an evaluation of how this element is related to the main article.
2. SameLeft: The count of other elements in the web page which has the same left edge with element i. Regular element like Content (C) or Navigation (N) link are always present a regular pattern, while elements like Form (F) are less likely to demonstrate such a pattern. In the same manner we also extract SameRight(i), SameTop(i), SameRight(i), SameBottom(i), SameSize(i)

In this way for each element we collect 28 features (22 internal features and 6 external features) to construct the feature vector for machine learning algorithm to construct the classifier.

## 4  Display Subsystem

We propose an innovative user interface to let user surf web page based on the functional classification algorithm. The basic idea is very simple: Instead of presenting the whole web page on small screen device, we classify the content in the web page into six categories and let user specify the content to delivered and rendered in his mobile device. The design goal of our method is to provide a user interface and let user freely read the specific content without being distracted by the irrelevant content.

Suppose the user is interested in a web page shown in Fig.4. The web page contains several different type of material: the navigation bar, the content and the advertisements. Before the page is delivered to user's mobile device, the middleware in the proxy server will decompose the page into elements and classify it into six categories, and generate six view of the web page. It will then deliver the Content (C) view, where only the content will be delivered. The user can read other type of content by clicking the category name on the top. That will leads to the wireless delivery that that content.

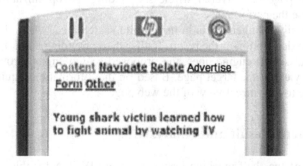

**Fig. 3.** The generated content view on a mobile device

The system allows the user to freely surf the exact the type of the material from mobile device, and simply click the category name to explore different material in the web page.

Related View

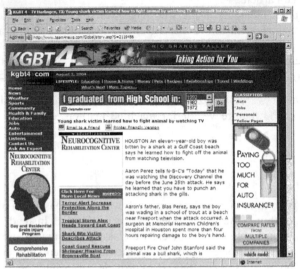

The original web page

Content View

**Fig. 4.** The web page and PDA view

Ideally, the algorithm will be loaded in a personal gateway, which can be our own desktop computer. A page retrieval request is sent to the personal gateway from the mobile device. It will retrieve and render the page in its memory on the behalf of the mobile device, and classified all the result into six categories and generate six web pages, and send the optimized page to the mobile device wirelessly. Normally optimization of a web page can be done within 1 second on a normal Pentium III computer. Because the desktop is connected to Internet with cable and only requested content of the page is delivered wirelessly, adding the optimization part will not greatly decrease the performance.

Another possible application is in the mobile search area, where the search engine returns the text description of the result. Rather than providing a link to the target web page as normal search engine does, the proposed search engine provide six links to the optimized view of the target page. Based on the information requirement the user can directly go to the correct view of the web page.

## 5   Experiment Result and Analysis

We used the off-the-shelf machine learning system WEKA for this experiment. The classifier was first trained and tested using 10-fold cross-validation. We follow the general practice of using precision (P), recall (R) and F-measure as the evaluation.

1. The precision (P) for certain category:
   Return= (number of element really belongs to certain category)/(number of elements classified as certain category)
2. The recall(R) for certain category:
   R= (number of element returned)/(all the elements that belong to certain category)
3. F-measure: 2PR/(P+R)

We created the data in the following manner. First, we randomly selected 150 websites from the Google directory as the training set, under the category of news. We developed a specialized annotation software tool. For each web site we chose 2 pages and manually label each of the 18,864 elements into one of the six categories. As we can see from the distribution of all the elements in the training set, only 7% of the elements are content.

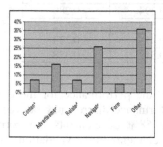

The distribution pattern

| | |
|---|---|
| Content | 7% |
| Advertisement | 16% |
| Related Link | 7% |
| Navigation | 26% |
| Form | 5% |
| Other | 36% |

**Fig. 5.** The distribution of category element

Not every element in a web page is equally valuable for classification. For example, a very small icon or the invisible images which is used to make up space like "space.gif" is not meaningful to classify, to remove them from the page will not affect the surfing experience. We design two simple rules to remove such elements from the list. First, the element with either width or height less than 5 pixels will be removed. Second, if same picture (decided by source image name) is used more than three times in a page, it will also be removed.

After preprocess, altogether 13,693 elements remained, we set aside the first 10,650 as training set and the rest as test set. Using the WEKA package J48 classifier, we first trained with the internal and external features using 10-fold cross-validation on the training set. The corresponding results are showed in Table 1.

**Table 1.** Experiment result with external features

| Class | Precision | Recall | F-Measure |
|---|---|---|---|
| Content | 0.882 | 0.877 | 0.879 |
| Advertisement | 0.739 | 0.753 | 0.746 |
| Relate | 0.870 | 0.865 | 0.867 |
| Navigation | 0.871 | 0.881 | 0.876 |
| Form | 0.660 | 0.629 | 0.644 |
| Other | 0.716 | 0.713 | 0.715 |

Of all the 10,650 elements, 8172 elements (82.13%) are correctly classified; the average F value for five categories is 0.78. The F-Measure for the Content is 0.88, which shows the features are quite effective and important for the classifier.

The second step of our experiment we use the training set to build the J48 classifier (we have tried many other classifiers provided by WEKA, the J48 performs the best), and evaluate 3,043 elements in the test set. The test set does not have any common web sites with the training set, and the elements are labeled with the same tool. The results are:

**Table 2.** Experiment result on test sets

| Class | Precision | Recall | F-Measure |
|-------|-----------|--------|-----------|
| Content | 0.798 | 0.792 | 0.795 |
| Advertisement | 0.475 | 0.597 | 0.529 |
| Related Link | 0.228 | 0.394 | 0.289 |
| Navigation | 0.678 | 0.575 | 0.622 |
| Form | 0.202 | 0.419 | 0.273 |
| Other | 0.347 | 0.312 | 0.328 |

As we can see from the Table 1 and Table 2, the performance for the content category is still reasonably good, but there is performance discrepancy between the cross validation result and the test result, only 58% of elements are correctly classified. This is probably due to the fact that the elements in the validation sets for cross validation belong to the same set of web pages as the training set but the elements from the test sets belong to different web sites altogether. There appears to be a lot of room for performance improvement, particularly on unseen web sites. In addition to that, all the sample websites are chosen from Google directory. It is likely that most of them are well organized and designed. More research work needs to be done in the future for the real world Internet where a lot of irregular web pages and misleading anchor text might exist.

## 6   Related Work

[9] proposed using vision based page segmentation algorithm to partition semantic blocks with a hierarchical structure, and extract spatial features ( position and size) as well as content features (number of image and links) to form feature vectors. A machine learning algorithm is used to train for block importance. The work is related to this paper in many ways. First, both works are trying to understand the role of each part after partitioning the page, [9] express it in three level of importance, while our research defines the roles as a specific functional category. Secondly, the two papers solve the problem at a different granularity. [9] solved the same problem from a block level. Consequently, how to define the block becomes a problem that will affect the accuracy. In our work the definition of the element is fixed.

In our previous work [15], we proposed to divide the web page in the same manner, and use a ranking algorithm similar to Google's PageRank algorithm to rank the content objects within a web page. This allows the extraction of only important parts of web pages for delivery to mobile devices. The limitation in that work is that it only selects the element based on importance. While many elements in a web page are not so important, removing them on the mobile device will hurt the surfing experience. For example, without navigation links, it will be very difficult to browse a web site on a mobile device.

The SmartView system in [8] is based on idea of "divide and view". It performs partitioning of HTML document content into logical sections that can further be selected by the user and viewed independently from the rest of the document. This paper presents the same idea of allowing the user to randomly access any website and gives the user full control of which content to be read. However, the system in [8] let user to control with logical section to read, so the user must first view a whole web page, and then decide which section to follow. Our system divide the webpage into functional category, and the user can select the type of content to view on from mobile device without downloading the whole page.

The web is not designed for mobile device. There are many commercial solutions to help people to surf web on mobile device. System like Web Clipping [13] and AvantGo [1] suggests building specialized website for different website to surf. We believe mobile Internet is an extension of existing Internet and we should develop systems that convert the content in the Internet to a format that is suitable for various small screen devices. The systems need to perform three functions, including scaling, manually authoring, transforming. The functions are summarized in [12]. For example, [4] and [3] use summaries of single or multiple pages to present to the user. [2] and [11] describe the process of manually extracting only the useful information from the existing web.

## 7  Conclusion

Our goal is to design a system that can classify the elements that make up a web page into six functional categories Content (C), Related Link (R), Navigation (N), Advertisement (A), Form (F) and Other (O). This allows the extraction of only certain type of content in a webpage to be delivered to a mobile device to fit user's specific needs, or to facilitate web information processes like web mining or mobile search. We achieve this by first dividing the web page into basic elements, and extracting internal and external features of the each element in a given web page. After the feature vector is obtained, we use an off-the-shelf machine learning package to classify each of them. Previous research in this area mainly focused on deciding the importance of a blocks or element in the web page, but we propose that it is possible to understand the functional category of each element. Based on a machine learning classification algorithm, we developed a system which extract certain type of content from original page and provide a different view for small screen devices.

With the current system, it is possible to select only certain category of content within a web page to be viewed on a mobile device. However, current work has been tested only on mainly news website, where the layout typically quite standard, we need further try different type of web page. Even though the result for 10-fold cross-validation is good, the precision and recall for categories other than Content (C) on an unseen website is still not satisfactory. Further work is required to improve the performance before the system is truly able to be used for surfing on mobile devices.

With the development of wireless technology and emergence of various mobile devices, people will not be limited to the desktop computer. We will access the Internet through all possible devices. Instead of building different webs for different devices, we strongly believe that the right direction is to convert and deliver the same content in different ways to different devices.

# References

1. AvantGo  http://www.avantgo .com
2. Bickmore, T., Schilit, B. Digester. Device Independent Access to the World Wide Web. In *Proceedings of the 6th International World Wide Web Conference*, 1997.
3. Buyukkokten, O., Garcia-Molina, H., Paepcke, A. Seeing the Whole in Parts: Text Summarization for Web Browsing on Handheld Devices. In *Proceedings of the $10^{th}$ World Wide Web Conference*, 2001.
4. Buyukkokten, O., Garcia-Molina, H., Paepcke, A., T. Winograd. Power Browser: Efficient Web Browsing for PDAs. In *Proceedings of the ACM Conference on Computers and Human Interaction*, 2000.
5. H. Bharadvaj, A. Joshi, and S. Auephanwiriyakul. An Active Transcoding Proxy to Support Mobile Web Access. In *Proceedings of 17th IEEE Symposium on Reliable Distributed Systems*, 1998.
6. Jinlin Chen , Baoyao Zhou , Jin Shi , Hongjiang Zhang , Qiu Fengwu. Function-based Object Model towards Website Adaptation. In *Proceedings of 10th Thirteenth International World Wide Web Conference*, 2001.
7. Lan Yi , Bing Liu , Xiaoli Li. Eliminating Noisy Information in Web Pages for Data Mining. In *Proceedings of the 9th ACM SIGKDD International Conference on Knowledge Discovery and Data Mining*, 2003.
8. Natasa Milic-Frayling, Ralph Sommerer. SmartView: Flexible Viewing of Web Page Contents. In *Proceedings of the 11th World Wide Web Conference*, 2002.
9. Ruihua Song, Haifeng Liu, Jirong Wen, Wei-Ying Ma. Learning Block Importance Models for Web Pages. In *Proceedings of 13th International World Wide Web Conference*, 2004.
10. Shipeng Yu, Deng Cai, Ji-Rong Wen, Wei-Ying Ma. Improving Pseudo-relevance Feedback in Web Information Retrieval Using Web Page Segmentation. In *Proceedings of the 11th World Wide Web Conference*, 2003.
11. Suhit Gupta, Gail Kaiser, David Neistadt, Peter Grimm. DOM-based Content Extraction of HTML Documents. In *Proceedings of the 12th International Conference on World Wide Web*, 2003.
12. Trevor, J. Hilbert, D.M., Schilit, B.N., Koh, T.K. From Desktop to Phone Top, a UI for Web Interaction on Very Small Devices. In *Proceedings of the $14^{th}$ annual ACM symposium on user interface software and technology,* 2001.
13. Web Clipping http://www.palmos.com/dev/tech/webclipping/
14. Xiao-Dong Gu, Jinlin Chen, Wei Ying Ma, Guo-Liang Chen. Visual Based Content Understanding towards Web Adaptation. In *Second International Conference on Adaptive Hypermedia and Adaptive Web-based Systems*, 2002.
15. Xinyi Yin, Wee Sun Lee. Using Link Analysis to Improve Layout on Mobile Devices. In *Proceedings of 13th International World Wide Web Conference*, 2004.
16. Yu Chen, Wei-Ying Ma, Hong-Jiang Zhang. Detecting Web Page Structure for Adaptive Viewing on Small Form Factor Devices. In *Proceedings of the 11th World Wide Web Conference*, 2003.
17. Yudong Yang, HongJiang Zhang. HTML Page Analysis Based on Visual Cues. In *7th International Conference on Document Analysis and Recognition*, 2001.

# Clustering-Based Navigation of Image Search Results on Mobile Devices

Hao Liu[2]*, Xing Xie[1], Xiaoou Tang[2], and Wei-Ying Ma[1]

[1] Microsoft Research Asia, 5F, Sigma building, No. 49, Zhichun Road, Beijing, 100080, P.R.China
[2] Department of Information Engineering, the Chinese University of Hong Kong, Shatin N.T., Hong Kong

**Abstract.** With the Internet information explosion as well as the rapid increasing of mobile users, web image search and browsing on mobile devices will become a big application in foreseeable future. Since mobile devices generally suffer from limited resources, like small screen factor and interaction facilities, etc, the interface provided by current image search engine is not suitable for search results navigation on mobile devices. In this paper, we propose a new method to increase user's experience on search results browsing and navigation. By clustering result images based on visual and semantic cues, user can catch the overview of the search results with only a few clicks; by using an optimal layout algorithm, the summary of each cluster can be presented on small screen with maximum scaling ratio; by employing the smart thumbnail of the presentation image, more information can be shown to users on limited screen. The navigation operations are designed to make the new interface easy to learn and simple to use. Experiment results demonstrate the effectiveness of the new presentation method.

## 1 Introduction

With the information explosion in the past decade, large collections of images have been made available on the Internet. Traditionally, people who were searching for particular types of images submit their queries to some trained librarians and waited for a few days for the results. This situation has been greatly improved since now we can simply type our query terms on an image search engine or the web interface of professional photo libraries, and the results will return within seconds.

Because of their portability and mobility, mobile devices with diverse capabilities are also undergoing a considerable progress in hardware and software. The rapid development makes it possible for users to access various kinds of services on Internet anywhere and anytime. Since visual contents such as images provide rich and vivid information preferred by users, mobile image search is becoming a big application in foreseeable future. Tom Yeh [23] demonstrated an example of

---

* This work was conducted at Microsoft Research Asia.

S. H. Myaeng et al. (Eds.): AIRS 2004, LNCS 3411, pp. 325–336, 2005.
© Springer-Verlag Berlin Heidelberg 2005

such application, in which mobile image search act as a director to help people in unfamiliar sites.

Browsing and navigation of large image collections is not a new topic too. It has been studied under different contexts, such as personal albums or professional libraries. However, none of them has taken web images into consideration. The characteristics of data sets can heavily affect the effectiveness of presentation approaches. For example, home photos would be best browsed in a chronicle order [17][18], by location [20], or by person [26]. As for professional photo libraries, they contain more accurate annotations and usually have been classified manually into different categories.

The problem of web image retrieval has already been studied for years. A number of approaches have been proposed to improve the retrieval performance, including Text-based method as Google [8] and Altavista [1], Content-based systems as [7] and [19], and Link-based method as PageRank [15] and PicASHOW [11]. However, the presentation of search results has been less studied. A simple ranking-based interface is still the most frequently used presentation in commercial systems. Unfortunately, since the performance of current image retrieval systems is far from satisfactory, a large number of irrelevant images are often retrieved along with relevant ones. So a great deal of manual page navigation is required to find the best match from the retrieval results. Only in very recently, the problem of web image search results navigation and browsing has been studied in [13][21]. It has shown that the new presentation method can help users to explore image search results more naturally and efficiently. However, web image search results navigation on mobile devices is still an open question.

In this paper, we intend to study new methods for web image search results presentations on mobile devices. The novel contributions of our work include: 1) We analyze user's information need and information accessing patterns in mobile search scenario; 2) a method is proposed to help user to explore the search results effectively; 3) the navigation operations are designed to make the new interface easy to learn and simple to use.

The rest of this paper is organized as follows: in Section 2, we briefly introduce several key concepts, which guides our proposed method in this paper. Section 3 discusses two image clustering strategies, i.e. clustering based on visual cues and clustering based on textual labels, used for search results navigation. Section 4 proposes a new high density image presentation method, which is designed specifically for small screen factors of mobile devices. Based on those techniques, we re-design navigation operations for smart phone system, a kind of poplar mobile devices in people's daily life, in Section 5. We give the experiment results in Section 6. Finally, concluding remarks and discussions are in Section 7.

## 2    Information Accessing on Mobile Devices

Since mobile devices generally suffer from limited resources, like small screen factor, interaction facilities and computational power, we can expect different navigation pattens between the users of mobile and desktop image search services. For example, for desktop users, user can input the query very quickly and get the results back with high speed network and powerful desktop machine in few seconds. Also user can refine the query easily to get further results until satisfied with the results. While, for mobile device users, it is more difficult to input a detail and exact query with small keyboard, waiting for a long list returned results is also annoying due to the bandwidth limitation on moving, small screen space makes it impossible for user to catch the whole results in a few clicks, etc. Current commercial image search engines only provide general search and browsing interfaces and do not take into account the special capability constrains of mobile devices in search results presentation design. For example, in those search interfaces, it needs too many navigation operations (scrolling and clicking) to get the target results, zooming in/out operations are required frequently in order to get a clear view of images of high dimension on small screen, etc.

*Information Foraging Theory* [16] analyzes trade-offs in the value of information gained against the costs of performing activity in human-computer interaction tasks. Cognitive systems engaged in information foraging will exhibit such adaptive tendencies, when feasible, to maximize gains of valuable information per unit cost. The theory has been applied to Web navigation [5] as well as the image browsing on mobile devices [12].

Image clustering is a feasible reorganization method to assist information foraging on search results. Clustering analysis can partition image collection into groups of image subsets such that the attributes of images in the same cluster are similar. The images of the same cluster are organized in a web page, one page for each cluster. At the same time, the images which stands for the profile of each cluster are extracted and presented to users as cues for page navigation. While browsing the search results, people can usually scan the representation images first, and decide which cluster the target images belongs to, and then move forward to the correspondent page and find target images. Therefore, clustering-based presentation can speed up search process by taking advantage of previewing representation images. It enables the user to navigate toward the interested subset images directly.

## 3   Search Results Clustering

Different from traditional CBIR systems, both the content features, which are visual presentation of the image, as well as the textual features, which are regard as the semantic meaning of the image, are available for web image search systems. Generally speaking, visual feature is the most intuitive cue for navigation. Clustering based on the visual features of image can provide good guidance for user to navigate among search results. On the other hand, it is an accurate way to reveal the semantic meaning of the image through those textual labels of the image. Therefore, two presentations can be generated for search results navigation. It is not an easy task to discover the optimal search results clustering for different queries. A solution to this problem is to provide multiple choices and let user decide their favorite clustering methods.

## 3.1    Visual Based Image Clustering

To represent the content information of each image, an extensible attention model proposed in [4][14] has been applied in many areas, such as mobile image browsing [12], progressive image delivery [9] as well as image retrieval [22]. Instead of treating an image as a whole, it detects each region-of-interest in the image separately. Since these regions can be regarded as the key regions which represent the semantic content of the image, clustering based on the key regions of the image can provide the visual cues as well as semantic cues for navigation.

Many low level features can be extracted from the key regions of the image, such as color features for example, correlogram, color moments, and texture features, for example, Gabor wavelet feature. Since color and texture features can reveal different aspects of images, their combination may be useful to represent the image. In our system, we employ Color Texture Moments (CTM) proposed by Yu et.al [24], which integrates the color and texture characteristics of the image in a compact form. CTM adopt local Fourier transform as a texture representation scheme and derive eight characteristic maps for describing different aspects of co-occurrence relations of image pixels in each channel of the color space. Then it calculates the first and second moments of these maps as a representation of the natural color image pixel distribution, resulting in a 48-dimensional feature vector. Also it would be important to note that the CTM features can be extracted in advance to speed up the clustering processing.

In our system, K-means clustering algorithm is employed on the visual feature vector space. Model selection is determined by a heuristic way. In order to reduce the number of operation at the first overview step, we set the number of clusters from 6 to 12, so users can catch the overview of search result collection with a few clicks. Figure 1 shows an example of saliency region based clustering using the key words "Hawaii".

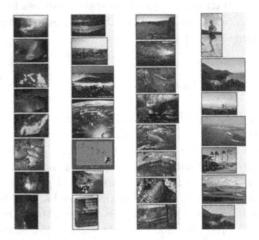

**Fig. 1.** Visual-based clustering

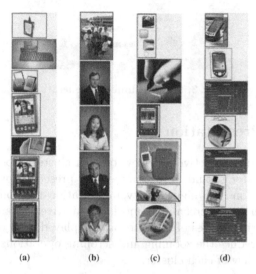

(a)            (b)            (c)            (d)

**Fig. 2.** Textual-based clustering

## 3.2    Textual Based Image Clustering

An alternative way is to cluster web image search results based on the textural labels of each image. These textual labels include text close to that image, image file name, URL of the image, alternate text of the image in web page source, and title of the web page.

Generally, there are three methods to extract the surrounding text: window based, DOM based and vision based. Widow based method treats html source as a text stream. For each image, it uses a fixed-length window to extract the text before and behind the image. DOM based method extracts the surrounding text from HTML DOM tree. In HTML DOM tree, an image is always a leaf node, DOM based method uses the text of the sibling nodes as the surrounding text of the image. VIsion-based Page Segmentation (VIPS) algorithm as proposed in [3] treats the web page from 2-D view, extract the semantic structure of a web page based on its visual presentation. Such semantic structure is a tree structure; each node in the tree corresponds to a block. Compared with DOM based methods, the segments obtained by VIPS are much more semantically aggregated. Moreover, each node in VIPS tree will be assigned a value to indicate how coherent of the content in the block based on visual perception. Thus, it is easy to decide which block should be the right image block according to the DoC value.

Once texture labels of image are extracted, clustering image search results becomes the same problem of web search results clustering [25]. We use linear regression method to extract the saliency phrase of each image cluster.

Figure 2 shows an example of search clustering based on saliency phase. User wants to find images of "PDA". The returned results are classified automatically into subcategories. The textual labels of the four categories shown in Figure 2 are: (a) palm, devices; (b) staff; (c) technology, design; and (d) tom hardware guide mobile.

**Fig. 3.** Smart thumbnail generation

# 4    Cluster Presentation

In order to generate a effective overview of each cluster, most important information of the cluster should be extracted and presented on the small screen space. In this section, two high density presentation methods are proposed to assist browsing on small screen. The most informative parts of each image and the most informative images of a cluster are displayed with optimal resolution. Therefore, it can reduce the zooming and scrolling operations required for users to get clear over view of each cluster.

## 4.1    Smart Thumbnail Generation

It is demonstrated that small degree of cropping will not affect the user understanding the content of an image [10]. Cropping-based thumbnail generation method proposed in [4] can be employed in our system to optimize the scaling ratio of each image shown on the screen. In our system, smart thumbnail generated base on attention model is applied to the the images presented to users. Figure 3 shows an example of smart thumbnail generated by attention model based cropping method.

## 4.2    Space Efficient Layout Algorithm

To help users navigate among different clusters efficiently, several images of each cluster are selected to represent the content of the cluster and serve as cues for navigation.

**Representation Images Selection.** For visual based image clustering method, we choose the distance between each image and the corresponding cluster center in visual vector space as the ranking function. For textual label based image clustering method, we choose the co-occurrence evaluation as well as the thumbnail size as the ranking criteria.

The number of images chosen to represent the cluster depends on the personal preference as well as the properties of the devices, such as screen size, aspect ratio, etc.

**High Density Layout Design.** In order to accommodate the represent images on limited screen space, a layout algorithm is employed in our system to optimize the scaling ratio of each image. A similar algorithm is designed under the ordered presentation constrain in [6]. This constrain comes from the facts the video shot

**Table 1.** The description of layout algorithm

---

Inputs:
- Screen resources $S = (w^s, h^s)$, where $x^s, y^s > 0$;
- An set of representation images: $R = r_i | r_i = (w_i^r, h_i^r), 0 < i < M$, i.e. $M$ images are chosen to represent the cluster;
- Corresponding smart thumbnails of the images: $T = t_i | t_i = (x_i^t, y_i^t)$;

Outputs:
- The coordinates of each thumbnail $t : C = c_i | c_i = (x_i^t, y_i^t)$; that maximize the scaling ratio $r_{max}$ so that the thumbnails can be packed in rows filling the screen.

---

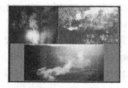 

(a) An example of 3 images     (b) An example of 4 images

**Fig. 4.** Presentation layout design

should be placed according to their time information in the original video. As for our problem, it is free from such constrain since no time information available for web image search results. Table 1 shows the description of layout algorithm.

Such layout algorithm optimized to get $r_{max}$, i.e. maximized scaling ratio, is NP problem. However, due to the limited screen size, we can not put too many images on the screen at the same time while user still can catch the content the images. When $M$ is small (less than 5), we can iteratively adjust the design to get the maximum scaling ratio. Therefore, the problem can be solved by finding line breaks in each permutation order of $M$ images.

In order to make each cluster looks more reasonable and speed up the optimal $r_{max}$ search process, we add an extra pre-processing step in advance by getting rid of those smart thumbnails with aspect ratio far from that of screen outline. Figure 4 shows examples of layout design for three and four representation images in (a) and (b) respectively.

## 5   Navigation Design

Smart phone system is a kind of poplar mobile devices in people's daily life. In this section, we design the image search results navigation operations for smart phone based on the methods discussed in the previous sections.

Figure 5 gives a walkthrough of web image search on mobile device using an example of query "PDA". The user's task is to find different design of PDA devices. The details of each step will be discussed in the following sub-sections.

## 5.1    The Start Page

A simple and easy to understand start page is very important for search engine [2]. This is especially true for mobile devices, which often suffers from the limited input capabilities. The start page is shown in Figure 5 (a), an input box and a button to trigger the search process is provided on this page. Suppose a user who wants to buy a new PDA is interested in the latest design of such mobile devices. S/he can enter the key word "PDA"at this page and the results will be returned to him as a list of clustered results, as shown in Figure 5 (b).

The only choice provided to users is the clustering method used in the result presentation page. The default presentation interface was set to be the visual based clustering presentation, because we believe it is more intuitial for user to navigate based on visual cues, especially for novice users. However, the textual based clustering can be generated by a simple switching operation. This provides another choose for expert users, who can make a judgment on query type and select the best presentation strategy. As for the user searching "PDA", clustering based on visual information may not be a good presentation method. So s/he can switch to textual based clustering presentation methods to quickly access the results in clusters labelled with "palm, devices"and "technology, design"as shown in Figure 5 (d).

## 5.2    The Results Presentation Pages

The returned images are presented in result pages. Two steps are provided to help users to understanding the search results, i.e. an overview page and detail view pages of each cluster.

The overview page provides the content summary of the search results as shown in Figure 5 (b) and (d). The cluster presentation is generated for users to get the information of each cluster and decide whether to continue to investigate the details of the cluster. This is an optimized one column view and only up-down scrolling is required to explore this page. For each up/down scrolling operation, a clustering box is moved to the center of the screen, so users can catch the entire results in few clicks. The layout relationship between the boxes of each cluster is determined by the importance of the cluster, that is, the ranking of the images within the cluster.

Once the user click a cluster in the overview page, the corresponding detail view page of the cluster will be generated and transmitted to the client, as shown in Figure 5 (c) and (e). Still, this is an optimized one column page and only up-down scrolling is required to explore it. The layout relationship of the image within the each cluster is determined by the importance of the image, that is, the ranking score of each image within that cluster. By clicking each image of the detail view page, user will navigate to the corresponding web page containing the image. A backwards operation by clicking the "backward"key on the keyboard is easy to employ to move back to the overview page.

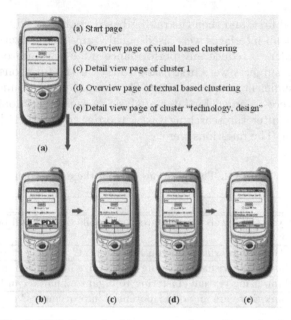

(a) Start page

(b) Overview page of visual based clustering

(c) Detail view page of cluster 1

(d) Overview page of textual based clustering

(e) Detail view page of cluster "technology, design"

(a)

(b)          (c)          (d)          (e)

**Fig. 5.** Walkthrough of image search results navigation

# 6    Experiment Results

We implemented the new browsing system using C# and ASP.Net on Windows operation system. In order to evaluate the new presentation method for search results navigation on mobile devices, an initial controlled user study experiment has been carried out. We compare our method with Google image search interface. Experiment results are presented and discussed in this section.

## 6.1    Subjects and Queries

There were four participants for our experiment. They were recruited from nearby universities. Three of them were master students and one of them is an undergraduate student. Before study, they were asked to performed image search tasks on desktop machine. Six queries were used in comparing different types of interfaces, which are PDA, Apple, Java, Hawaii, Orange, and Pie. Among them, the query "PDA" was provided to allow subjects to practice and get familiar with the interfaces.

## 6.2    Tasks

Each user was asked to complete all the six queries in the same order. For each query, they tried to find a few images most relevant to the query terms. The interface presentation order was varied for each user: two users were given the ranking-based, then the clustering-based approach; the other two saw the clustering-based first, and then the raking-based approach. This balanced ordering meant that all interfaces were used equally any query-interface biases were reduced. In order to reduce any performance influence due to familiarity and experience, the subjects were first asked to try all the functions of different interfaces for a sufficient amount of time. The user interactions and corresponding time stamps were recorded for later analysis. A small questionnaire was given to the subjects after the searching task, in order to get their feedback on the interface design. The questions were:

**Table 2.** The questions used in the experiment

| ID | Question |
|----|----------|
| 1  | When do you think web image search on mobile devices will be useful? Give a few scenarios. |
| 2  | Which of the two interfaces do you like the best? |
| 3  | Do you think using visual clustering to organize images can help searching? |
| 4  | Do you think using textual clustering to organize images can help searching? |
| 5  | Do you think there are any other potential improvements? |

### 6.3    Results

Results show that the subjects thought that web image search on mobile devices would be useful for those circumstances that desktop machine is not available to access, for example travelling to unfamiliar sites, shopping, or waiting friends at coffee house. Some of them also searched images just for fun, for example, searching for photos related to the latest movie or some famous people.

The experiment results were shown in Figure 6. On average, users spent 217 seconds for each query using the clustering-based interface and 268 seconds for the raking-based interface. As we expected, the new approach outperformed the raking-based interface by reducing 19% of the total search time.

All of the four subjects prefer clustering-based approach for search results browsing on mobile devices. They said that visual based clustering was in good accordance with human perception, therefore, was helpful to narrow down the

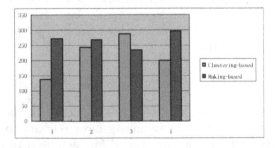

**Fig. 6.** The average search time for subjects/interface combination

searching process. While, it is not very convenient to find images of interest by textual based clustering, this is partly because the textual label is not as intuitive as visual cues. The subjects thought the presentation image selection is a key factor related to the browsing performance and should be improved in future.

# 7    Conclusions and Future Work

In this paper, we proposed a web image search results navigation strategy based on image clustering to increase the usability of image search on mobile devices. By clustering result images based on visual and semantic cues, user can catch the overview of the search results with only a few clicks so as to speed up the navigation process. Initial experiment results show the effectiveness of the new method.

There are several issues to be investigated in current design. It is an interesting topic to find an optimal clustering methods which can use both the visual and textual cues for search results re-organization. Also, an long term user study should be carried out to validate the usability of the new design. We will continue to investigate these directions in future work.

# References

1. AltaVista Image Search. http://www.altavista.com/image
2. Baeza-Yates, R. and Ribeiro-Neto, B., Modern Information Retrieval, Addison Wesley Longman 1999.
3. Cai, D., Yu, S., Wen, J. R., and Ma, W. Y., VIPS: a visionbased page segmentation algorithm, Microsoft Technical Report, MSR-TR-2003-79, 2003.
4. Chen, L. Q., Xie, X., Fan, X., Ma, W. Y., Zhang, H. J., and Zhou, H. Q., A Visual Attention Model for Adapting Images on Small Displays, ACM Multimedia Systems Journal, Vol. 9, No.4, pp353-364, 2003.
5. Chi, E. H., Pirolli, P., Chen, K., and Pitkow, J., Using information scent to model user information needs and actions on the Web, ACM CHI 2001, Seattle, Washington, Mar. 2001.
6. Chiu, P., Girgensohn, A., and Liu, Q., Stained-Glass Visualization for Highly Condensed Video Summaries, ICME 2004, Taipei, Taiwan, June 2004.
7. Flickner, M., Sawhney, H., Niblack, W., Ashley, J., Huang, Q., Dom, B., Gorkani, M., Hafner, J., Petkovic, D., Lee. D., Steele, D., and Yanke, P., the QBIC system, IEEE Computer, Vol. 28, No. 9, pp23-32, 1995.
8. Google Image Search. http://images.google.com
9. Hu, Y.S., Xie, X., Chen, Z.H., and Ma, W.Y., Attention Model Based Progressive Image Transmission, ICME 2004, Taipei, Taiwan, June 2004.
10. Intraub, H., Gottesman, C. V., Willey, E. V., and Zuk, I. J., Boundary Extension Briefly Glimpsed Photographs: Do Common Perceptual Processes Result in Unexpected Memory Distortions? J. of Memory and Language, Vol. 35, No. 2, pp118-134, 1996.
11. Lempel, R. and Soffer, A., PicASHOW: Pictorial Authority Search by Hyperlinks on the Web, ACM Trans. on Information Systems, Vol. 20, No. 1, pp1-24, Jan. 2002.

12. Liu, H., Xie, X., Ma, W. Y., and Zhang, H. J., Automatic Browsing of Large Pictures on Mobile Devices, ACM Multimedia 2003, Berkeley, CA, USA, Nov. 2003.

13. Liu, H., Xie, X., Tang, X., Li, Z., and Ma, W. Y., Effective Browsing of Web Image Search Results, ACM Multimedia Workshop on Information Retrieval, New York, NY, USA, Oct. 2004.

14. Ma, Y. F. and Zhang, H. J., Contrast-based Image Attention Analysis by Using Fuzzy Growing, ACM Multimedia 2003, Berkeley, CA, USA, Nov. 2003.

15. Page, L., Brin, S., Motwani, R., and Winograd, T., the PageRank Citation Ranking: Bringing Order to the Web, Technical Report, Computer Science Dept., Stanford University, 1998.

16. Pirolli, P. and Card, S.K., Information Foraging, Psychological Review, Vol. 106, No. 4, pp643-675, 1999.

17. Platt, J. C., Czerwinski, M., and Field, B. A., PhotoTOC: Automatic Clustering for Browsing Personal Photographs, Microsoft Technical Report, MSR-TR-2002-17, Feb. 2002.

18. Rodden, K. and Wood, K., How Do People Manage Their Digital Photographs? ACM CHI 2003, Fort Lauderdale, USA, April 2003.

19. Rui, Y., Huang, T., and Chang, S., Image Retrieval: Past, Present, and Future, J. of Vis. Com. and Image Rep. Vol. 10, No. 4, pp39-62, Apr. 1999.

20. Toyama, K., Lofan, R., and Roseway, A., Geographic Location Tags on Digital Images, ACM Multimedia 2003, Berkeley, CA, USA, Nov. 2003.

21. Wang, X. J., Ma, W. Y., He, Q. C., and Li, X., Grouping Web Image Search Result, ACM Multimedia 2004, New York, USA, Oct. 2004.

22. Wang, X. J., Ma, W. Y., and Li, X., Data-Driven Approach for Bridging the Cognitive Gap in Image Retrieval, ICME 2004, Taipei, Taiwan, June 2004.

23. Yeh, T., Tollmar, K., and Darrell, T., Searching the Web with Mobile Images for Location Recognition, CVPR 2004, Washington DC, USA, Jun. 2004.

24. Yu, H., Li, M., Zhang, H.J., and Feng, J., Color Texture Moments for Content-based Image Retrieval. ICIP 2002, Rochester, N.Y., USA, Sept. 2002.

25. Zeng, H. J., He, Q. C., Chen, Z., and Ma, W. Y., Learning to Cluster Search Results. SIGIR 2004, Sheffield, UK, July 2004.

26. Zhang, L., Chen, L. B., Li, M. J., and Zhang, H. J., Automated Annotation of Human Faces in Family Albums, ACM Multimedia 2003, Berkeley, CA, USA, Nov. 2003.

# Author Index

# Lecture Notes in Computer Science

For information about Vols. 1–3309

please contact your bookseller or Springer

Vol. 3357: H. Handschuh, M.A. Hasan (Eds.), Selected Areas in Cryptography. XI, 354 pages. 2004.

Vol. 3356: G. Das, V.P. Gulati (Eds.), Intelligent Information Technology. XII, 428 pages. 2004.

Vol. 3355: R. Murray-Smith, R. Shorten (Eds.), Switching and Learning in Feedback Systems. X, 343 pages. 2005.

Vol. 3353: J. Hromkovič, M. Nagl, B. Westfechtel (Eds.), Graph-Theoretic Concepts in Computer Science. XI, 404 pages. 2004.

Vol. 3352: C. Blundo, S. Cimato (Eds.), Security in Communication Networks. XI, 381 pages. 2005.

Vol. 3351: G. Persiano, R. Solis-Oba (Eds.), Approximation and Online Algorithms. VIII, 295 pages. 2005.

Vol. 3350: M. Hermenegildo, D. Cabeza (Eds.), Practical Aspects of Declarative Languages. VIII, 269 pages. 2005.

Vol. 3349: B.M. Chapman (Ed.), Shared Memory Parallel Programming with Open MP. X, 149 pages. 2005.

Vol. 3348: A. Canteaut, K. Viswanathan (Eds.), Progress in Cryptology - INDOCRYPT 2004. XIV, 431 pages. 2004.

Vol. 3347: R.K. Ghosh, H. Mohanty (Eds.), Distributed Computing and Internet Technology. XX, 472 pages. 2004.

Vol. 3346: R.H. Bordini, M. Dastani, J. Dix, A.E.F. Seghrouchni (Eds.), Programming Multi-Agent Systems. XIV, 249 pages. 2005. (Subseries LNAI).

Vol. 3345: Y. Cai (Ed.), Ambient Intelligence for Scientific Discovery. XII, 311 pages. 2005. (Subseries LNAI).

Vol. 3344: J. Malenfant, B.M. Østvold (Eds.), Object-Oriented Technology. ECOOP 2004 Workshop Reader. VIII, 215 pages. 2005.

Vol. 3342: E. Şahin, W.M. Spears (Eds.), Swarm Robotics. IX, 175 pages. 2005.

Vol. 3341: R. Fleischer, G. Trippen (Eds.), Algorithms and Computation. XVII, 935 pages. 2004.

Vol. 3340: C.S. Calude, E. Calude, M.J. Dinneen (Eds.), Developments in Language Theory. XI, 431 pages. 2004.

Vol. 3339: G.I. Webb, X. Yu (Eds.), AI 2004: Advances in Artificial Intelligence. XXII, 1272 pages. 2004. (Subseries LNAI).

Vol. 3338: S.Z. Li, J. Lai, T. Tan, G. Feng, Y. Wang (Eds.), Advances in Biometric Person Authentication. XVIII, 699 pages. 2004.

Vol. 3337: J.M. Barreiro, F. Martin-Sanchez, V. Maojo, F. Sanz (Eds.), Biological and Medical Data Analysis. XI, 508 pages. 2004.

Vol. 3336: D. Karagiannis, U. Reimer (Eds.), Practical Aspects of Knowledge Management. X, 523 pages. 2004. (Subseries LNAI).

Vol. 3335: M. Malek, M. Reitenspieß, J. Kaiser (Eds.), Service Availability. X, 213 pages. 2005.

Vol. 3334: Z. Chen, H. Chen, Q. Miao, Y. Fu, E. Fox, E.-p. Lim (Eds.), Digital Libraries: International Collaboration and Cross-Fertilization. XX, 690 pages. 2004.

Vol. 3333: K. Aizawa, Y. Nakamura, S. Satoh (Eds.), Advances in Multimedia Information Processing - PCM 2004, Part III. XXXV, 785 pages. 2004.

Vol. 3332: K. Aizawa, Y. Nakamura, S. Satoh (Eds.), Advances in Multimedia Information Processing - PCM 2004, Part II. XXXVI, 1051 pages. 2004.

Vol. 3331: K. Aizawa, Y. Nakamura, S. Satoh (Eds.), Advances in Multimedia Information Processing - PCM 2004, Part I. XXXVI, 667 pages. 2004.

Vol. 3330: J. Akiyama, E.T. Baskoro, M. Kano (Eds.), Combinatorial Geometry and Graph Theory. VIII, 227 pages. 2005.

Vol. 3329: P.J. Lee (Ed.), Advances in Cryptology - ASI-ACRYPT 2004. XVI, 546 pages. 2004.

Vol. 3328: K. Lodaya, M. Mahajan (Eds.), FSTTCS 2004: Foundations of Software Technology and Theoretical Computer Science. XVI, 532 pages. 2004.

Vol. 3327: Y. Shi, W. Xu, Z. Chen (Eds.), Data Mining and Knowledge Management. XIII, 263 pages. 2005. (Subseries LNAI).

Vol. 3326: A. Sen, N. Das, S.K. Das, B.P. Sinha (Eds.), Distributed Computing - IWDC 2004. XIX, 546 pages. 2004.

Vol. 3325: C.H. Lim, M. Yung (Eds.), Information Security Applications. XI, 472 pages. 2005.

Vol. 3323: G. Antoniou, H. Boley (Eds.), Rules and Rule Markup Languages for the Semantic Web. X, 215 pages. 2004.

Vol. 3322: R. Klette, J. Žunić (Eds.), Combinatorial Image Analysis. XII, 760 pages. 2004.

Vol. 3321: M.J. Maher (Ed.), Advances in Computer Science - ASIAN 2004. Higher-Level Decision Making. XII, 510 pages. 2004.

Vol. 3320: K.-M. Liew, H. Shen, S. See, W. Cai (Eds.), Parallel and Distributed Computing: Applications and Technologies. XXIV, 891 pages. 2004.

Vol. 3319: D. Amyot, A.W. Williams (Eds.), System Analysis and Modeling. XII, 301 pages. 2005.

Vol. 3318: E. Eskin, C. Workman (Eds.), Regulatory Genomics. VII, 115 pages. 2005. (Subseries LNBI).

Vol. 3317: M. Domaratzki, A. Okhotin, K. Salomaa, S. Yu (Eds.), Implementation and Application of Automata. XII, 336 pages. 2005.

Vol. 3316: N.R. Pal, N.K. Kasabov, R.K. Mudi, S. Pal, S.K. Parui (Eds.), Neural Information Processing. XXX, 1368 pages. 2004.

Vol. 3315: C. Lemaître, C.A. Reyes, J.A. González (Eds.), Advances in Artificial Intelligence - IBERAMIA 2004. XX, 987 pages. 2004. (Subseries LNAI).

Vol. 3314: J. Zhang, J.-H. He, Y. Fu (Eds.), Computational and Information Science. XXIV, 1259 pages. 2004.

Vol. 3313: C. Castelluccia, H. Hartenstein, C. Paar, D. Westhoff (Eds.), Security in Ad-hoc and Sensor Networks. VIII, 231 pages. 2005.

Vol. 3312: A.J. Hu, A.K. Martin (Eds.), Formal Methods in Computer-Aided Design. XI, 445 pages. 2004.

Vol. 3311: V. Roca, F. Rousseau (Eds.), Interactive Multimedia and Next Generation Networks. XIII, 287 pages. 2004.

Vol. 3310: U.K. Wiil (Ed.), Computer Music Modeling and Retrieval. XI, 371 pages. 2005.